Programmable Logic Controllers

Programmable Logic Controllers

Principles and Applications

Fourth Edition

John W. Webb

Northcentral Technical College
Wausau, Wisconsin

Ronald A. Reis

Los Angeles Valley College
Van Nuys, California

PRENTICE HALL
Upper Saddle River, New Jersey
Columbus, Ohio

Library of Congress Cataloging-in-Publication Data

Webb, John W.
Programmable logic controllers : principles and applications / John W. Webb, Ronald A. Reis. — 4th ed.
p. cm.
Includes bibliographical references and index.
ISBN 0-13-679408-4
1. Programmable controllers. I. Reis, Ronald A. II. Title.
TJ223.P76W43 1999
629.8′9—dc21 98-13424
CIP

Cover photo: © H. Armstrong Roberts
Editor: Charles E. Stewart Jr.
Production Editor: Alexandrina Benedicto Wolf
Cover Designer: Lisa Stark
Design Coordinator: Karrie M. Converse
Production Manager: Deidra M. Schwartz
Editorial/production supervision: Tally Morgan, WordCrafters Editorial Services, Inc.
Marketing Manager: Ben Leonard

This book was set in Times Roman by Maryland Composition Co., Inc., and was printed and bound by Courier/ Kendallville, Inc. The cover was printed by Phoenix Color Corp.

Printed in the United States of America

10 9 8 7 6 5 4 3 2

ISBN: 0-13-679408-4

Prentice-Hall International (UK) Limited, *London*
Prentice-Hall of Australia Pty. Limited, *Sydney*
Prentice-Hall Canada Inc., *Toronto*
Prentice-Hall Hispanoamericana, S.A., *Mexico*
Prentice-Hall of India Private Limited, *New Delhi*
Prentice-Hall of Japan, Inc., *Tokyo*
Simon & Schuster Asia Pte. Ltd., *Singapore*
Editora Prentice-Hall do Brazil, Ltda., *Rio de Janiero*

To our wives:
Thelma Webb and Karen Reis

And to the up-and-coming generation:
Evie, Nora, and Tom
Christine
Paul, Adam, and Karina
Alex and Samantha
Ben and Matthew
Austin

Preface

Programmable logic controllers (PLCs) are being used increasingly worldwide. Originally, in the 1980s, PLCs were primarily single machine control devices. Today, they are used in tandem with each other, with other control devices, and in complex network system control. Training, retraining, and upgrading of skills in PLC use are necessary with their increased popularity and complexity. The purpose of this book is to fulfill these training requirements.

The fourth edition has been revised to include a number of new features:

- Equipment illustrations have been updated.
- Additional functions are covered in various chapters.
- Most chapters have been expanded to include recent developments, such as new program languages, new manufacturers and models, and enhanced functions.
- Example problems have been added.
- Troubleshooting exercises have been added to appropriate chapters.
- The glossary and bibliography have been expanded.
- The chapters have been rearranged for better "flow" of instruction.
- The text includes programming formats and functions for the major manufacturers in as generic a format as possible.
- A matrix listing of major function designations by manufacturer is included for cross-reference purposes.

The book is divided into eight sections. The sections may be used to match units in courses and training programs. Unit one might cover the basics in Sections I through III. Unit two could be Sections IV and V, intermediate and data handling functions. Sections VI and VII are for inclusion in a longer course or in a second course as appropriate. Section VIII can be a "stand-alone" section. Alternatively, chapters of Section VII can be inserted anywhere in the training course. For example, Chapter 26, PLC Auxiliary Commands and Functions, could be included in unit one. Chapter 25, Alternate Programming Languages, and 27, PLC Installation, Troubleshooting and Maintenance, could be included in an earlier unit, depending on the needs of the instructional program.

Each chapter includes learning objectives, an introduction, explanations, examples and, in some cases, troubleshooting problems. At the end of each chapter, chapter exercises are included in the order of the chapter sequence. An instructor's manual which includes worked-out answers for every exercise is available.

Acknowledgments

We would like to thank the reviewers for their helpful comments:

Marvin Anderl, John Tyler Community College; Donald Huskey, Bainbridge Junior College; James Knack, Henry Ford Community College; Dennis Lindgren, Range Technical College; William Maxwell, Nashville State Technical Institute; and Paul Weingartner, Cincinnati Technical College.

We would also like to thank the following companies for their support and cooperation in preparing the fourth edition of *Programmable Logic Controllers:* ABB Robotics; Allen-Bradley/Rockwell Automation; Amatrol; Best Power Technology; Bussmann Division/Cooper Industries; Control Technology Company; Cuttler Hammer Products; Eaton Corporation; GE Fanuc; General Electric; Giddings and Lewis; I & CS Magazine of Chilton Publishing Company; IDEC, Inc.; IDEL Corporation; ITT Robotics; Mitsubishi Electronics; Modicon/Schneider Automation; Square D/Schneider Automation; and Westinghouse Corporation

A special thanks to Curtis Clifton of L & S Electric of Schofield, Wisconsin for his technical input into the fourth edition.

Contents

I

PLC Basics

1

An Overall Look at Programmable Logic Controllers

OUTLINE

1–1 Introduction □ **1–2** Definition and History of the PLC □ **1–3** PLC Advantages and Disadvantages □ **1–4** Overall PLC System □ **1–5** CPUs and Programmer/Monitors □ **1–6** PLC Input and Output Modules □ **1–7** Printing PLC Information

OBJECTIVES

At the end of this chapter, you will be able to

- □ Discuss the history of the PLC.
- □ List and discuss advantages and disadvantages of the PLC.
- □ Evaluate knowledge levels needed for PLC programming and operating.
- □ List the four major parts of a PLC system.
- □ Describe the function of each of the four parts.
- □ Describe how the parts of the system are connected electrically.
- □ List the major classifications of input/output (I/O) modules.
- □ Outline the major precautions to follow when connecting I/O modules.
- □ Explain baud rates and how they are set.

1–1 INTRODUCTION

In this introductory chapter, we describe a programmable logic controller (PLC). We then discuss the evolution of relay logic and computer systems into the present-day PLC. We also list and discuss some advantages and disadvantages of using a PLC over other control systems. Finally, the knowledge level required for PLC programming and operating is evaluated.

We also describe the components and modules that make up a PLC control system. A simple PLC system is housed in one or possibly two enclosures, each of which includes multiple functions. A more complex PLC, controlling a large process, may have three to five or more separate interconnected enclosures containing the PLC subsystems.

Illustrations of the various subparts of a PLC are shown, as are general connection paths. The electrical interconnections of the various PLC parts are described in general terms. Details of the connections are discussed in chapter 27.

Most PLC electrical connecting is easily done with single cables between units. However, connecting the input/output (I/O) modules to the outside world can be fairly complicated. I/O module connections to the processes are discussed in this chapter and throughout the book. The proper setting of module switches is also described in this chapter.

Today, the personal computer is available to carry out PLC programming and, in some cases, take the place of a PLC. Used as a sophisticated programming device, the computer must be able to run PLC software that allows it to operate as a PLC. When functioning as a full-fledged PLC, the computer has to have, of course, some way to receive information from sensors and transducers and, in turn, to actuate relays, coils, lights, and motors. These personal computer systems are also discussed herein.

PLC systems operate at different computer rates. The rate, commonly called the *baud rate*, depends on what parts of the PLC system are communicating. We discuss baud rates later in the chapter.

1–2 DEFINITION AND HISTORY OF THE PLC

Originally, the PLC was represented by the acronym PC. There was some confusion with using this acronym as it is commonly accepted to represent *personal computer*. Therefore, PLC is now commonly accepted to mean *programmable logic controller*.

A PLC is a user-friendly, microprocessor-based specialized computer that carries out control functions of many types and levels of complexity. Its purpose is to monitor crucial process parameters and adjust process operations accordingly. It can be programmed, controlled, and operated by a person unskilled in operating computers. Essentially, a PLC's operator draws the lines and devices of ladder diagrams with a keyboard onto a display screen. The resulting drawing is converted into computer machine language and run as a user program.

The computer takes the place of much of the external wiring required for control of a process. The PLC will operate any system that has output devices that go on and off (known as discrete, or digital, outputs). It can also operate any system with variable (analog) outputs. The PLC can be operated on the input side by on–off devices (discrete, or digital) or by variable (analog) input devices.

Today, the big unit growth in the PLC industry is at the low end—where small keeps getting smaller. When a few years ago the micro PLC entered the market, some thought that these devices had "bottomed out." Now, nano PLCs—generally defined as those with 16 or fewer I/O—are spreading. Some can fit into your shirt pocket, being no larger than a deck of cards and at the time of this writing, PLC*Direct* plans to introduce a PLC the size of a box of Tic-Tac candy that will include many features of current micro models.

The first PLC systems evolved from conventional computers in the late 1960s and early 1970s. These first PLCs were installed primarily in automotive plants. Traditionally, the auto plants had to be shut down for up to a month at model changeover time. The early PLCs were used along with other new automation techniques to shorten the changeover time. One of the major time-consuming changeover procedures had been the wiring of new or revised relay and control panels. The PLC keyboard reprogramming procedure replaced the rewiring of a panel full of wires, relays, timers, and other components. The new PLCs helped reduce changeover time to a matter of a few days.

There was a major problem with these early 1970s computer/PLC reprogramming procedures. The programs were complicated and required a highly trained programmer to make the changes. Through the late 1970s, improvements were made in PLC programs to make them somewhat more user friendly; in 1978, the introduction of the microprocessor chip increased computer power for all kinds of automation systems and lowered the computing cost. Robotics, automation devices, and computers of all types, including the PLC, consequently underwent many improvements. PLC programs, written in high-level language, became more understandable to more people, and PLCs became more affordable.

In the 1980s, with more computer power per dollar available, the PLC came into exponentially increasing use. Some large electronics and computer companies and some diverse corporate electronics divisions found that the PLC had become their greatest volume product. The market for PLCs grew from a volume of $80 million in 1978 to $1 billion per year by 1990 and is still growing. Even the machine tool industry, where computer numerical controls (CNCs) have been used in the past, is using PLCs. PLCs are also used extensively in building energy and security control systems. Other nontraditional uses of PLCs, such as in the home and in medical equipment, have exploded in the 1990s and will increase as we enter the new millennium.

A person knowledgeable in relay logic systems can master the major PLC functions in a few hours. These functions might include coils, contacts, timers, and counters. The same is true for a person with a digital logic background. For persons unfamiliar with ladder diagrams or digital principles, however, the learning process takes more time.

A person knowledgeable in relay logic can master advanced PLC functions in a few days with proper instruction. Company schools and operating manuals are very helpful in mastering these advanced functions. Advanced functions in order of learning might include sequence/drum controller, register bit use, and move functions.

Figure 1–1 shows an older relay-type control panel used in process control. It is large and contains lots of wiring, interconnections, and relays, which can have maintenance problems. Figure 1–2 shows a typical PLC, which replaces the relay panel and performs the

FIGURE 1–1
Relay Panel for Logic Control

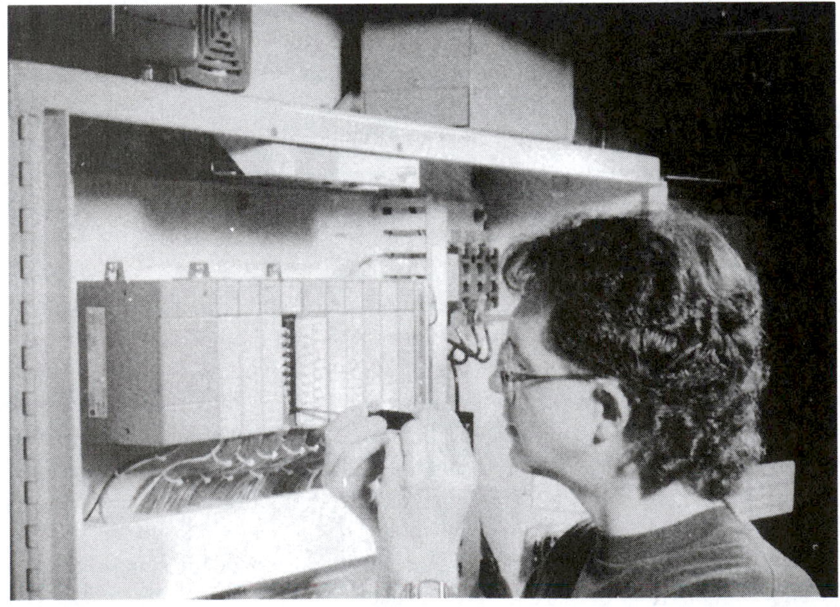

FIGURE 1–2
PLC for Logic Control (Courtesy of Allen-Bradley)

same logic control task. Of course, the wiring to the process is the same for each type of control. However, the PLC is smaller and more reliable. For control system logic changes, the relay panel must be rewired, whereas the PLC is quickly reprogrammed for any changes.

1–3 PLC ADVANTAGES AND DISADVANTAGES

Following are 13 major advantages of using a programmable controller:

Flexibility. In the past, each different electronically controlled production machine required its own controller; 15 machines might require 15 different controllers. Now it is possible to use just one model of a PLC to run any one of the 15 machines. Furthermore, you would probably need fewer than 15 controllers, because one PLC can easily run many machines. Each of the 15 machines under PLC control would have its own distinct program.

Implementing Changes and Correcting Errors. With a wired relay-type panel, any program alterations require time for rewiring of panels and devices. When a PLC program circuit or sequence design change is made, the PLC program can be changed from a keyboard sequence in a matter of minutes. No rewiring is required for a PLC-controlled system. Also, if a programming error has to be corrected in a PLC control ladder diagram, a change can be typed in quickly.

Large Quantities of Contacts. The PLC has a large number of contacts for each coil available in its programming. Suppose that a panel-wired relay has four contacts and all are in use when a design change requiring three more contacts is made. Time would have to be taken to procure and install a new relay or relay contact block. Using a PLC, however, only three more contacts would be typed in. The three contacts would be automatically available in the PLC. Indeed, a hundred contacts can be used from one relay—if sufficient computer memory is available.

Lower Cost. Increased technology makes it possible to condense more functions into smaller and less expensive packages. Now you can purchase a PLC with numerous relays, timers, and counters, a sequencer, and other functions for a few hundred dollars.

Pilot Running. A PLC programmed circuit can be prerun and evaluated in the office or lab. The program can be typed in, tested, observed, and modified if needed, saving valuable factory time. In contrast, conventional relay systems have been best tested on the factory floor, which can be very time consuming.

Visual Observation. A PLC circuit's operation can be seen during operation directly on a CRT screen. The operation or misoperation of a circuit can be observed as it happens. Logic paths light up on the screen as they are energized. Troubleshooting can be done more quickly during visual observation.

In advanced PLC systems, an operator message can be programmed for each possible malfunction. The malfunction description appears on the screen when the malfunction is detected by the PLC logic (for example, "MOTOR #7 IS OVERLOADED"). Advanced PLC systems also may have descriptions of the function of each circuit component. For example, input #1 on the diagram could have "CONVEYOR LIMIT SWITCH" on the diagram as a description.

Speed of Operation. Relays can take an unacceptable amount of time to actuate. The operational speed for the PLC program is very fast. The speed for the PLC logic operation is determined by scan time, which is a matter of milliseconds.

Ladder or Boolean Programming Method. The PLC programming can be accomplished in the ladder mode by an electrician or technician. Alternatively, a PLC programmer who works in digital or Boolean control systems can also easily perform PLC programming.

Reliability and Maintainability. Solid-state devices are more reliable, in general, than mechanical systems or relays and timers. The PLC is made of solid-state components with very high reliability rates. Consequently, the control system maintenance costs are low and downtime is minimal.

Simplicity of Ordering Control System Components. A PLC is one device with one delivery date. When the PLC arrives, all the counters, relays, and other components also arrive. In designing a relay panel, however, you may have 20 different relays and timers from 12 different suppliers. Obtaining the parts on time involves various delivery dates and availabilities. With a PLC you have one product and one lead time for delivery. In a relay system, forgetting to buy one component would mean delaying the startup of the control system until that component arrives. With the PLC, one more relay is always available—provided that you ordered a PLC with enough extra computing power.

Documentation. An immediate printout of the true PLC circuit is available in minutes, if required. There is no need to look for the blueprint of the circuit in remote files. The PLC prints out the actual circuit in operation at a given moment. Often, the file prints for relay panels are not properly kept up to date. A PLC printout is the circuit at the present time; no wire tracing is needed for verification.

Security. A PLC program change cannot be made unless the PLC is properly unlocked and programmed. Relay panels tend to undergo undocumented changes. People on late shifts do not always record panel alterations made when the office area is locked up for the night.

Ease of Changes by Reprogramming. Since the PLC can be reprogrammed quickly, mixed production processing can be accomplished. For example, if part B comes down the assembly line while part A is still being processed, a program for part B's processing can be reprogrammed into the production machinery in a matter of seconds.

These 13 items are some of the advantages of using a programmable logic controller. There will, of course, be other advantages in individual applications and industries.

Following are some of the disadvantages of, or perhaps precautions involved in, using PLCs:

Newer Technology. It is difficult to change the thinking of some personnel from ladders and relays to the PLC computer concept. Although today, with the pervasive use of computers not only at home and in the office but on the factory floor, acceptance of the computer as a powerful and reliable productivity-enhancing tool is, if not universal, almost so. Electricians and technicians are lining up to take courses on PLCs because they know that doing so contributes to job security and advancement.

Fixed Program Applications. Some applications are single-function applications. It does not pay to use a PLC that includes multiple programming capabilities if they are not needed. One example is in the use of drum controller/sequencers. Some equipment manufacturers still use a mechanical drum with pegs at an overall cost advantage. Their operational sequence is seldom or never changed, so the reprogramming available with the PLC would not be necessary.

Environmental Considerations. Certain process environments, such as high heat and vibration, interfere with the electronic devices in PLCs, which limit their use.

Fail-Safe Operation. In relay systems, the stop button electrically disconnects the circuit; if the power fails, the system stops. Furthermore, the relay system does not automatically restart when power is restored. This, of course, can be programmed into the PLC; however, in some PLC programs, you may have to apply an input voltage to cause a device to stop. These systems are not fail-safe. This disadvantage can be overcome by adding safety relays to a PLC system, as shown later in this text.

Fixed-Circuit Operation. If the circuit in operation is never altered, a fixed control system (such as a mechanical drum) might be less costly than a PLC. The PLC is most effective when periodic changes in operation are made.

1–4 OVERALL PLC SYSTEM

Figure 1–3 shows, in block form, the four major units of a PLC system and how they are interconnected. The four major parts, each of which is described later in detail, are

1. *Central Processing Unit (CPU).* The "brain" of the system, which has three subparts:
   a. Microprocessor. The computer center that carries out mathematic and logic operations.
   b. Memory. The area of the CPU in which data and information is stored and retrieved. Holds the system software and user program.
   c. Power supply. The electrical supply that converts AC line voltage to various operational DC values. In the process, the power supply filters and regulates the DC voltages to ensure proper computer operation.
2. *Programmer/Monitor.* The programmer/monitor (PM) is a device used to communicate with the circuits of the PLC. Hand-held terminals, industrial terminals, and the personal computer exist as PM devices. In a hand-held unit, input takes place through a membrane keypad and the display is usually a liquid crystal

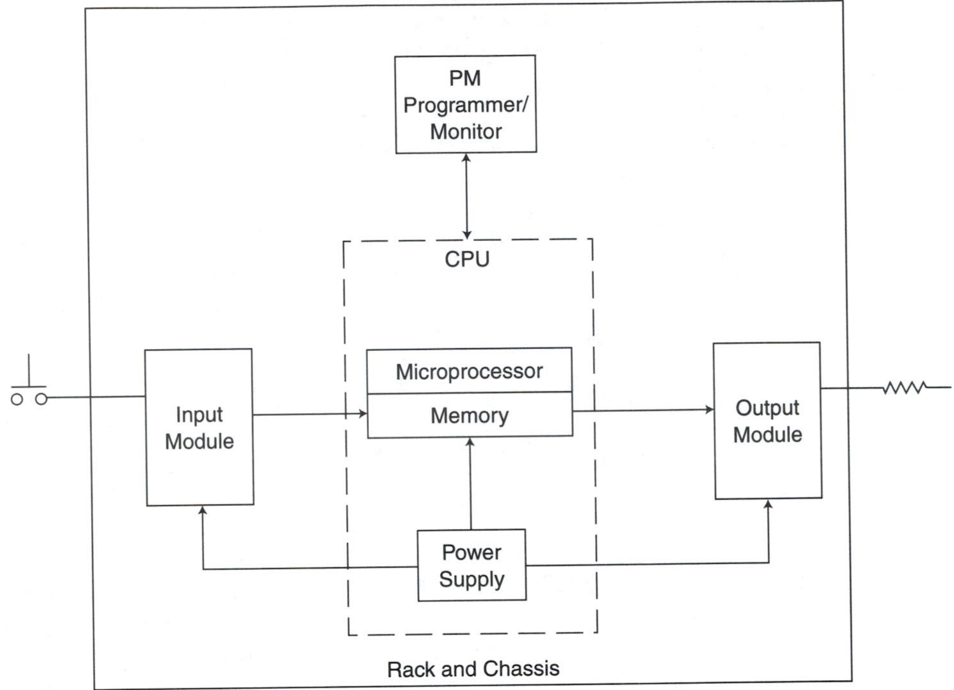

FIGURE 1–3
PLC System Layout and Connection

(LCD). With the industrial terminal or personal computer, more complex, typewriter-type keyboards and cathode ray tubes (CRTs) are employed.

3. *I/O Modules.* The input module has terminals into which outside process electrical signals, generated by sensors or transducers, are entered. An electronic system for connecting I/O modules to remote locations can be added if needed. The actual operating process under PLC control can be thousands of feet from the CPU and its I/O modules.
4. *Racks and Chassis.* The racks on which the PLC parts are mounted and the enclosures on which the CPU, PM, and I/O modules are mounted.

Optional units often a part of the PLC system are

- □ *Printer.* A device on which the program in the CPU may be printed. In addition, operating information may be printed upon command.
- □ *Program Recorder/Player.* Some older PLC systems use tape to provide secondary storage for CPU programs. Today, PLCs use floppy disks, with hard disks for secondary storage. The stored programs provide backup and a way to download programs written off-line from the PLC process system.

For large operations, a master computer is often used to coordinate many individual, interconnected PLCs. In such systems, the interconnecting electrical buses are sometimes referred to as data highways (more on this topic in chapters 24 and 27).

1–5 CPUs AND PROGRAMMER/MONITORS

The CPU is the heart of the PLC system. A typical CPU is shown in figure 1–4. The CPU you use may be smaller or larger than the one shown, depending on the size of the process to be controlled. It is important to size the system CPU according to the internal memory needed to run the process. Controlling a small operation requires only a small PLC unit with limited memory; controlling a larger system would require a larger unit with more memory and functions.

Some CPUs can have additional memory easily added at a later date; others cannot be added to or expanded. Advanced planning with the manufacturer is required to match present and future needs with the size of the system being purchased.

The CPU contains various electrical receptacles for connecting the cables that go to the other PLC units. It is important to connect the proper receptacles with the correct cables supplied by the manufacturer.

Many CPUs contain backup batteries that keep the user process control ladder program in storage in the event of a plant power failure. Typical retentive backup time is one month to one year. The basic operating system is stored permanently in the CPU, in read-only memory (ROM), and is not lost when input power is lost. However, the user process control ladder program, being in random access memory (RAM), is not stored permanently. Battery backup power enables the CPU to retain the user program in the event of power loss. Only the user program can be lost or erased when PLC CPU power is lost.

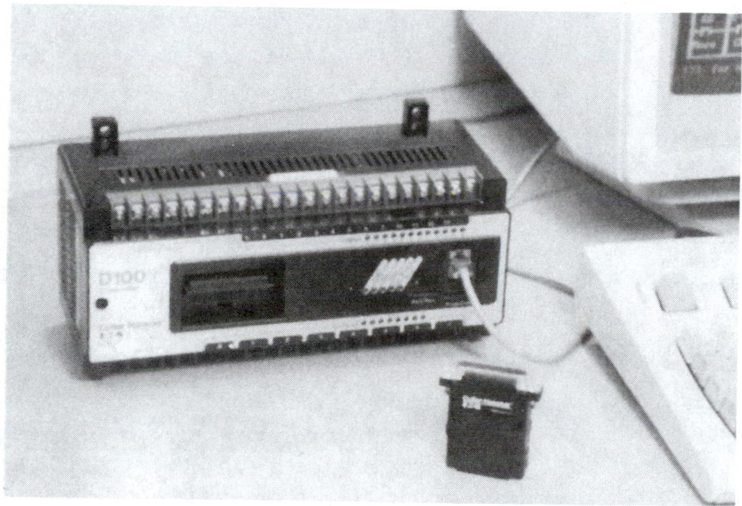

FIGURE 1–4
Central Processing Unit (Courtesy of Eaton Corp.)

The CPUs all have operational switches, some of which require a key to prevent unauthorized personnel from running a turned-off process. The key-type switch also can prevent unauthorized alterations to the operating system program. The switch positions vary from manufacturer to manufacturer, but are similar. Typical positions are

- □ Off. System cannot be run or programmed.
- □ Run. Allows the system to run, but no program alterations can be made.
- □ Disable. Turns all outputs off or sets them to the inoperable state.
- □ Monitor. Turns on screen that displays operating information.
- □ Run/Program. System can run, and program modifications can be made to it while it is running. This mode must be used with caution. In this mode, the program cannot be completely erased (for safety) but can only be modified. To delete an entire program, the key must be in the *disable* position.
- □ Off/Program or Program. System cannot run, but can be programmed or reprogrammed.

Some manufacturers' programmers may have other special key positions in addition to these.

Figure 1–5 shows some typical large programmer/monitors with large cathode ray-tube screens. Figure 1–6 shows some typical, small, hand-held programmers with small display windows. The difference in display size is directly related to cost. The units in figure 1–5 cost considerably more, but give more information on the screen. A large monitor screen shows an entire circuit. The smaller hand-held screen display shows only one part of the circuit at a time. With the smaller unit, you may have to go through two or three steps to see all of just one ladder rung. A less costly laptop program monitor is commonly in use. A system is shown in figure 1–7.

FIGURE 1–5
Large-Screen Programmer/Monitors (Courtesy of Giddings and Lewis and General Electric)

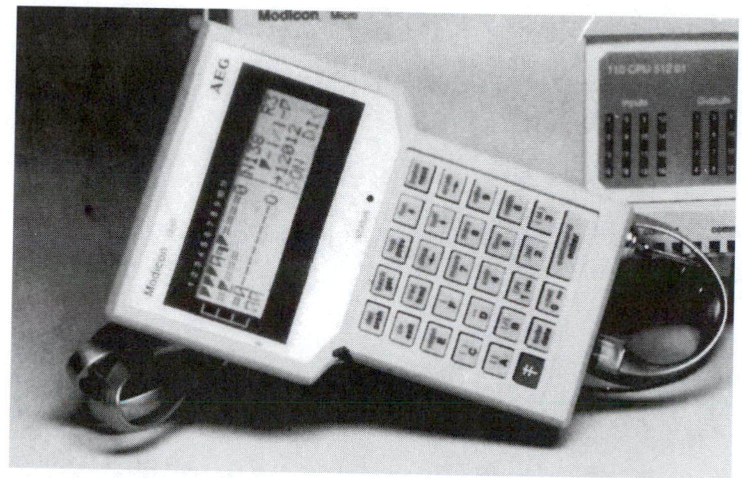

A

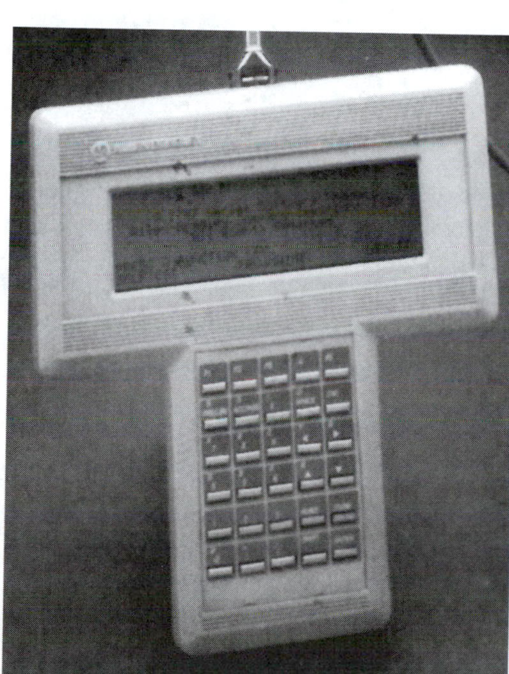

B

C

FIGURE 1–6
Hand-Held PMs [Courtesy of (a) Modicon/Schneider Automation, (b) Allen-Bradley, and (c) Eaton Corp.]

FIGURE 1–7
Laptop PM System

The programmer/monitor (PM) is connected to the CPU by a cable. After the CPU has been programmed, the PM is no longer required for CPU and process operation and can be disconnected and removed. Therefore, you may need only one PM for a number of operational CPUs. The PM may be moved about in the plant as needed. The same PM can even be used in the office or lab to pretest programs. PM keyboard and screen operations are discussed in detail in subsequent chapters.

1–6 PLC INPUT AND OUTPUT MODULES

We get information in and out of the PLC through the use of input and output modules. The input module terminals receive signals from wires connected to input sensors and transducers. The output module terminals provide output voltages to energize actuators and indicating devices.

There are typically 4, 8, 12, or 16 terminals per module. There may be the same number of terminals for a PLC's I/O modules, but often there are different numbers of terminals for input and output; for example, a system may have 12 inputs and 8 outputs. A typical module is shown in figure 1-8.

In smaller systems, the input and output terminals may be included on the same frame as the CPU. Figure 1–9 shows two such units. In other, larger PLC systems, the input and output modules are separate units. In these larger systems, modules are placed in groups on racks, as shown in figure 1–10. The racks are connected to the CPU via appropriate connector multiconductor cables.

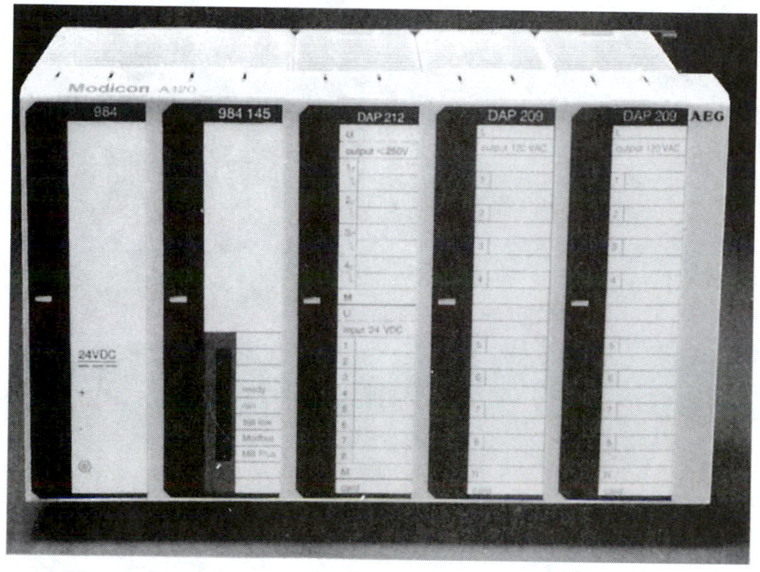

FIGURE 1–8
I/O Modules (Courtesy of Modicon/Schneider Automation)

FIGURE 1–9
I/O Terminals Combined with a CPU (Courtesy of Modicon/Schneider Automation)

FIGURE 1–10
Rack Mounts for I/O Modules
(Courtesy of Giddings and Lewis)

Typically, up to 256 terminals may be controlled using only 9 to 24 interconnecting wires. The exact number of wires is determined by the type of computer configuration used for terminal-to-CPU information interchange. The electrical controlling signals from the CPU to the I/O terminals are coded and decoded electronically, making 256 wires for 256 terminals unnecessary.

For multiple modules in a rack, it is necessary to set module switches for each individual module. These settings specify each module's operational number series. Again, for 256 inputs and 256 outputs on a rack, there are 9 to 24 wires in the cable connected to the CPU. Each rack group knows what numbers it should respond to by the system of single in-line package (SIP) switch settings.

Some PLC systems use programming instead of switches to configure I/O module settings. Some small systems require no address settings on the I/O modules. The order in which you plug them in determines the address number for these small systems. Other, larger systems set the address numbers by following a programming procedure on the PM.

Various parts of a PLC system require different computer operational rates for proper operation. These rates are called *baud rates.* A PLC CPU computer may "converse" with its keyboard at a rate of 4800 baud. For remote operation, it might use 2400 baud. Two peripheral devices might use rates of 600 and 1200 baud. The baud rates vary for each manufacturer and its individual PLC device. Each device's baud rate is set automatically when the PLC is turned on. The baud rates may have to be reset for certain modes of PLC operation. If, for example, you attempt to print a ladder diagram and get an unreadable result, it may be that the baud rate is set incorrectly. Refer to and follow the manufacturer's manual program section on setting peripheral baud rates.

A most important consideration for an I/O module is the module's voltage and current rating. Both voltage and current must match the electrical requirements of the system to which it is connected. An input module rated at 24 volts DC will not work on 120 volts AC and may even be damaged if the module fuse does not act quickly. An output device requiring 4.5 amperes cannot be turned on by a 2-ampere output module; the module fuse would blow. PLC manufacturers have a wide variety of input and output modules available. Module ratings are chosen by the manufacturers to cover the most common applications of their customers. Typical ratings available from manufacturers are shown in figure 1–11.

Sometimes the processes to be controlled by a PLC are a long distance from the CPU or from each other. The normal input and output electrical signals will be reduced to a value too low for module recognition due to long interconnecting wires. Remote amplifier units are available for cases such as these. A typical remote setup is shown in figure 1–12. The input and output signals from the CPU are coded by an adjacent coding unit into digital electrical pulses. The pulses are transmitted over two wires, or by a fiber optics system, to the remote location. At the remote location, a matching station decodes the digital signals. The digital pulses are decoded back into the separate signals that feed the remote modules. The signals originally leaving the CPU are, therefore, exactly duplicated at the remote modules—a module a mile away will operate as if it were 10 feet away. Other communication systems include telemetering and radio continuous-wave communication.

The early chapters of this text discuss the most common type of module, the discrete, or digital, type. Inputs in the discrete type of module are either on or off and the outputs are either energized or deenergized. Chapter 22 covers the basic principles of a different type of module—the analog. These analog modules work with variable signals with varying values.

Many newer modules have an internal computer for faster process control operation. For example, a process might have a critical input that must be acted upon immediately for the safety of process personnel. Sending signals to the CPU, CPU analysis, and return-signal time takes too long. The module can do the analysis continuously and quickly, and can take action immediately.

There is one major precaution to be considered with PLC output modules. In relay operation, when a relay contact is open, there is no current flow in the associated controlled circuit. However, PLC output modules, when turned off, are not strictly off. A small leakage current from the output terminal to the output module still exists, even though the output module is turned off. The output current of each module terminal comes from the output of a thyristor semiconductor called a *triac*. When not turned on, the triac still puts out a small amount of current. The leakage current is a matter of a few milliamperes and is often of no consequence; however, the leakage current may have to be considered in some applications. For example, a PLC output terminal might supply a neon bulb that indicates that the output is on; the neon will glow dimly when the module is off due to the leakage current. It might be necessary to add an amplifier or shunting resistors in the electrical output system to bleed off this leakage current.

VOLTAGE LEVEL	MODULE	CATALOG NUMBER IC600BF	CIRCUIT QUANTITY	UNIT OF I/O LOAD
115 Vac/dc	Input	804	8	2
115 Vac, 2 amp	Output	904	8	9
115 Vac/dc, Isolated	Input	810	6	2
115 Vac, 3 1/2 amp, Isolated	Output	910	6	8
115 Vac, 4 amp, Protected	Output	930	4	8
220 Vac/dc	Input	805	8	2
220 Vac, 2 amp	Output	905	8	9
220 Vac/dc, Isolated	Input	812	6	2
220 Vac, 3 1/2 amp, Isolated	Output	912	6	8
12 Vac/dc	Input	806	8	2
12 Vdc, Sink	Output	906	8	7
12 Vdc, Source	Output	907	8	7
24–48 Vac/dc	Input	802	8	2
24 Vdc, Sink	Output	902	8	7
24 Vdc, Source	Output	908	8	7
48 Vdc, Sink	Output	903	8	7
48 Vdc, Source	Output	909	8	7
120 Vdc, 1 1/2 amp	Output	924	8	5
5 VTTL/10–50Vdc w/o Lights	Input	811	32	4
5 VTTL w/o Lights	Output	911	32	3
10–50 Vdc, Sink, w/o Lights	Output	913	32	3
10–50 Vdc, Source, w/o Lights	Output	919	32	3
5 VTTL/10–50 Vdc with Lights	Input	831	32	4
5 VTTL with Lights	Output	921	32	3
10–50 Vdc, Sink, with Lights	Output	923	32	3
10–50 Vdc, Source, with Lights	Output	929	32	3
100 VA Reeds (NO/NC)	Output	914	6	17
0–10 Vdc Analog	Input	841	8	29
−10 to +10 Vdc Analog	Input	842	8	29
4–20ma/1–5Vdc Analog	Input	843	8	29
0–10 Vdc Analog	Output	941	4	29
−10 to +10 Vdc Analog	Output	942	4	29
4–20ma Analog	Output	943	4	29
Thermocouple Type J	Input	813	8	9
Thermocouple Type K	Input	814	8	9
Thermocouple Type S	Input	815	8	9
Thermocouple Type T	Input	816	8	9
Axis Position, Type 1	Output	915	1	42
High Speed Counter	I/O	827	1	19
Interrupt	Input	808	8	3
I/O Local Receiver		800		9
I/O Local Transmitter		900		34
I/O Remote Receiver		801		42
I/O Remote Driver		901		38

FIGURE 1–11
Typical Available Module Ratings (Courtesy of General Electric)

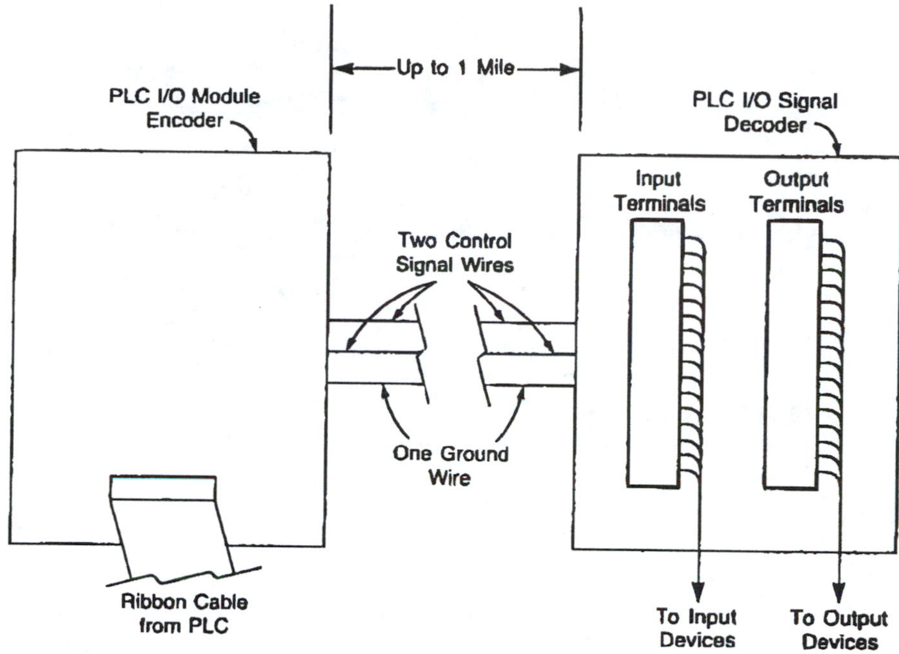

FIGURE 1–12
Remote PLC Operation

1–7 PRINTING PLC INFORMATION

A typical PLC printer is shown in figure 1–13. Printers are used to record information from the CPU for visual analysis. Lengthy ladder programs cannot be completely shown on a screen; typically, a screen shows only one to five rungs. A printout on a continuous paper roll can show ladder diagrams and programs of any length. In industrial settings, the complete diagram can be used to analyze the complete circuit. In educational settings, printouts may be used for written assignments to check for correct program construction.

There are many different types of PLC information that can be printed out:

- Ladder diagrams (which may include coil/contact cross-references)
- Status of registers
- Status and listing of forced conditions
- Timing diagrams of contacts
- Timing diagrams of registers
- Other special diagrams or information

In Chapter 26 we describe in detail how to use the printer and list the possible benefits of the various printouts.

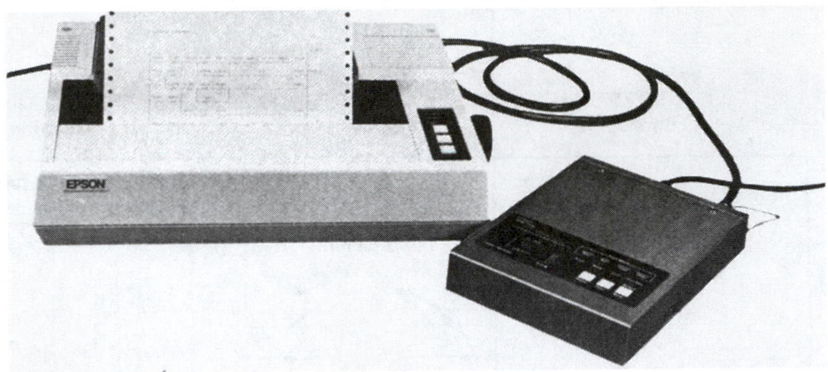

FIGURE 1–13
A PLC Printer (Courtesy of Eaton Corp.)

EXERCISES

1. Discuss the evolution of relay logic and the computer evolution into the PLC.
2. List a dozen advantages of using a PLC. Use items discussed in the text plus some of your own ideas.
3. List a few disadvantages of using a PLC. Again, use the text ideas plus your own.
4. Based on your own knowledge and skill level in the areas of relay and digital logic, evaluate the level of difficulty you will have in learning PLC programming and operations. Do the same for two or three typical factory workers, electricians, or technicians.
5. Obtain the manuals from one or two different PLC models. List the various types of I/O modules available for each model.
6. From the manuals in exercise 5, determine and list the baud rates at which the models operate, including the peripheral baud rates and other relevant details.

2

The PLC: A Look Inside

OUTLINE

2–1 Introduction □ **2–2** The PLC as a Computer □ **2–3** The Central Processing Unit □ **2–4** Solid-State Memory □ **2–5** The Processor □ **2–6** I/O Modules (Interfaces) □ **2–7** Power Supplies

OBJECTIVES

At the end of this chapter, you will be able to

- □ Describe the difference between a data processing and a process control computer.
- □ List and define the functions of each of the major sections of a PLC CPU.
- □ List and describe the various types of solid-state memory used in a PLC CPU.
- □ Describe, using diagrams, how a CPU processes information internally.
- □ Describe the operation of a typical input module.
- □ Describe the operation of a typical output module.
- □ Describe how an AC-in/DC-out power supply functions.

2–1 INTRODUCTION

In chapter 1 we introduced the programmable logic controller (PLC). We saw how it has evolved, and we examined its many advantages while noting a few disadvantages. We looked at the overall system and briefly discussed the CPU, P/M, I/O modules, and a peripheral, the printer. In this chapter we look inside the PLC and investigate its internal operation.

In the first part of chapter 2 we discuss the PLC as a process control computer. We also examine the CPU as it exists in small to large PLCs. In the second part of the chapter we investigate the four main elements comprising all PLCs. We look at solid-state *memory*, both volatile and nonvolatile. We examine what the *processor* consists of and what it does. We analyze the makeup of discrete *input* and *output modules (interfaces)*, seeing how they link real-world sensors and actuators to the PLC. Finally, we take a look at the *power supply*, a circuit that supplies the DC voltages necessary to power a PLC. Examining what goes on inside the PLC will enhance our understanding and appreciation for the many topics to come.

2–2 THE PLC AS A COMPUTER

A PLC is a computer, but a different type from the one you are probably used to seeing and working with. Most people are familiar with *data processing computers*, especially microcomputers such as those from Apple and IBM (figure 2–1). These machines sit on your

FIGURE 2–1
Data Processing Computer (Courtesy of Tandy)

desk or lap and have powerful systems and applications software that let you play games, do word processing, create computer-aided design (CAD) drawings, lay out spreadsheets, and explore the wonders of the Nile River. Such computers process reams of data, which is why they are called data processing machines. Their input peripherals are the keyboard and mouse; their output peripherals, the video display terminal (VDT), printer, and plotter.

There is another type of computer, however, known as a *process control computer*. Although it, of course, processes data, its main function is to control manufacturing and industrial processes (machinery, robots, assembly lines, etc.). Such computers are said to be event driven. Although they may have a keyboard input peripheral, their control inputs are switches and sensors, and although output peripherals such as VDTs and printers may be attached, the process control computer primarily controls such devices as motors, solenoids, lights, and heaters (figure 2–2). Such process control computers, which number in the millions, are the control element in virtually all modern factory operations.

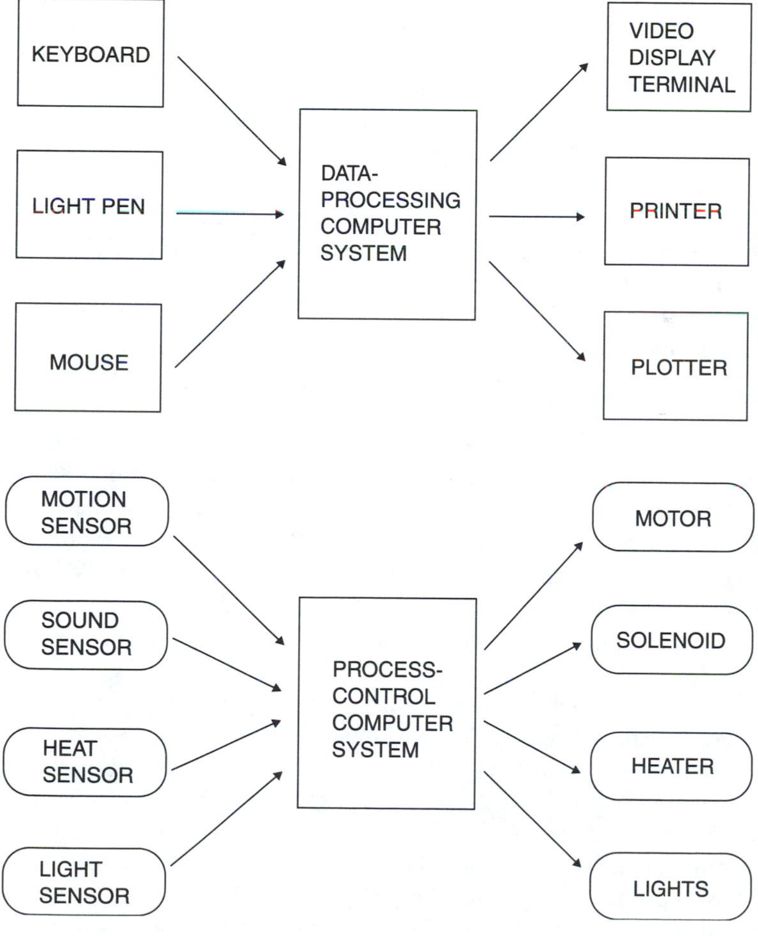

FIGURE 2–2
Data Processing and Process Control Computers

PLCs are a type of process control computer: small, relatively inexpensive, environmentally hardened, and easy to program, operate, maintain, and repair. They are often installed close to the machinery or process they control and are thus seen as an extension of industrial equipment. At the *Los Angeles Times* Olympic printing plant, for example, 72 separately operating small PLCs control the various processes necessary to print 1 million newspapers in an evening shift. In other plants, such as the Foster Farms dairy in Modesto, California, one large, powerful PLC system, with 222 I/O terminals, practically runs the whole operation. Either way, dedicated control or systems management, the PLC is the heart of today's manufacturing and process operations.

2–3 THE CENTRAL PROCESSING UNIT

In small PLCs, the processor, solid-state memory, I/O modules, and power supply are housed in a single compact unit. The programming device, usually a hand-held unit with a keypad and LCD display, is separate but tethered to the main unit with a cable. In larger PLCs, the processor and memory are in one unit, the power supply in a second unit, and the I/O interfaces (modules) in additional units. The programming device, which may be a personal computer (PC), is, of course, a separate, tethered item.

Regardless of PLC size (small, medium, or large), the processor and memory are always in the same unit. This unit is called the *central processing unit* (CPU). In larger PLCs, the CPU contains just the processor and memory. In small PLCs, the CPU also consists of the I/O interfaces and power supply. It is also possible for the CPU to contain the processor, memory, and power supply, with the I/O interfaces placed in external modules. Such a scheme is shown in figure 2–3. As we see, the fixed memory contains the program set by the manufacturer. This operating system program, which has the same function as a DOS program in your PC, is set into special IC chips called read-only memory (ROM). The fixed program in ROM cannot be altered or erased during the CPU's operation. The program in this nonvolatile memory is retained when power is removed from the CPU.

The alterable memory contains many sections, which are outlined later in the chapter. Its information is stored on IC chips that can be programmed, altered, and erased by the programmer/user. The alterable memory is stored mainly in random access memory (RAM) chips. Information can be written into or read from a RAM chip. RAM is often called read/write memory. The typical RAM chip will lose any information it has stored when input power is lost. It is therefore a volatile device; that is, its memory is erased when power is lost.

Note that as stated in chapter 1, there is battery backup power in most CPUs. If input power fails and the power supply can no longer deliver voltage to the system, power backup preserves any program that has been inserted into the CPU RAM.

As illustrated in figure 2–3, the processor section has computer flow connections to other subsections of the CPU and to outside devices. The processor is the controller that keeps information going from one place to another. It responds to programmed instructions stored in memory, causing output devices to be energized and deenergized in response to the on–off status of input devices. Before examining the processor in more detail, we first turn our attention to a more in-depth look at solid-state memory.

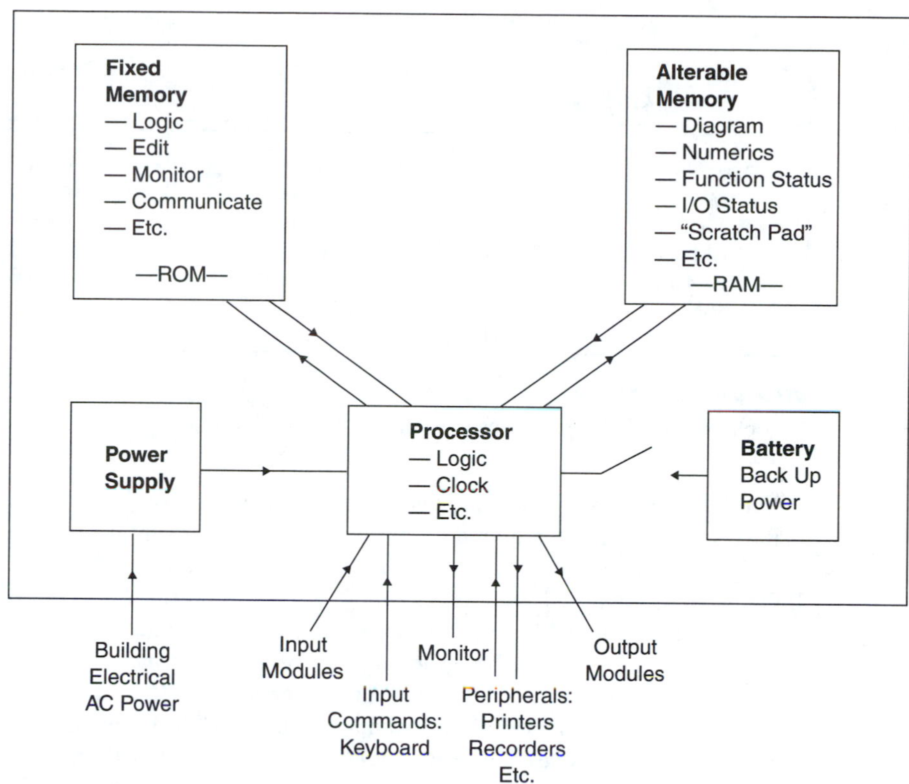

FIGURE 2–3
Operational Sections of a PLC CPU

2–4 SOLID-STATE MEMORY

In section 2–3 we discussed the use of ROM and RAM IC chips. The other major types of solid-state memory chips used in PLC CPUs are PROM, EPROM, EEPROM, and NOVRAM. Figure 2–4 shows a summary of the operational characteristics of these chips.

The *programmable read-only memory* (PROM) chip is similar to the ROM except that it may be programmed once, and once only, by the user/programmer. In other words, the manufacturer furnishes the chip in an unprogrammed or semiprogrammed state. The user then programs the chip to his or her requirements. No erasures are possible. To change the program in a programmed PROM, you throw it away and replace it with a new, unprogrammed PROM. The PROM is seldom used because it requires special programming circuits. It does, however, have the advantage of being an unalterable backup to a ROM.

The *erasable programmable read-only memory* (EPROM) is a PROM that can be erased. The EPROM is erased by subjecting a window in its top to ultraviolet (UV) light for a few minutes. Thus it is also called a UVPROM.

When exposed to UV light, the chip's memory bits are reset to 0. The chip's window is covered during normal use to prevent unwanted erasure. The advantage of the EPROM

CHIP	FIXED (F) OR ALTERABLE (A)	APPLICATION	ERASABLE BY
ROM	F	Fixed Operating Memory	No
RAM	A	User Program	No
PROM	F	User Program	No
EPROM	A	User Program	UV Light
EEPROM	A	User Program	Electrical Signals
NOVRAM	A	User Program	Electrical Signals

FIGURE 2–4
Major Types of IC Memory Chips Used in PLC CPUs

is that it can be reused. There are two major disadvantages of the EPROM, however. One is the downtime interval required for its reprogramming. Downtime includes removal time, UV light exposure time, and reinsertion time. Two, when the EPROM is exposed to UV light, all of its memory locations are erased. The EPROM must then be completely reprogrammed, even if only one or two memory slots required updating.

The *electrically erasable programmable read-only memory* (EEPROM) is similar to the EPROM. Instead of UV light exposure for erasure, though, an electrical signal is applied to the chip. The EEPROM's advantage over the EPROM is the ease and speed with which it is reset and erased. The EEPROM is used in place of RAM when you want fast erasure without using time for individual reprogramming of each part of the chip's memory. Today, the EEPROM is the memory of choice for storing, backing up, or transferring PLC programs.

The *nonvolatile random access memory* (NOVRAM) is a combination chip. It is a combination of an EEPROM and a RAM. When the power is about to go off, the contents of the RAM memory are quickly stored in the EEPROM. The stored data can then be read into the RAM memory when the power is again restored. The NOVRAM chip combines the flexibility of RAM memory with the nonvolatility of EEPROM.

Whether solid-state memory is volatile or nonvolatile, its chips are classified according to bit (or cell) size. A bit is a 0 or a 1 (low or high voltage) that occupies a given cell. Cells are arranged in slots, usually 8 or 16 bits wide. When bits are thus combined, they are referred to as *words*. An 8-bit word is called a byte. Two bytes are often arranged side by side to form a 16-bit word. In figure 2–5a we see the arrangement for a 1-kilobyte (1 KB) memory. It has 1K (actually 1024) slot locations, each 8 bits, or 1 byte, wide. Figure 2–5b shows a typical 2 KB memory. It has 1K slot locations, each 16 bits (or 2 bytes) wide. Today's PLCs contain anywhere from 1 to 256 KB of solid-state memory, most of which is RAM.

Obviously, the more processes you want to control, the more memory the PLC requires. The amount of memory you need is described in individual manufacturer's specification manuals. You need more memory for analog control than for discrete operation of a comparable process. As memory size increases, the cost of the CPU unit also increases. It

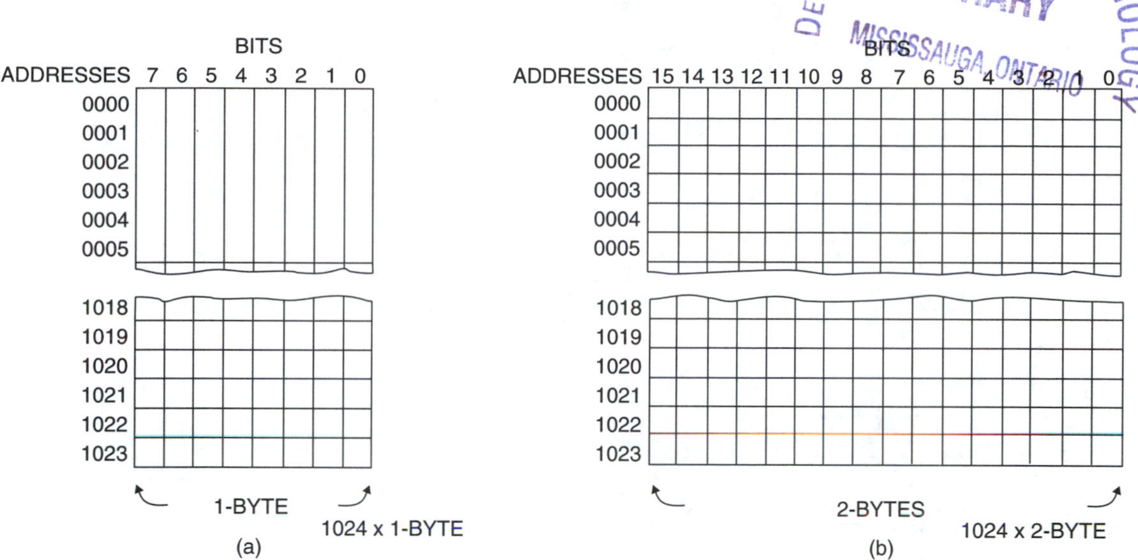

FIGURE 2–5
Memory Size

is possible to buy too much memory if your needs are not calculated properly. Yet with memory so inexpensive today, overbuying is not the issue it used to be.

When an application is matched to a PLC, the memory required depends on the number of inputs, the number of outputs, and the complexity of the control diagram. A most important feature of a PLC as these factors increase is expandability of memory. Some PLC models do not have memory expansion capabilities and have to be completely replaced if more memory for bigger tasks becomes necessary. However, many PLC models can have memory modules added to the existing CPU. Adding a new memory module or two is much less costly than replacing the entire PLC system. It is wise to consider memory expandability when purchasing a PLC.

How is solid-state memory organized within the PLC? To find out, we draw what is known as a memory map for a given PLC. Such a map might look like the one shown in figure 2–6. It can be divided into two broad categories: user memory and storage memory. The former contains the ladder logic program. The latter stores information needed to carry out the user program: the status of discrete input and output devices, the preset and accumulated values of counters and timers, numerical values, sequencer patterns, internal I/O relay equivalents, and so on. The user memory occupies the greater portion of total memory, often 75 percent or more. A PLC with 16 KB of memory will often devote 12 KB or more to the ladder logic program, leaving 4 KB or less of memory for data storage.

Taking a closer look at the memory map of figure 2–6, we see first that all addresses are given in octal. (For a review of the octal numbering system, see chapter 13.) Second, we note that each word in memory is 16 bits wide.

Concentrating on storage memory, we find the first section, consisting of eight words, is the *input image status* area (addresses 110_8–117_8). It is here that the status of discrete, real-world inputs is stored.

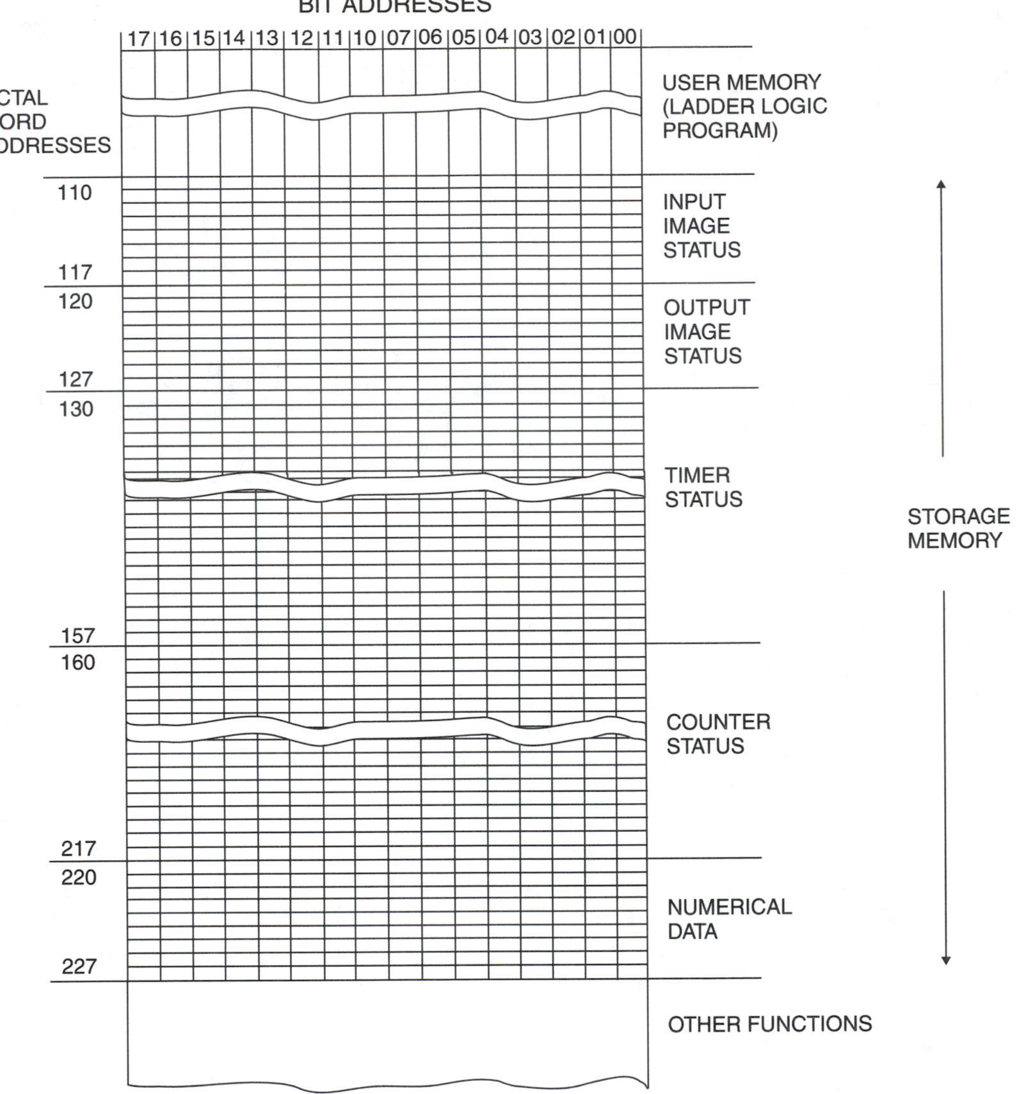

FIGURE 2–6
Memory Map

The second section, also consisting of eight words, is the *output image status* area (addresses 120_8–127_8). This is where the binary data (0's or 1's) that will activate real-world outputs are stored.

Timer status, accumulated values and preset values, is stored in the third section of storage memory, using a total of 24 words from address 130_8 to address 157_8.

Counter status, accumulated values and preset values, is stored in the fourth section, also using a total of 24 words. These words are located at addresses 160_8 to 217_8.

The fifth section, *numerical data*, is used for number system conversion. In this section, eight 16-bit words have been reserved, from address 220_8 to address 227_8.

Other functions can be continued at the bottom of the map as needed. The additional memory size required depends on the total number of different functions included in the CPU.

2–5 THE PROCESSOR

All computer processors are designed to carry out arithmetic and logic operations. Since the early 1970s, when Intel engineers were able to cram the complex circuitry necessary to do these functions onto a single chip, processors have been known as *microprocessors*. Such devices, which are the "brains" of every computer, have a unique characteristic—they are programmable, which means they are "told" what to do by a set of instructions, compiled to form a program. When the microprocessor is to carry out a different task, a new program is written and fed to it.

Microprocessors are classified as to how powerful they are. Two factors determine power: bit size and clock speed. There are 4-, 8-, 16-, and 32-bit microprocessors, which manipulate data 4, 8, 16, or 32 bits at a time, respectively. The larger the bit size, the more powerful the computer. Clock speed determines how quickly a microprocessor executes instructions. Clock speeds range from a low of 1 MHz to over 300 megahertz (MHz). The faster the clock speed, the more powerful the computer.

Intel, the inventor of the microprocessor, has continued to develop ever more powerful processors. As shown in figure 2–7, beginning with the 8085, bit size and average clock speed have kept on increasing. Today, a Pentium chip manipulates instructions and data in 32- and 64-bit chunks at an average clock speed of 300 MHz.

Although some high-end, large PLCs use the Pentium chip, most are content to operate with less powerful microprocessors. Some small PLCs function quite nicely with 8-bit microprocessors running at 4 MHz. The average small PLC, however, is a 16-bit machine operating at 33 MHz.

As we have seen, the microprocessor is the part of the PLC CPU that receives, analyzes, processes, and sends data. The data, in digital pulse form, is sent and received as shown in figure 2–8. Let's examine the figure more closely.

To begin with, a ROM with the fixed operating system program interfaces to the control section. This unalterable program manages the operation of the PLC. Whatever the logic scan (user) program is asking the PLC to do, the operating system program is there to do housekeeping chores for the PLC.

FIGURE 2–7
Microprocessors: Bit Size and Speed

Microprocessor	Bit Size	Clock Speed
8085	8-bit	1 MHz
8086	16-bit	4.77 MHz
80186	16-bit	8 MHz
80286	16-bit	12.5 MHz
80386	32-bit	33 MHz
80486	32-bit	50 MHz
Pentium	32-bit/64-bit	200 MHz

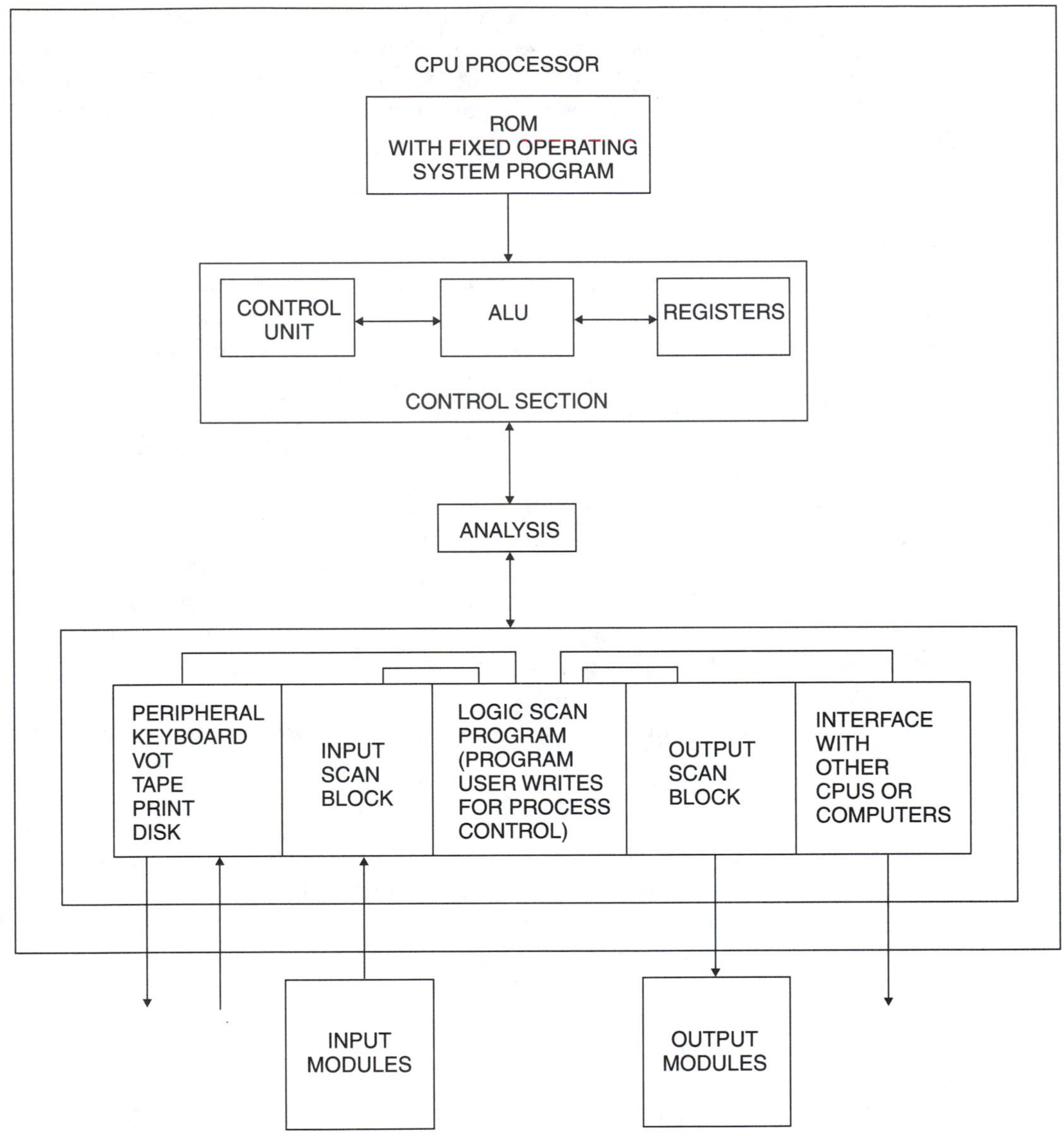

FIGURE 2–8
CPU Processor

The control section, the heart of the microprocessor, consists of a control unit with a clock, an arithmetic/logic unit (ALU), and a few internal temporary storage registers. The control section determines which operating sections are to be functional, in what order, and for how long.

The input scan block, when called upon to operate, scans the inputs and places the individual input statuses in RAM memory. After analysis, the logic scan (user ladder logic program) updates the output scan block to the appropriate state. Next, the outputs are scanned and updated. The output statuses are changed or left alone, depending on logic analysis. Output status depends on the output status signals of the CPU.

Other typical functions carried out by the microprocessor are also shown in the figure. The keyboard and video display peripherals take action based on any keyboard operation that occurs. The video display is then appropriately updated. Other peripherals, such as the tape drive, disk drive, or printer, may also become involved.

On the far right of figure 2–8 is an optional interfacing section. This section is required if the PLC is part of a larger system. This section carries out communication with other PLC CPUs and a master computer, if one is used.

2–6 I/O MODULES (INTERFACES)

The input module performs four tasks electronically. First, it senses the presence or absence of an input signal at each of its input terminals. The input signal tells what switch, sensor, or other signal is on or off in the process being controlled. Second, it converts the input signal for on, or high, to a DC level usable by the module's electronic circuit. For a low, or off, input signal, no signal is converted, indicating off. Third, the input module carries out electronic isolation by electronically isolating the input module output from its input. Finally, its electronic circuit must produce an output to be sensed by the PLC CPU. All these functions are illustrated by the module layout in figure 2–9.

A typical input module has 4, 6, 8, 12, 16, or 32 terminals, plus common and safety ground terminals. The figure shows the circuit for only one terminal. All terminals in a given module have identical circuits. The first block receives the input signal from the switch, sensor, and so on. For AC voltage inputs, the DC converter consists of rectifiers and

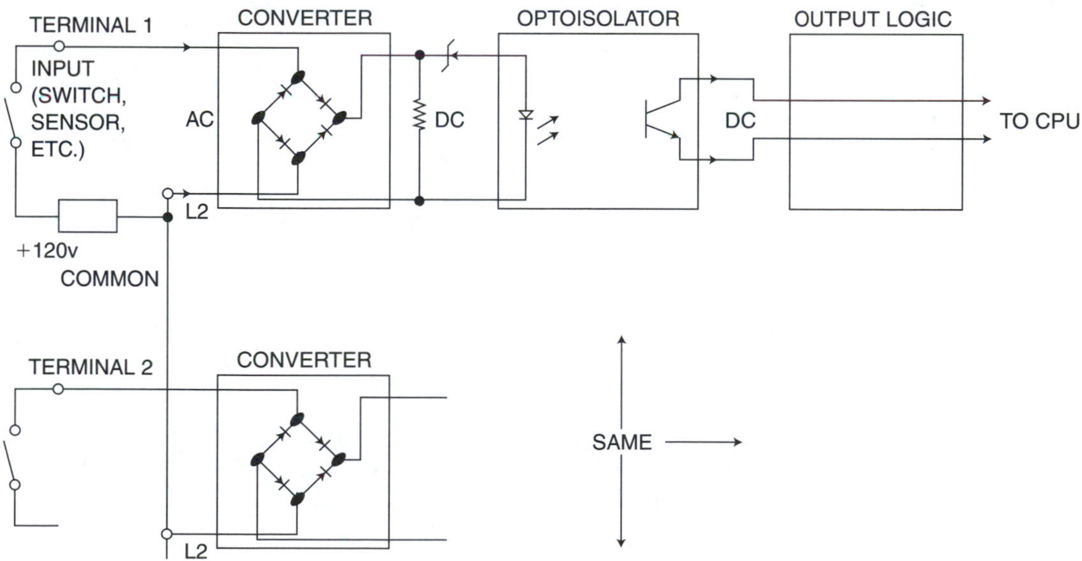

FIGURE 2–9
PLC Input Module Layout

a means to step the voltage down to a usable level, usually with a zener diode. For input DC voltages, some type of DC-to-DC conversion within the converter block is required.

The output of the converter is not directly connected to the CPU. If it were, an input surge or circuit malfunction could reach the CPU. For example, if a rectifier in the converter should open or short out, you could have 120 volts AC fed to the CPU. Because most CPUs work on only 5 volts DC, they would be damaged. The isolation block protects the CPU from this type of damage.

The isolation is usually accomplished by an optoisolator, as shown. The on-off signal is carried on a light beam (produced by an LED) in one direction. Electrical surges will not pass through the optoisolator in either direction.

When its input is on, the isolator sends a signal to the CPU via the output logic block. When the isolator's output is on, it is sensed by a coded signal from the CPU. Each module is assigned a coded series of numbers by its SIP or DIP switch settings. Each terminal number of the module is assigned a number in consecutive order. The on-off status for each number is checked on each sweep of the input scan. The result, on or off, is placed in RAM memory, as previously discussed.

The output module operates in the opposite manner from the input module. A DC signal from the CPU is converted through each module section (terminal) to a usable output voltage, either AC or DC. A block diagram of the output module is shown in figure 2–10.

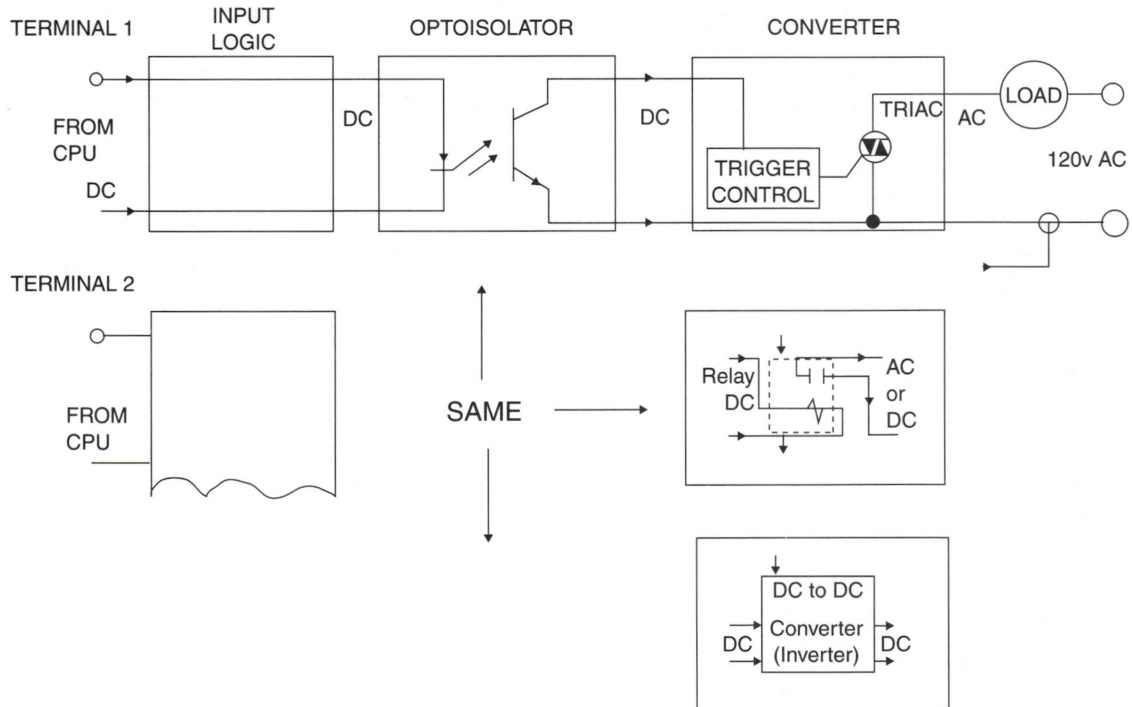

FIGURE 2–10
PLC Output Module Layout

A signal from the CPU is received by the output module logic, once for each scan. If the CPU signal code matches the assigned number of the module, the module section is turned on. The identification numbers of the module are again determined by the setting of the module SIP switches. As with input modules, there are 4, 6, 8, 12, 16, and even 32 terminals or sections. If no matching signal is received by a terminal during the output scan, the module terminal is not energized.

The matching CPU signal, if received, goes through an isolation stage. Again, isolation is necessary so that any erratic voltage surge from the output device does not get back into the CPU and cause damage. The isolator output is then transmitted to switching circuitry or an output relay. AC switching is usually accomplished by turning on a triac. The output of a module section may be through a relay, or a DC or AC output. All three types are shown in figure 2–10.

All terminals of a single module have the same output system. In other words, an 8-terminal module would not have some AC and some DC outputs or voltages of differing values. All would be the same.

An actual block diagram for a small PLC input/output module (interface) is shown in figure 2–11. At the input end is an internal DC power supply (24 volts) that supplies voltage for up to eight input switches or sensors (terminals 0 to 7). When a switch or sensor is

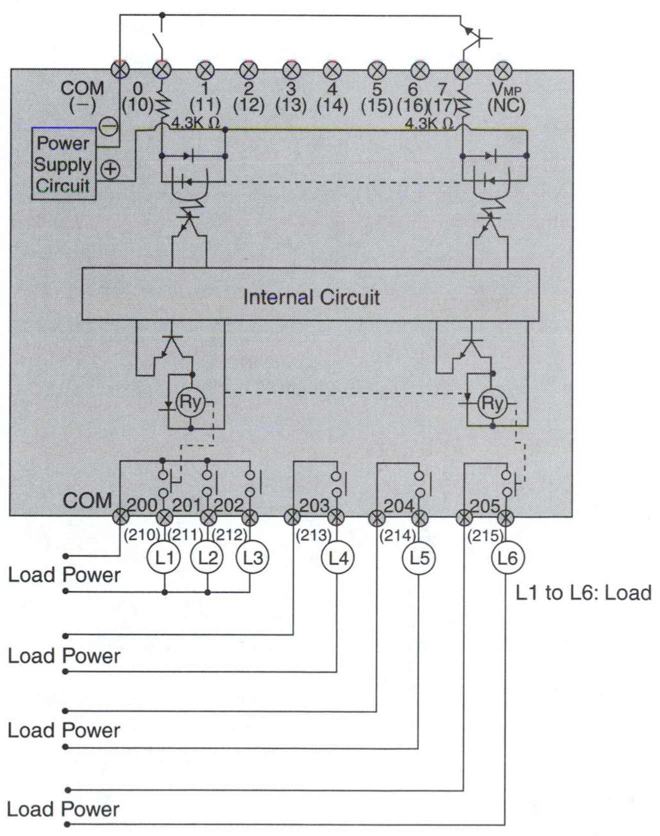

FIGURE 2–11
PLC I/O Module (Courtesy of IDEL Corporation)

closed, a current path is completed through the LED of an optoisolator, a phototransistor conducts, and as a result, a signal is received by the internal circuitry.

At the output end, six relays open or close respective contacts when "told" to do so by the internal circuitry. Output terminals 200 to 202 share a common return to the load power. Terminals 203, 204, and 205 are completely independent of each other. Each has its own connection to load power.

2–7 POWER SUPPLIES

The power available in most plants is 120 volts alternating current (AC) at 60 Hz (cycles per second). Most PLCs operate on +5 and −5 volts DC. Therefore, the PLC CPU must contain circuitry to convert the 120-volt AC input to the required 5-volt DC values. The conversion is accomplished by a built-in voltage-converting power supply. Figure 2–12 includes the makeup of a typical power supply in block diagram form. The figure also shows voltage waveforms versus time at various points in the power supply.

Four parts are shown in the diagram, plus a switching system for the battery backup system. The first block on the left, the AC conditioning block, could be included in the PLC CPU. More often it is a separate external unit that is sized according to the CPU current rating. The AC conditioner purifies the AC waveform. The input waveform is normally a perfect sine wave, but it can be distorted at times by two external factors. First, the power company's generating system sine wave might be distorted during system switching or by

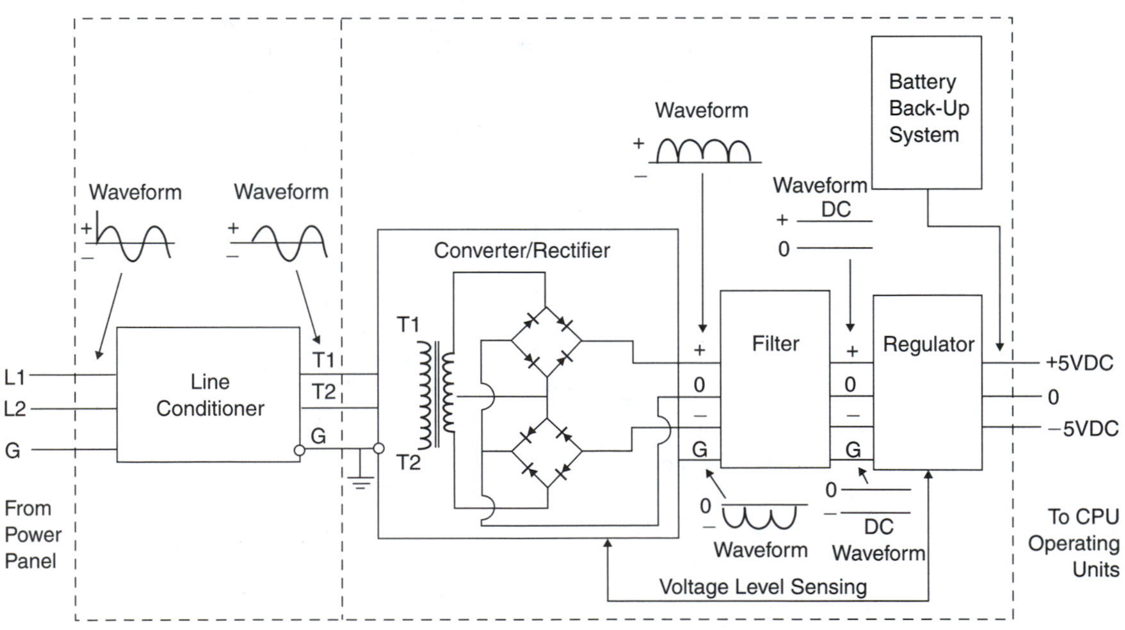

FIGURE 2–12
PLC CPU Power Supply

generation problems; second, equipment in your plant may cause electrical back surges that affect the purity of the electrical sine wave.

The second block in the diagram is the converter/rectifier. It changes the bidirectional AC to a pulsating, unidirectional DC waveform. Internally, a transformer steps the voltage down to an appropriate level. Then bridge rectifiers produce pulsating DC (PDC) outputs. One output is +5 volts; the other is −5 volts. This dual voltage is required to operate many of the IC chips in the CPU.

A computer needs a constant (not pulsating) input DC voltage for correct operation. Therefore, a means of smoothing out the PDC is required. The third block in the diagram is the filter section, which accomplishes the required smoothing. The filter consists of internal circuitry, including capacitors and resistors or inductors. Alternatively, the filtering may be accomplished electronically by this filtering block. A fourth block shown in the diagram, the regulator, is always included. A regulator keeps the voltages at or near the 5-volt levels regardless of load (CPU) demands.

The battery backup switch is shown on the upper right of the diagram. The switch (not shown) can transfer the output from power supply to battery. The switch is set to switch the output from power supply to battery backup power quickly and automatically if the input power supply ceases. Normal power supply voltage ceases if the CPU plug is disconnected from its socket. It also ceases when building power fails. Continuity of power voltage keeps the user program from being lost, as previously discussed. Note that there is some circuitry included in the CPU to convert battery DC (for example, 24 volts) to the two 5-volt, DC-required levels. The conversion system is not shown but is given in the operating manual.

EXERCISES

1. Describe how the following sequence is carried out by a PLC CPU:
   a. A program on a disk is placed in the program memory. The program recalled from the disk includes a line where switch 34 causes output 54 to go on.
   b. Switch 34, which is connected to input 34, is turned on.
   c. The CPU recognizes that the switch is on.
   d. The CPU logic turns internal output 54 on.
   e. The 54 output status is conveyed through the output module, terminal 54, to an external light.
   f. Next, two alterations are made in the ladder program.
   g. The entire logic program is then recorded on a printer.
2. Obtain one or two manufacturers' manuals. Determine the following from the manual or manuals:
   a. Size of memory.
   b. Memory map. Which sections of memory are used for what?
   c. Input module system and output module system of operation.
3. List five microprocessors used in PLC CPUs. Which is the least powerful? Which is the most powerful? Why?
4. Describe the feasibility of the following chips: EPROM, EEPROM, NOVRAM.

3

General PLC Programming Procedures

OUTLINE

3–1 Introduction □ **3–2** Programming Equipment □ **3–3** Programming Formats □ **3–4** Proper Construction of PLC Ladder Diagrams □ **3–5** Process Scanning Considerations □ **3–6** PLC Operational Faults

OBJECTIVES

At the end of this chapter, you will be able to

- □ Describe a typical PLC keyboard layout and its operational procedures.
- □ Describe a typical PLC display for hand-held and full-size units.
- □ Describe the difference between legal (proper) and illegal (improper) PLC ladder programming layouts.
- □ List the important considerations of program scanning rate and sequence, and their effects on system operation.
- □ Describe what action to take when a PLC operational fault occurs.

3–1 INTRODUCTION

We have seen what a PLC is, examined its system hardware, and peeked inside to explore the CPU, I/O interfaces, and power supply. It is now time to investigate general programming procedures.

We begin by surveying programming equipment: programmer/monitors (PMs) and PLC software for the personal computer. Next we look at programming formats and study a typical data entry sequence, then proceed to the proper construction of PLC ladder diagrams with examples of construction limitations. A discussion of process scanning in general, and specific considerations in particular, follows. Finally, we review how the PLC examines operational faults: how it tells us what's wrong through error messages and panel-mounted LEDs. When you have completed this chapter you will be ready to begin specific programming procedures, starting with chapter 5.

3–2 PROGRAMMING EQUIPMENT

PLC programming equipment exists to allow you to write, edit, and monitor a program, as well as perform various diagnostic procedures. In most cases the programming device, the PM, must be connected to the CPU while programs are written. Other PMs, however, allow you to program offline and then download the program to the PLC CPU. The programs are usually written in ladder logic, although alternative programming languages are available (see chapter 25).

Three types of PMs, also referred to as program loaders, are in common use. At the low end are the hand-held, palm-size units with dual-function keypads and a liquid-crystal display (LCD) or light-emitting diode (LED) window. At a more user-friendly level are the full-size keyboards, accompanied by a large LCD display or cathode-ray tube (CRT) screen. A third programming option exists with software that allows programs to be developed on IBM-compatible PCs. Let's take a moment to explore each type of PM in a bit more detail.

Hand-held units have come a long way in recent years. With the Shift key function, which works like a second function key on calculators, a relatively full keypad is available. Device symbols, function indicators, numeric keys, program editing and entry buttons, and cursor movement keys are right at your fingertip. A typical keypad is color-coded on the basis of function and uses membrane keys that provide an audio feedback.

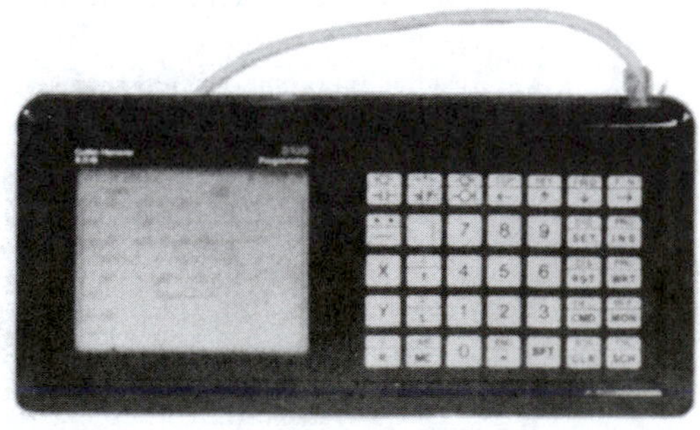

FIGURE 3–1
Hand-Held Programmer (Courtesy of Cutler-Hammer)

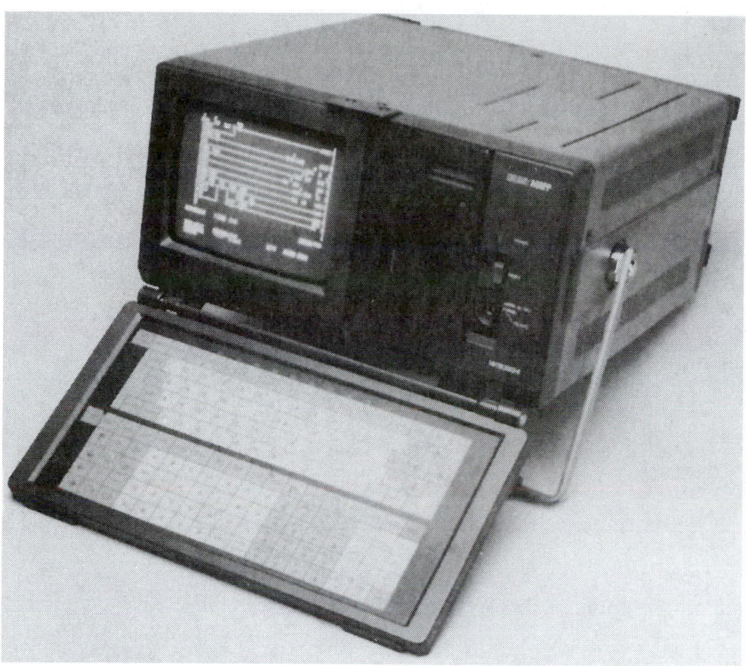

FIGURE 3–2
Full-Size Programmer/Monitor (Courtesy of Mitsubishi Electronics)

The display for these palm-sized units has also been expanded and improved upon. The LCD shown in figure 3–1 is capable of indicating eight rungs of a ladder diagram at one time—with each rung containing up to nine elements (contacts) and a coil function. In addition, written messages, in full alphanumerics, appear on a message line. Furthermore, when the unit is placed in the monitor mode, the operation of devices can be observed, not only on the message line, but on the ladder diagram as well. For example, when a timer program is run, the message line will count down the time, while various elements appear shaded as power is being passed. Their small size aside, some of today's hand-held displays are almost as "revealing" as their big brother monitors.

Full-size PMs give you a complete keyboard and a large monitor, the latter either LCD-, plasma-, or CRT-based (figure 3–2). The keyboard usually contains all the ASCII symbols (typical computer keyboard) plus a host of function-keys dedicated to PLC programming. No Shift key is required to bring up a second function as is the case with the smaller hand-held units. Because of its larger size, the monitor display can present a considerable amount of information at one time. A typical full-size monitor display is shown in figure 3–3.

In addition to the use of dedicated hand-held or full-size programmers, powerful software for programming PLCs is available to run on IBM-compatible machines. Once actual

```
 SCR 1     X000  X107  R000  R177  L000  L177  T000  T077  C000  Y000
 PALET    |-] [-+-]/[-+-] [-+-]/[-+-] [-+-]/[-+-] [-+-]/[-+-] [-+-(  )+
  #1      |            |        |
 CNTRL    | C077  S000 |        |S377  Y000                        T020  |
          |-]/[-+-] [-+         +-] [-+-]/[-+-----+-----+-----+-( T )-|
 DEV      |     |                   |                              0285
 FEED     |     |X001  R001  L001   |T001                          R177  |
 LIMIT    |     +-]/[-+-] [-+-]/[-+-] [-+-----+-----+-----+-(   )-|
 SWITCH   |            |        |           |
          | Y057  C001 |        |S001       |
 [  82]   |-] [-+-]/[-+         +-] [-+-----+
FUNC. KEYS|            |
 F1  F2   |            |X106  R176  X002                         C035  |
 WRT INS  |            +-] [-+-]/[-+-] [-+-----+-----+-----+-( C )-|
 F3  F4   |                                                       1234
 DEL SCH  | L176  T076  C076  S250  Y021  X062  R057  L046         C035  |
 F5  F6   |-] [-+-]/[-+-]/[-+-]/[-+-] [-+-]/[-+-] [-+-]/[-+-----+-(RC)-|
 CLR D-CMT|     |                               |     |              1234
 F7  F8   |     |T051  C030  S322               |     |
 GO  S-CMT|     +-] [-+-]/[-+-] [-+-----+       +     +
 F9  F10  |            |                              |
 More P-CMT|           |Y012  X043              R007  |              L115  |
          |            +-] [-+-]/[-+-----+-----+-] [-+-----+-----+-(   )-|
[?] FOR HELP
```

FIGURE 3–3
Typical Full-Size Monitor Display (Courtesy of Cutler-Hammer)

programming is complete, the program is downloaded to the PLC. All such PLC software programs are now menu driven (figure 3-4). Throughout the program, you may step from menu to menu by entering the indicated selection number. When further information is required, the program requests an entry from you. Such an entry can be made by pressing the appropriate key or, in some programs, by using a mouse.

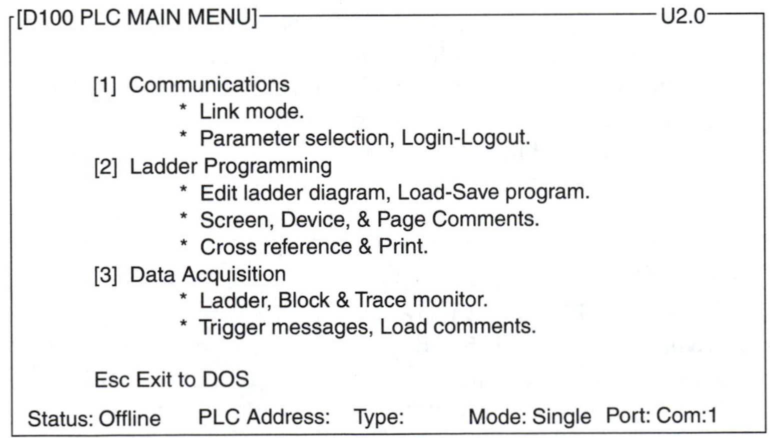

FIGURE 3–4
Main Menu Screen (Courtesy of Cutler-Hammer)

3–3 PROGRAMMING FORMATS

In certain chapters throughout the book we show different manufacturers' format approaches to controlling processes. We use a general format like those of companies having a major share of the PLC market at present. Experience has shown that when a person learns to program one type of PLC, he or she can easily master other PLC systems, even though the formats differ somewhat.

Some of the factors that vary between formats are nomenclature, numbering schemes, and screen appearance. Nomenclature descriptions are covered in examples in individual chapters. Another format variation is in the numbering formats for contacts, outputs, and registers. These formats include letters, numbers, or a combination of both. Individual PLC operating manuals explain the various systems of designating functions and registers.

A typical hand-held keypad sequence for a three-wire holding circuit is shown in figure 3-5a. In the circuit (see figure 3-5b), output Y0 can be turned on and off through the operation of the two inputs X0 and X1. X0 and X1 are the two NO (normally open) pushbuttons connected to the controller input. The sequence is as follows:

Clear RAM Memory

1. Turn PLC on.
2. Clear RAM memory.
3. Clear the screen. Programming can now begin.

Program First Screen (This program uses only one screen.)

4. Press contact device symbol (normally open).
5. Press function; X for input.
6. Assign contact number (0) by pressing numerical keys.
7. Press WRT to enter contact.
8. Press contact device symbol (normally closed).
9. Press function; X for input.
10. Assign contact number (1) by pressing numerical keys.
11. Press WRT to enter contact.
12. Press coil device symbol.
13. Press function; Y for output.
14. Assign coil number (0) by pressing numerical keys.
15. Press WRT to enter coil.
16. Return to left of display, one line down.
17. Press contact device symbol (normally open).
18. Press vertical connection symbol key (1) used to tie a device to line above it on the ladder diagram.
19. Press function; Y for output.
20. Assign contact number (0) by pressing numerical keys.
21. Press WRT to enter contact.

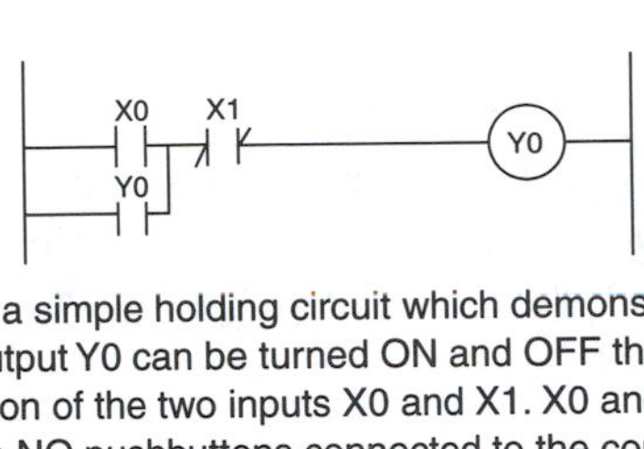

This is a simple holding circuit which demonstrates how output Y0 can be turned ON and OFF through the operation of the two inputs X0 and X1. X0 and X1 are the two NO pushbuttons connected to the controller input. The following figure shows the programmer keystrokes and the resultant screen display.

Clear Ram Memory

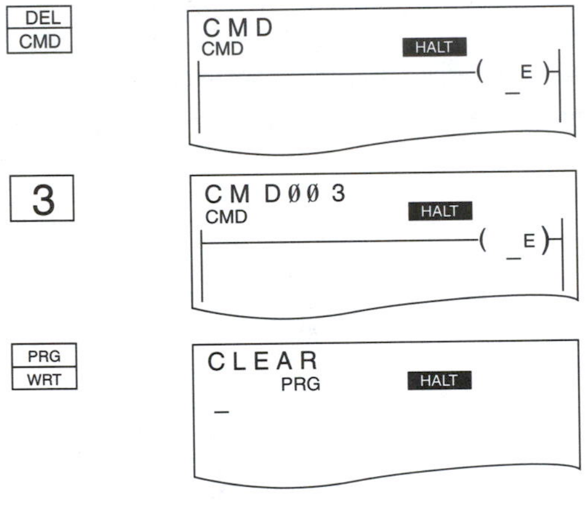

Program First Screen

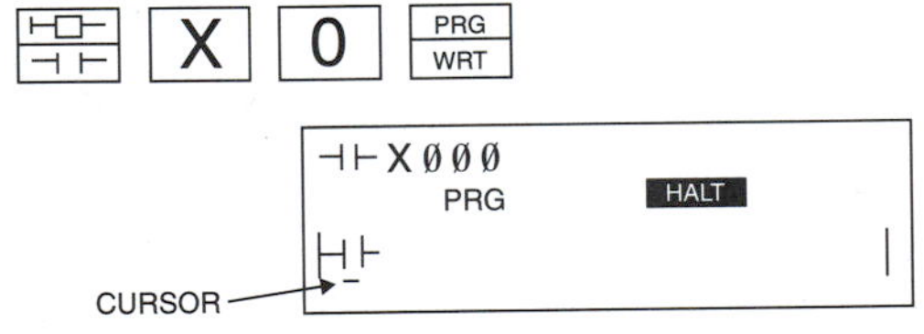

X Ø Ø Ø
PRG
HALT
CURSOR

(a)

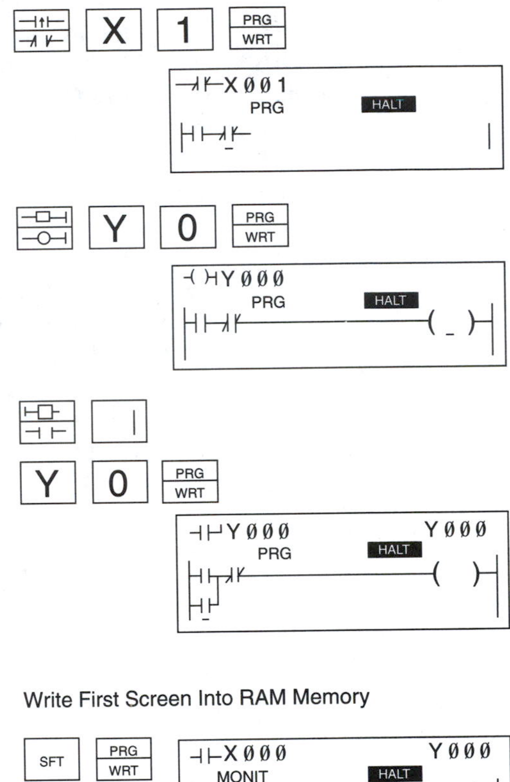

Write First Screen Into RAM Memory

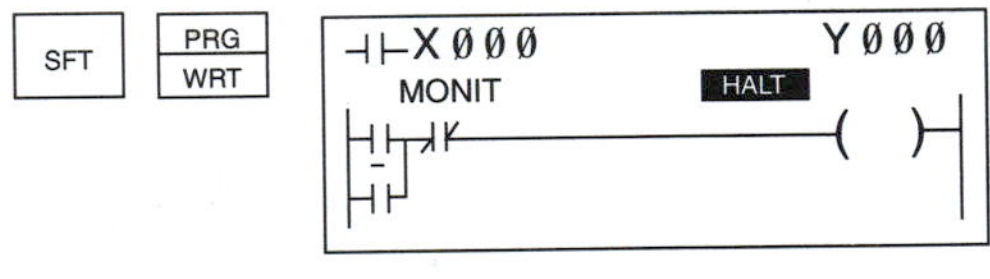

Halt/Run

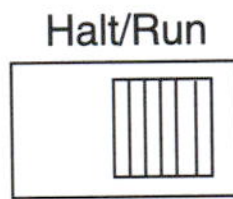

FIGURE 3–5
Keypad Sequence (Courtesy of Cutler-Hammer)

Write First Screen into RAM Memory

22. Write first screen (program) into RAM by pressing SFT (shift) and PRG (program).

The PLC is now switched from halt to run mode. When the input X0 pushbutton is pressed, the Y0 output will energize and remain energized after the button is released. This is because the Y0 contact (II) is closed when the Y0 coil (c) is energized. The Y0 contact is said to latch the Y0 coil on. Pressing input X1 pushbutton causes Y0 to deenergize.

3–4 PROPER CONSTRUCTION OF PLC LADDER DIAGRAMS

A PLC programming format's limitations must be observed when programming a PLC ladder diagram. Otherwise, the PLC CPU will not accept the screen-programmed ladder diagram into its memory. In some cases, when incorrectly formatted ladder diagrams are not received, an error message appears on the screen showing that the program was not entered and why. Why might the ladder diagrams be incorrect for a PLC? Because various ladder construction limitations were probably not observed. Here are examples of such limitations for a typical PLC:

1. A contact must always be inserted in slot 1 in the upper left (see figures 3–6 through 3–10).

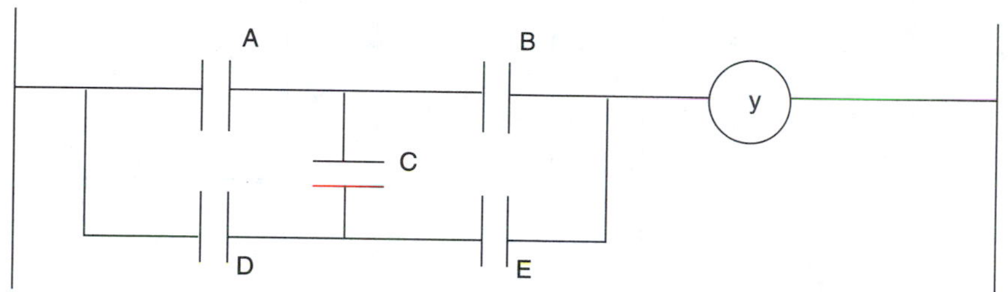

(a)

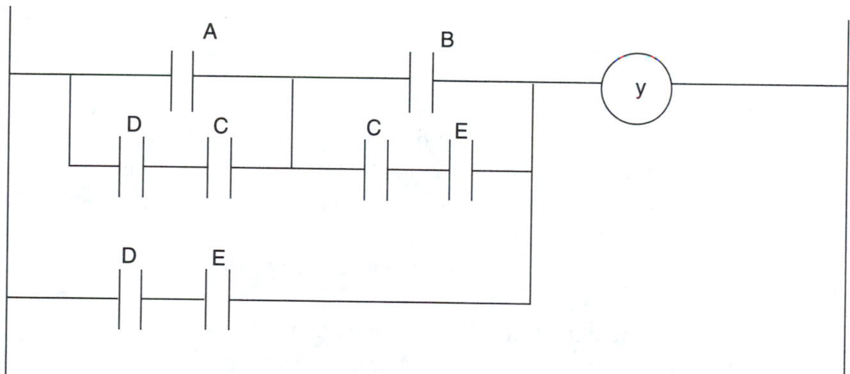

(b)

FIGURE 3–6
Proper PLC Ladder Diagrams

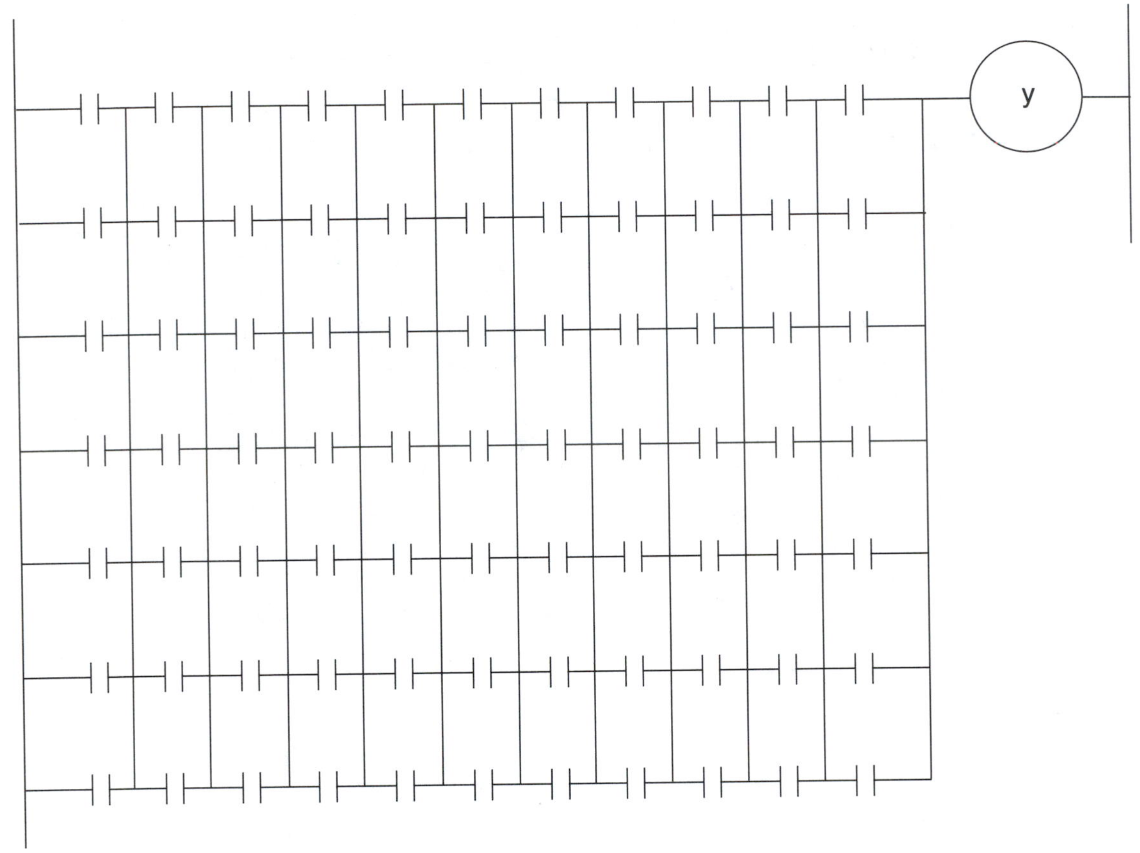

FIGURE 3–7
Contact Matrix

2. A coil must be inserted at the end of a rung (see figure 3-6b).
3. All contacts must run horizontally. No vertically oriented contacts are allowed. In Figure 3–6a, contact C is programmed incorrectly. The ladder diagram in figure 3–6b represents one solution to the problem.
4. The number of contacts per matrix (network) is limited—for example, 11 across by 7 down (see figure 3–7).
5. Only one output may be connected to a group of contacts (see figure 3–7).
6. Contacts must be "nested" (a branch circuit programmed within a branch circuit) properly or, in some PLCs, not at all. Figure 3–8 shows one manufacturer's required format.
7. Flow must be from left to right (see figure 3–9).
8. Contact progression should be straight across (see figure 3–10).

Again, the individual operational manuals contain information on the proper programming of a given PLC system.

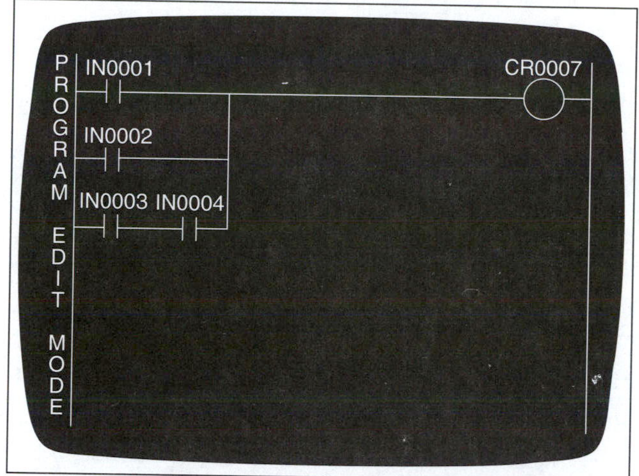

INCORRECT

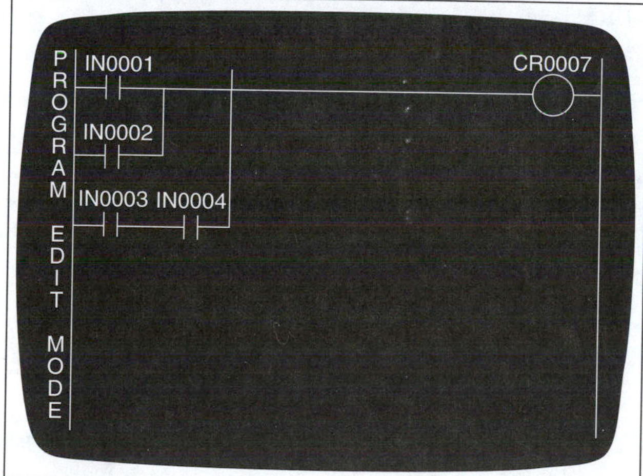

FIGURE 3–8
Proper Diagram Nesting Required Orientation

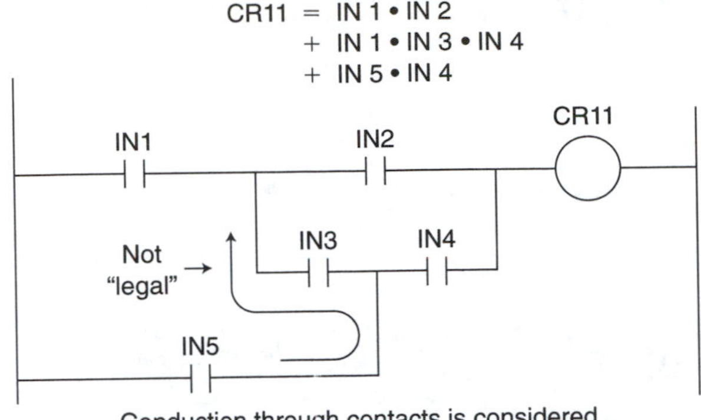

$$CR11 = IN1 \bullet IN2 + IN1 \bullet IN3 \bullet IN4 + IN5 \bullet IN4$$

CR11
IN1
IN2
IN3
IN4
Not "legal" →
IN5
Conduction through contacts is considered to occur from left to right...
Will not let current flow to left through 3.
CR11
IN1
IN2
IN5
IN3
IN3
IN4
IN5
Note: This type of "nested branching" is illegal in some systems, which require 2 to 3 ladder lines feeding the same coil repeated.
The addition of 2 contacts (IN5 and IN3 on the left) *adds* the path IN5, IN3, IN2.

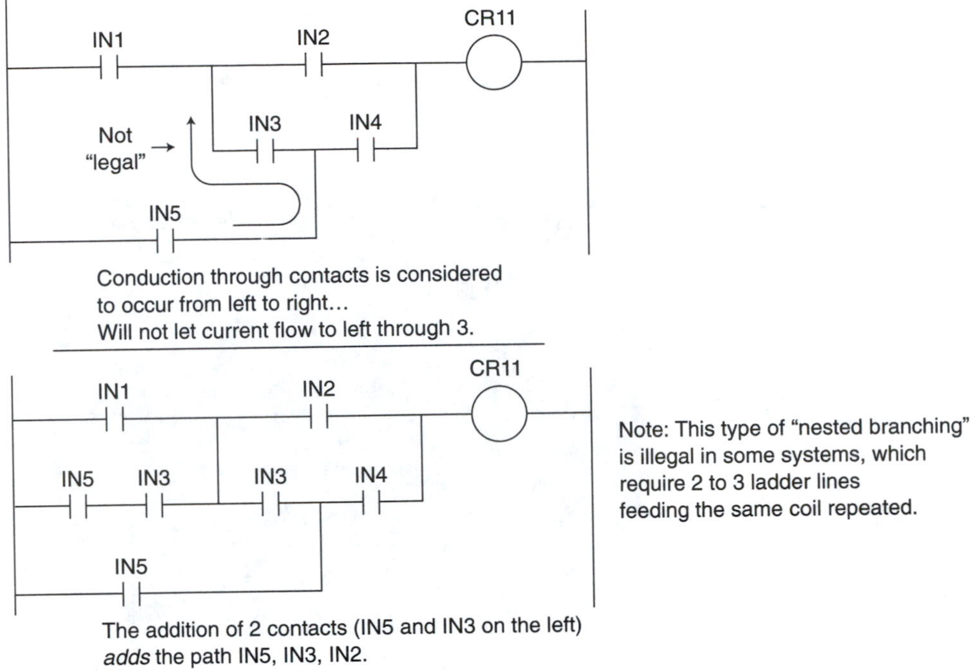

FIGURE 3–9
Proper Diagram Flow Orientation

FIGURE 3–10
Proper "Straight Across" Orientation for Contact Insertion

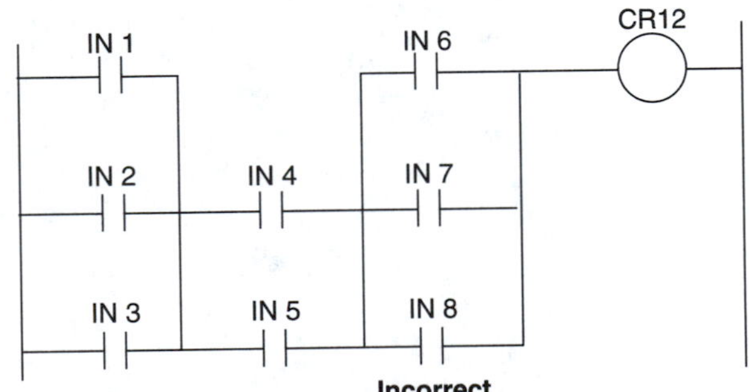

IN 1
IN 6
CR12
IN 2
IN 4
IN 7
IN 3
IN 5
IN 8
Incorrect
Use the topmost available junctions
IN 1
IN 4
IN 6
CR12
IN 2
IN 5
IN 7
IN 3
IN 8
Correct

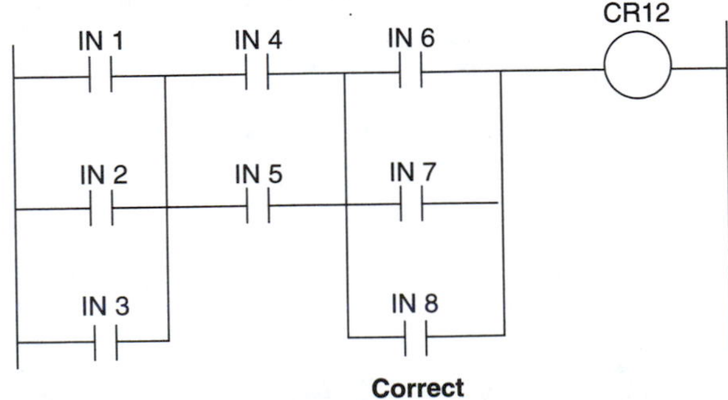

3–5 PROCESS SCANNING CONSIDERATIONS

PLCs function by scanning their operational programs. Each PLC operational cycle is made up of three separate parts: (1) input scan, (2) program scan, and (3) output scan (see figure 3–11a). The total time for one complete program scan is a function of processor speed and length of user program. With a high-speed processor and a short program hundreds of complete scans can take place in 1 second.

During the *input scan*, input terminals are read and the input status table is updated accordingly (see figure 3–11b).

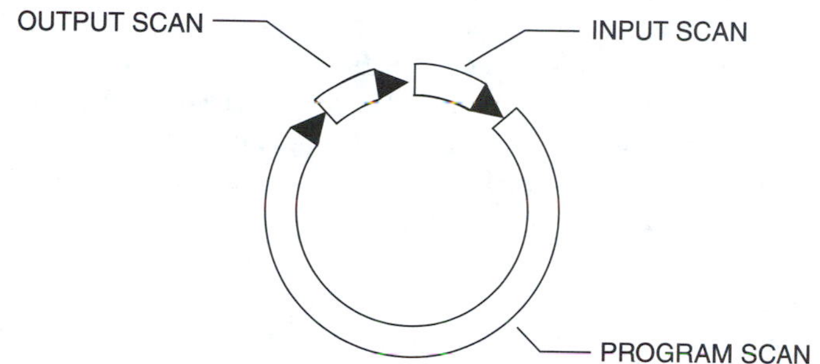

(a)

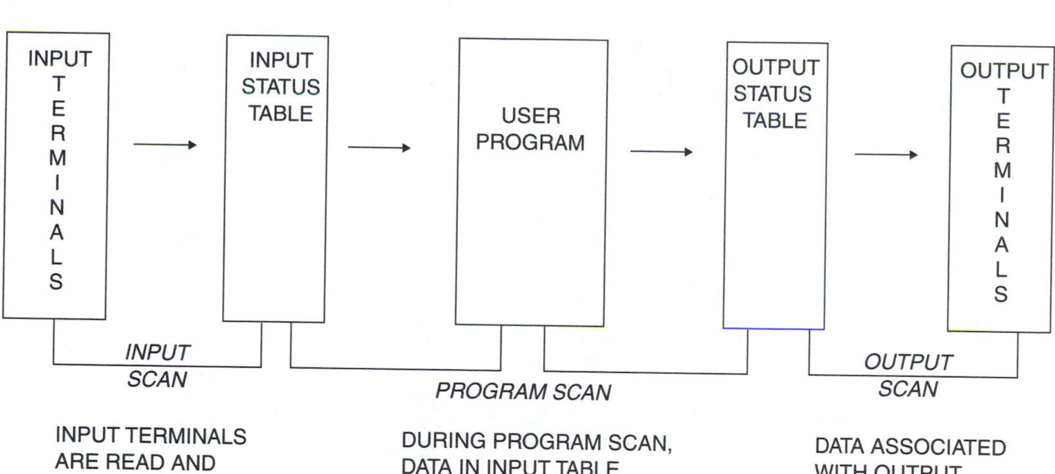

(b)

FIGURE 3–11
PLC Scanning

During the *program scan*, data in the input status table is applied to the user program, the program is executed (instructions carried out in sequence), and the output status table is updated appropriately.

During the *output scan*, data associated with the output status table is transferred to output terminals.

It is important to understand that the input, program, and output scans are separate, independent functions. Hence, any changes in the status of input devices during the program or output scan are not recognized until the next input scan. Furthermore, data changes in the output table are not transferred to the output terminal during the input and program scans. The transfer affecting the output devices takes place only during the output scan.

With all PLCs, there are special processing considerations to note. First, as we've indicated, all PLCs take a specific amount of time to scan their operational programs completely. Typically, the program scanning takes place left to right across each rung and from the top to bottom rungs, in order. Usually, the complete ladder scan time is a few milliseconds. Early computers took as long as a few seconds to make a complete scan. Although the present-day microprocessor-based PLC scans much faster, its speed must often be considered. For example, we might have a critical safety point in the diagram that must be monitored twice per millisecond. Suppose that scan time is 5 milliseconds. The critical safety point is, therefore, only checked out once every 5 milliseconds, not the required once every 0.5 millisecond. There are advanced techniques to handle the programming problem.

Another scanning consideration involves proper operational sequencing of events. An output might not go on immediately in sequence as it would in a relay logic system. In a relay logic system, an event occurring anywhere in the ladder control system results in immediate action. In a PLC ladder control diagram, however, no effect takes place until the rung is scanned. In most cases, the PLC logic delay effect is inconsequential. However, in

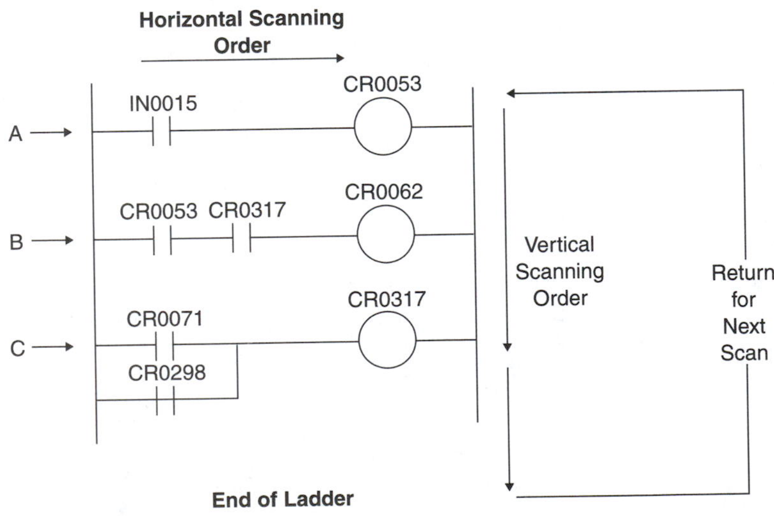

FIGURE 3–12
PLC Scanning Sequence Example

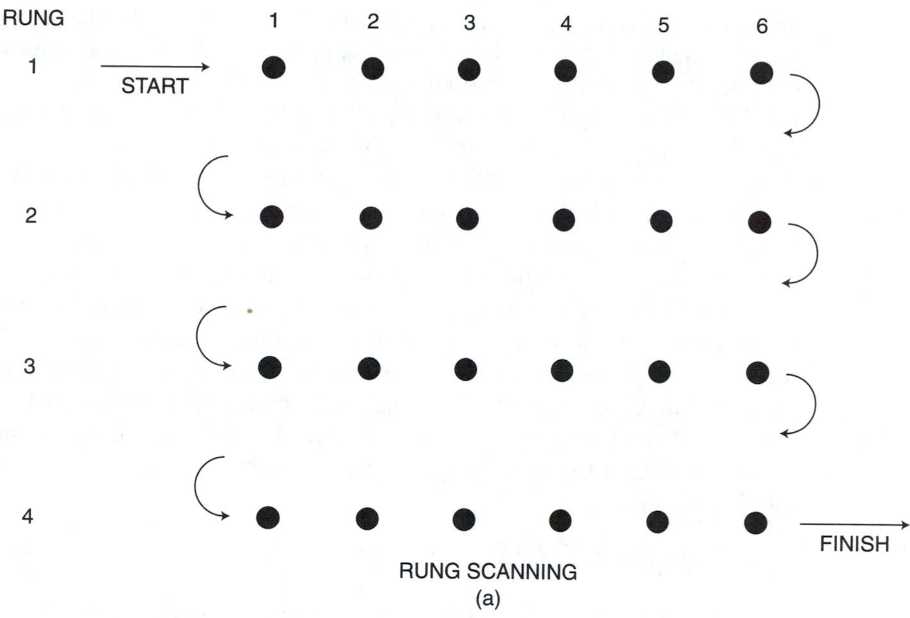

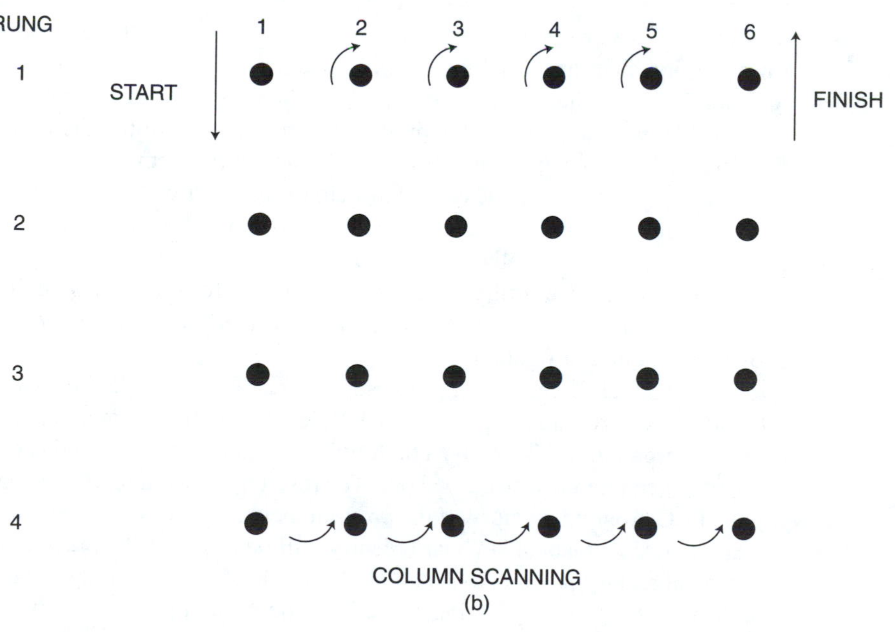

FIGURE 3–13
Rung and Column Scanning

fast-acting, interlocked, or rapidly sequenced PLC programs, the elapsed time required for scanning must be considered. For example, in figure 3–12 we see that the on–off status of the output on line B is identified as CR0062. CR0062 is controlled by two contacts, CR0053 and assigned input number CR0317. The input switch connected to IN0015 is closed. CR0053 on line A is turned on by the contact IN0015. Then, the CR0053 contact closes on line B. If output CR0317 is energized by one of its two contacts just after we go past B, output CR0053 will not go on immediately. The CR0317 contact on line B will therefore not close until we go to B on the next scan.

As mentioned a moment ago, program scanning typically takes place left to right across each rung and from the top to bottom rungs, in order. Known as *rung scanning*, it is the method used, for example, by Allen-Bradley and illustrated in figure 3–13a. Another method, used by Modicon/Schneider Automation known as *column scanning*, is shown in figure 3–13b. Here the processor "looks" at the first contact at the top left corner and reads the first column from top to bottom. It next reads the second column from top to bottom, and so on. Either method, rung or column, is appropriate.

3–6 PLC OPERATIONAL FAULTS

Every PLC has error codes for identifying incorrect programming and misoperation. The codes appear on the monitor, usually in code form in small systems or in user-friendly language in larger systems, when something is incorrect.

For example, in one small system, the error code is displayed using one or two digits in hexadecimal (0–F). Each digit of the error code indicates a different set of conditions requiring attention, as shown in the charts of figure 3–14. To illustrate, an error code might read out "24." As a result, two conditions require attention: "Program Sum Check Error" from the first chart and "Memory Pack Replacement" from the second chart. An "80" error code indicates that a "Programming Error" exists (8 is the error code display digit on the left; 0, the error code display digit on the right). You would then refer to your operations manual for an explanation of how to locate the problem.

In many systems, in the case of system misconnection, or poor connections, you will get a message such as "communication error." For other problems, different messages will appear on the screen, usually at the bottom.

Most PLCs have further diagnostic aids in the form of LEDs on the controller front panel. One manufacturer, for instance, uses five LEDs to indicate the various conditions shown in the chart of figure 3–15.

The fault LED, indicating processor failure, is of particular interest. Such a fault usually indicates a programming "hang-up." Reference to the operating manual is required for correct interpretation. Typically, a fault light going on at the CPU indicates that a memory-clearing procedure must be carried out. The resetting procedure involves completely clearing the PLC program memory. If the program being used has not been previously recorded on tape or disk, a manual keyboard reentry will be required. To prevent the time-consuming manual reentry process, it is a good idea to have each operating program saved in case a fault occurs. After the clearing procedure, the saved program can be quickly reentered into the CPU.

Error Code Display: Digit on the Left	Error Codes			
	TIM/CNT Present Value CRC Error	Program Sum Check Error	Keeping Data Sum Check Error	Programming Error
No display (No error)				
1	●			
2		●		
3	●	●		
4			●	
5	●		●	
6		●	●	
7	●	●	●	
8				●
9	●			●
A		●		●
B	●	●		●
C			●	●
D	●		●	●
E		●	●	●
F	●	●	●	●
ERROR LED (ON or OFF)	OFF	ON	OFF	ON

Error Code Display: Digit on the Right	Error Items			
	Power Failure or Memory Pack Removal	WDT Error	Memory Pack Replacement	User Memory CRC Error
No error 0				
1	●			
2		●		
3	●	●		
4			●	
5	●		●	
6		●	●	
7	●	●	●	
8				●
9	●			●
A		●		●
B	●	●		●
C			●	●
D	●		●	●
E		●	●	●
F	●	●	●	●
ERROR LED (ON or OFF)	OFF	OFF	OFF	ON

FIGURE 3–14
Error Message Charts (Courtesy of IDEC Corp.)

FIGURE 3–15
LED Fault Indicators (Courtesy of Cutler-Hammer)

LED Indicators on Controllers

LED	Indication
POWER	Internal power supply is functioning.
RUN	The processor is scanning the program and controlling output. LED flashes when PC is STOPPED – outputs are retained. LED is off when PC is halted – outputs are disabled.
FAULT	Processor failure has been detected.
LOW BATTERY	Battery is below low limit.
PROM	PROM module is plugged in and operating.

EXERCISES

Obtain two or more manufacturers' manuals for different PLCs to use as reference for exercises 1 through 6.

1. Compare keyboard layouts and functions for two or more PLC models. How are they alike? What are their major differences?

2. Compare programming formats for coils and contacts for two or more models. How do the procedure sequences differ? What reference numbers and letters must be used for each system?

3. Compare other function formats such as timers, counters, and sequences in the manner of exercise 2. What are the major differences and similarities?

4. What are the programming ladder arrangement rules for one of the models chosen for analysis? What format arrangements will not be accepted by the CPU?

5. What is the scan rate for the units chosen?

6. What corrective procedures are to be taken when a CPU fault light goes on? If there is more than one fault light, what are the corrective procedures for each?

4

Selecting a PLC

OUTLINE

4–1 Introduction □ **4–2** Industrial Control and the Rise of the PLC □ **4–3** The PLC versus the PC □ **4–4** Factors to Consider in Selecting a PLC □ **4–5** PLC Manufacturers

OBJECTIVES

At the end of this chapter, you will be able to

- □ Identify the precursors of the PLC.
- □ Explain the roles of the PLC and PC in industrial control.
- □ List some of the factors to consider in selecting a PLC.
- □ Identify the major PLC manufacturers.

4–1 INTRODUCTION

We should now have a good idea as to what a PLC is, how it is configured, how its CPU, I/O interfaces, and power supply interact, and the programming equipment and procedures necessary to get such a device to carry out machine and process control. In this chapter we turn our attention to what is involved in selecting a PLC.

To begin, we look at the development of machine and process control and the rise of the PLC. Just how is it that PLCs have become the workhorses of today's high-tech manufacturing? Next, we examine the PLC versus PC issue. Will the low-cost PC, hardened for industrial use, hasten the demise of the PLC? Third, we investigate key factors to keep in mind when selecting a PLC. How do such particulars as cost, serviceability, and training impinge on selection? Finally, we look at PLC manufacturers. To aid in making the appropriate choice, we provide a wide-ranging guide to the current PLC market.

4–2 INDUSTRIAL CONTROL AND THE RISE OF THE PLC

Industrial automation, with its machine and process control, had its origins in the 1920s with the advent of rudimentary *pneumatic controllers*. These devices, using compressed air, were flexible, economical, and safe. They created no spark hazard in an explosive atmosphere and could be used under wet conditions without electrical shock hazard. It was easy to connect one device to another with tubing, pipe, or flexible hose. Actions were controlled by a simple manipulation of valves, which in turn were controlled by relays and switches. The air system provided great flexibility in speed and motion control. Because there were few moving parts, reliability and low maintenance costs were evident. Even today, pneumatic control is widely used in conjunction with PLCs and industrial computers in all sorts of machine and process control applications.

Even though pneumatics had, and still has, its advantages, especially with the advent of digital-logic pneumatic-control components, as an actual controlling device it began to be replaced by discrete solid-state controllers in the 1960s. These new controllers, consisting first of transistors and later of individual low- and medium-scale integrated circuits, provided reliable, small, low-power digital logic control. The age of electronic control had definitely arrived.

The big leap into true automation and industrial control, however, took place with the arrival of the microprocessor in the early 1970s. Now it was the program, or software, that modeled and emulated the modulating control that had been previously achieved by hard-wiring discrete gates, encoders/decoders, counters, timers, flip-flops, and similar digital circuitry. Furthermore, with the tiny microprocessor controller, *distributive intelligence* became a reality. The controller was brought closer to the part of the process to be controlled.

Although the microprocessor controller (often called a *microcontroller*) has found its place in machine and process control, buried (embedded) as it is within the controlled device, it is the PLC that has come to dominate industrial automation. With its dedicated I/O (analog and digital), hardened hardware, scanner processing, and ladder logic programming, the PLC today represents the optimum way to achieve industrial control. Or does it? There are those who believe that PLCs will in turn be replaced by industrial PCs, as the latter go beyond their traditional data processing function to usurp the PLC's process control domain. Let us see if this is indeed to be the case.

4–3 THE PLC VERSUS THE PC

In the late 1980s, many in the industry began to predict the demise of the PLC. They felt that as the power of PCs increased, such machines would eventually kill off the PLC. One would just hang some I/O capability off the back of a PC, and presto—a powerful, fully functional PLC! Before we accept this scenario, let us examine more closely the difference between the PC and the PLC and see why we are likely to witness the converging, even merging, of the two technologies in the years to come.

The industrial PC differs from the PLC mainly in its packaging and software. Although the PC has membrane-type keyboards to protect against moisture, grease, dirt, and general industrial grime, and may have an enclosed, shielded video display as well, it is less protected than the more rugged PLC. We could say that the industrial PC, used on the factory floor, lies midway between the relatively pristine PC in the air-conditioned office and the PLC on the assembly line.

Nonetheless, today's industrial computer is a powerful hardware package. Here are the specs for a high-end Honeywell industrial computer:

> The Pentium-based HCC-5000 industrial control computer features a 10.4-in. TFT VGA flat-panel color monitor or STN passive color display, a 52-position sealed-membrane keypad on the front panel, and a NEMA 4/12 rating. It includes a PCI video adapter with 2 Mbytes of VRAM that can drive a simultaneous display on the flat panel and an external CRT, 16 Mbytes of DRAM (expandable to 128 Mbytes), a 1-Gbyte hard drive, a floppy drive, three ISA slots, three PCI slots, and a shared ISA/PCI slot.

When it comes to software, PC-based control software has arrived. As an example, Windows NT, with its OpenControl, from NemaSoft, is a suite of software solutions and development tools that provides impressive control applications. The software includes a family of visual programming languages for control, the 32-bit FloPro for the Windows NT control engine, and servers for device networks and PLC systems.

The PLC, on the other hand, is built for a specific purpose. It has unique components and architecture to make it better for control. It also has software designed for a particular job. Furthermore, because it has cyclical scans that can be monitored easily, the PLC is relatively easy to diagnose and troubleshoot. PCs don't manage failure in the same way and are thus more difficult to bring back online when problems do occur.

So, what's it to be—PC or PLC? The answer, of course, is both, depending on the application. Perhaps Bill Cummings, senior product marketing specialist, Omron Electronics, Inc., puts it best in the April issue of *I&CS:* "A PLC is still a cost-effective solution and has a reliability that most customers need for production. If the customer doesn't really need total real-time control, like a batch process or something, they might be able to use a PC, or if their training needs are such, they might be going to a PC. But the PLC is still very cost effective, and you don't have to interface to the I/O; you're already connected to the I/O through the backplane."

4–4 FACTORS TO CONSIDER IN SELECTING A PLC

In the billion-dollar PLC industry, with over a hundred PLC manufacturers producing a thousand or more individual models, how do you choose the right machine for your needs? If you're a *systems integrator*, working as a consultant or for a large manufacturing company, it's your full-time job to know when, where, and what to acquire. But even if you are an electrician or electronics technician, hovering on the factory floor, you should have at least a rudimentary familiarity with the key factors that must be considered in selecting a PLC. Specifically, you should be cognizant of such issues as *cost, serviceability/support, flexibility/expandability*, and *training/documentation*. Not only might you be called upon to give advice on such matters, you might also become directly involved in implementing any purchasing decisions based on these factors.

When you consider cost, two factors are at work. First, you must determine the crossover point, where it becomes economically advantageous to go with a PLC as opposed to another (particularly a hard-wired) solution. Generally, given the high cost of quality, long-lasting industrial relays, if your application involves a half-dozen or more of these electromagnetic components, it may be time to consider a PLC instead. With the latter below $200 in some cases, the PLC is becoming ever more cost-effective. This leads to the second consideration—overall cost. It is important to remember, especially when installing networked PLC systems, to weigh not only the initial cost of the PLCs, but installation, maintenance, and training costs as well.

In addition to cost (or, in fact, related to it) are the serviceability/support concerns that must be considered when purchasing a PLC. Of course, one wants the most reliable PLC for the money, but if failure occurs, is your PLC equipped with adequate self-diagnosis? And once the problem is found, can it be corrected with minimal effort and time? When it comes to vendor support, will the company that sold you the PLC be there with replacement parts for a quick turnaround? Keep such factors in mind when making any PLC selection decision.

Flexibility and expandability are also factors to consider in selecting a PLC. Your PLC system must be able to grow with your company's needs. Fortunately, most PLCs, even the "shoebox" variety, are designed with these two factors in mind. Memory, I/O, and system expansion, along with the communications infrastructure that goes with them, should not be taken for granted, however. Plan for the future by carefully analyzing your PLC system capabilities and limitations at the outset.

Finally, don't forget training and documentation when deciding what PLC vendor to deal with. At least five factors should be considered with regard to training: (1) Is training supplied at all? (2) Is training provided on site, at your plant? (3) If training does not take place at your plant, what will it cost for employees to be away for training at a regional center or company home office? (4) Is training included when system upgrades are involved? (5) Is the training conducted by competent, industrially experienced instructors who know how to communicate with electricians and electronics technicians?

Documentation—user manuals, software support, and the like—is also extremely important and must accompany any hardware product. Unfortunately, too often documentation is an afterthought, something thrown together hurriedly as the product is being packaged for delivery. Actually, although that is a bit of an exaggeration, it is best to take the

I&CS Guide to programmable controllers

Manufacturer & Model	Memory: Type	Memory: Size	I/O Capability: No. Total I/O	Types: Analog	Types: AC	Types: DC	Types: H.S. Counter	Types: Positioning	Types: PID	Types: ASCII	Languages: Ladder	Languages: Boolean	Languages: Grafcet/sfc	Languages: Other	Programming Devices: Manual	Programming Devices: CRT	Programming Devices: Tape	Programming Devices: Computer	Programming Devices: Other	Networking: Remote I/O	Networking: Host Comp.	Networking: PLC-to-PLC	Networking: Data Highway	Networking: MAP	Other: Math	Other: Diagnostics	Other: Documentation	Other: Color Graphics	Other: Multiple CPUs	Reader Service No.
ABB Process Automation Inc. MasterPiece 90	FLASH PROM BB RAM	128K 256K	256	•	•	•	•	•	•	•			•			•		•			•	•	•		•	•	•	•	•	241
MasterPiece 51	BB RAM	4K	32		•	•							•			•		•			•		•			•	•			
ABB Kent Taylor Batchpac	Hard disk	Multi-Meg	Unlim	•	•	•			•					•				•		•	•		•		•		•	•	•	242
Modcell 2000 Multiloop Cont.		64K		•	•	•			•					•				•			•	•	•		•	•	•	•	•	
Active Systems Group The Stepladder PLC	EEPROM	512B	24	•	•	•				•	•	•						•		•	•	•			•	•	•			243
Adatek 90-30 SLP State Logic Processor FOR GE 90-30 PLCs	RAM/PROM EEPROM	46K	512	•	•	•			•	•				•				•	•		•	•			•	•	•		•	244
90-70 SLP State Logic Processor for GE90-70 PLCs	CMOS RAM	21K-512K	1024	•	•	•	•	•	•	•				•				•	•	•	•	•		•	•	•	•		•	
DOS/State Engine Turns PC into PLC	PC RAM		3340	•	•	•	•	•	•	•				•				•	•	•	•	•			•	•	•			
ESE 2000 State Engine Computer supports GE Genius Adapter, I/O	NVRAM	32K	3840	•	•	•	•	•	•	•				•				•	•	•	•	•			•	•	•			
CO10E	NVRAM	32K	1272	•	•	•			•	•				•				•	•		•	•			•	•	•			
Allen-Bradley SLC 100	BBRAM EEPROM	885 words	24	•	•	•					•				•			•		•	•					•				245
SLC 150	BBRAM EEPROM	1200 words	72	•	•	•	•	•			•				•			•		•	•				•	•				
SLC 500	BBRAM	1K	104	•	•	•	•	•		•	•							•	•		•		•		•	•	•			
SLC 5/01	BBRAM	1K or 4K	256	•	•	•	•	•		•	•							•	•		•		•		•	•	•			
SLC 5/02	BBRAMM	4K	480	•	•	•	•	•	•	•	•							•	•		•	•	•		•	•	•			
SLC 5/03	BBRAM	12K	960	•	•	•	•	•	•	•	•							•		•	•	•	•		•	•	•	•		
PLC-5/10	BBRAM EEPROM	6K	512	•	•	•	•	•	•	•	•		•					•			•	•	•	•	•	•	•	•		
PLC-5/11	BBRAM EEPROM	8K	640	•	•	•	•	•	•	•	•		•	•				•		•	•	•	•	•	•	•	•	•		
PLC-5/12	BBRAM EEPROM	6K	512	•	•	•	•	•	•	•	•		•					•			•	•	•	•	•	•	•	•		
PLC-5/15	BBRAM EEPROM	14K	896	•	•	•	•	•	•	•	•		•					•		•	•	•	•	•	•	•	•	•		
PLC-5/20	BBRAM EEPROM	16K	896	•	•	•	•	•	•	•	•		•	•				•		•	•	•	•	•	•	•	•	•		
PLC-5/25	BBRAM EEPROM	21K	1920	•	•	•	•	•	•	•	•		•					•		•	•	•	•	•	•	•	•	•		
PLC-5/30	BBRAM EEPROM	32K	1920	•	•	•	•	•	•	•	•		•	•				•		•	•	•	•	•	•	•	•	•		
PLC-5/40	BBRAM EEPROM	48K	4224	•	•	•	•	•	•	•	•		•	•				•		•	•	•	•	•	•	•	•	•		
PLC-5/40L	BBRAM EEPROM	48K	4224	•	•	•	•	•	•	•	•		•	•				•			•	•	•	•	•	•	•	•		
PLC-5/60	BBRAM EEPROM	64K	3072	•	•	•	•	•	•	•	•		•					•		•	•	•	•	•	•	•	•	•		
PLC-5/60L	BBRAM EEPROM	64K	3072	•	•	•	•	•	•	•	•		•					•			•	•	•	•	•	•	•	•		
PLC-5/250	BBRAM	8 3M	4096	•	•	•	•	•	•	•	•		•					•		•	•	•	•	•	•	•	•	•	•	
PLC-5/VME	BBRAM EEPROM	14K	512	•	•	•	•	•	•	•	•		•				•	•		•	•	•	•	•	•	•	•	•	•	
PLC-5/40BV	BBRAM EEPROM	48K	1920	•	•	•	•	•	•	•	•		•					•		•	•	•	•	•	•	•	•	•	•	
PLC-5/40LV	BBRAM EEPROM	48K	1920	•	•	•	•	•	•	•	•		•					•			•	•	•	•	•	•	•	•	•	
PLC-3	BB EDC RAM	3 84M	8192	•	•	•	•	•	•	•	•			•		•	•	•		•	•	•	•	•	•	•	•	•		
PLC-3/10	BB EDC RAM	128K	8192	•	•	•	•	•	•	•	•			•		•	•	•		•	•	•	•	•	•	•	•	•		
Analogic Corporation DCS 9200	RAM, EPROM EEPROM, SRAM	16M	320	•	•	•	•	•	•	•				•				•	•	•	•	•	•		•	•	•			246
Aromat M2R	RAM, E/EPROM	2 5K	54		•	•	•				•	•		•	•			•							•	•	•			247
M1T	RAM, E/EPROM	2 5K	192	•		•	•				•	•		•	•			•							•	•	•			
FP1-C16	RAM, E/EPROM	900	12	•	•	•	•	•			•	•			•			•			•	•	•		•	•	•			
FP1-C24	RAM, E/EPROM	2 7K	120	•	•	•	•	•		•	•	•			•			•			•	•	•		•	•	•	•		
FP1-C40	RAM, E/EPROM	15K	2048	•	•	•	•	•		•	•	•			•			•			•	•	•		•	•	•	•		
FP3	RAM, E/EPROM	15K	2048	•	•	•	•	•		•	•	•		•	•			•		•	•	•	•		•	•	•	•	•	

FIGURE 4–1

Guide to Programmable Controller Hardware. (*I&CS*, April 1997 issue, with permission of Chilton Publishing Co.)

I&CS Guide to programmable controllers

Manufacturer & Model	Memory		I/O Capability								Languages				Programming Devices					Networking					Other					Reader Service No.
	Type	Size	No.	Types																										
			Total I/O	Analog	AC	DC	H.S. Counter	Positioning	PID	ASCII	Ladder	Boolean	Grafcet/sfc	Other	Manual	CRT	Tape	Computer	Other	Remote I/O	Host Comp.	PLC-to-PLC	Data Highway	MAP	Math	Diagnostics	Documentation	Color Graphics	Multiple CPUs	
ACS Systems																														248
ASC/86	RAM, EPROM	54K & Up	128	•	•	•	•	•	•	•	•	•	•	•	•	•		•	•	•	•	•	•	•	•	•	•	•	•	
ASC/486	BBRAM EEPROM	256K & Up	512	•	•	•	•	•	•	•	•	•	•	•	•	•	•	•	•	•	•	•	•	•	•	•	•	•	•	
Automatic Timing & Controls																														249
ATCOM 64	RAM, EEPROM	8K	64	•	•	•	•			•				•	•			•			•				•	•	•			
Autotech Controls																														250
M1500 Pc• PLS	BB RAM	2K	20/20			•	•	•			•						•	•								•		•		
Bailey Controls																														251
SLC01	EPROM, RAM NVRAM	128K	13	•		•	•	•	•	•	•	•		•							•	•	•		•	•	•	•	•	
CLS03/04	NVRAM	2KB	13	•		•		•	•		•	•		•							•	•	•		•	•	•	•	•	
CBC01	EPROM, RAM NVRAM	128KB	14 16KB	•	•	•	•	•	•	•	•	•		•						•	•	•	•		•	•	•	•	•	
CSC01	EPROM, RAM NVRAM	128KB	28 16KB		•	•		•	•	•	•	•		•						•	•	•	•		•	•	•	•	•	
CPC01	EPROM NVRAM	128KB	6 8 KB	•	•	•	•	•	•			•		•											•	•	•			
LMM02	EPROM, BBRAM	32K, 8K	1024		•	•	•		•		•	•		•	•	•		•	•				•		•	•	•			
MPC01	EPROM, BBRAM	128K, 20K	1024		•	•	•		•		•	•		•	•	•		•	•				•		•	•	•			
MFP01	ROM, NVRAM	256K, 64K	10000	•	•	•	•		•		•	•		•	•	•		•	•	•		•	•	•	•	•	•	•		
MFP02	EPROM	256K	10000	•	•	•	•		•		•	•		•	•	•		•	•	•		•	•	•	•	•	•	•		
MFP03	EPROM, BBRAM	2000K, 512	10000	•	•	•	•		•		•	•		•	•	•		•	•	•		•	•	•	•	•	•	•		
Basicon, Inc.																														252
SBC-64	RAM EEROM EEROM	6K 8K 8K	64	•	•	•	•			•				•				•		•	•						•			
Blue Earth																														253
Micro-440e	EPROM, SRAM	32K, 32K	21	•		•	•			•				•				•		•	•	•			•		•			
Micro-485	EPROM, SRAM EEPROM	64K, 128K 1K	31	•		•	•			•				•				•		•	•	•			•		•			
ALC-51	EPROM FLASH SRAM EEPROM	64K 128K 128K 1K	14			•	•			•				•				•		•	•	•			•		•			
Robert Bosch Corp.																														254
Bosch CL500	RAM, EEPROM	4 x 64KW	4Kx	•	•	•	•	•	•	•	•	•				•		•		•	•	•		•	•	•	•	•	•	
Bosch CL300	RAM/EPROM/ EEPROM	32, 32 16K	4K	•	•	•	•	•	•	•	•	•				•		•		•	•	•		•	•	•	•	•		
Bristol Babcock																														255
DPC 3335	RAM/PROM	256K	400	•	•	•	•	•	•	•		•		•				•		•	•	•	•		•	•	•	•		
DPC 3330	RAM/PROM	384K	400	•	•	•	•	•	•	•		•		•				•		•	•	•	•		•	•	•	•		
RTU 3310	RAM/PROM	384K	400	•	•	•	•	•	•	•		•		•				•		•	•	•	•		•	•	•	•		
B&R Industrial Automation																														256
MINICONTROL	EEPROM, RAM	16K-32K	192	•		•	•	•	•	•	•	•		•		•		•	•		•	•	•		•	•	•	•		
MIDICONTROL	EEPROM, RAM FLASH	16K-84K	192	•	•	•	•	•	•	•	•	•		•		•		•	•		•	•	•		•	•	•	•	•	
MULTICONTROL	EEPROM, RAM FLASH	16K, 384K	1536	•	•	•	•	•	•	•	•	•		•		•		•	•		•	•	•		•	•	•	•	•	
M264	EEPROM, RAM FLASH	16K-384K	264	•	•	•	•	•	•	•	•	•		•		•		•	•		•	•	•		•	•	•	•		
Cegeic Automation Inc.																														257
GEM80/131	EPROM	15000 inst	512	•	•	•	•		•	•	•			•		•	•	•			•	•	•		•		•			
GEM80/164	EPROM	25000 inst	8192	•	•	•	•		•	•	•			•		•	•	•		•	•	•	•		•		•	•		
GEM80/165	EPROM	25000 inst	8192	•	•	•	•		•	•	•			•		•	•	•		•	•	•	•		•		•	•		
GEM80/166	EPROM	25000 inst	8192	•	•	•	•		•	•	•			•		•	•	•		•	•	•	•		•		•	•		
GEM80/312	EPROM	36000 inst	8192	•	•	•	•		•	•	•			•		•	•	•		•	•	•	•		•		•	•		
GEM80/163	EPROM	25000	8192	•	•	•	•		•	•	•			•		•	•	•		•	•	•	•		•		•	•		
Cincinnati Milacron																														258
APC-500 Relay	CMOS RAM	32K-6128K	512	•	•	•	•	•	•	•	•				•	•	•	•		•	•	•	•		•	•	•	•		
APC-500MCL	CMOS RAM	32K-64K	2048	•	•	•	•	•	•	•				•	•	•	•	•		•	•	•	•		•	•	•	•		
Control Technology Corp.																														259
2200XM	NVRAM	10K	160	•	•	•	•	•	•	•				•				•			•	•			•	•	•		•	
2600XM	NVRAM	24K	160	•	•	•	•	•	•	•				•				•			•	•			•	•	•		•	
2600XM-10	NVRAM	24K	320	•	•	•	•	•	•	•				•				•			•	•			•	•	•		•	
2700	NVRAM	64K	512	•	•	•	•	•	•	•				•				•			•	•			•	•	•		•	
Digitronics Sixnet																														260
IOMUX	RAM, EPROM	1M	512	•	•	•	•	•	•	•		•		•				•		•	•	•	•		•	•	•	•	•	
Versamux RTU	RAM, EPROM	1M	512	•	•	•	•	•	•	•		•		•				•		•	•	•	•		•	•	•	•	•	

FIGURE 4–1 (continued)

I&CS Guide to programmable controllers

Manufacturer & Model	Memory: Type	Memory: Size	I/O Capability: No.I: Total I/O	I/O Types: Analog	I/O Types: AC	I/O Types: DC	I/O Types: H.S. Counter	I/O Types: Positioning	I/O Types: PID	I/O Types: ASCII	Languages: Ladder	Languages: Boolean	Languages: Grafcet/sfc	Languages: Other	Programming Devices: Manual	Programming Devices: CRT	Programming Devices: Tape	Programming Devices: Computer	Programming Devices: Other	Networking: Remote I/O	Networking: Host Comp.	Networking: PLC-to-PLC	Networking: Data Highway	Networking: MAP	Other: Math	Other: Diagnostics	Other: Documentation	Other: Color Graphics	Other: Multiple CPUs	Reader Service No.
Divelbiss																														261
Baby Bear Bones	EPROM	4K	58		•	•	•				•				•		•	•		•						•	•			
Boss Bear	E/EEPROM SRAM	128K, 8K 128K	256	•	•	•	•	•	•	•				•	•			•		•	•	•			•					
HD Bear Bones	EPROM	16K	249		•	•	•				•				•		•	•		•						•	•			
Bear Bones Plus	EPROM	16K	249		•	•	•				•				•		•	•		•						•	•			
Eagle Signal Controls																														262
Micro 190	BBRAM, EPROM, UVPROM	32K	128		•	•	•				•	•			•			•			•	•			•	•	•	•		
Micro 190 +	BBRAM, UVPROM	32K	128	•	•	•	•		•		•	•			•			•			•	•			•	•	•	•		
Eptak 225	BBRAM	16K	128		•	•					•	•			•			•			•	•			•	•	•	•		
Eptak 245	BBRAM	16K	128	•	•	•			•		•	•		•	•			•			•	•			•	•	•	•		
Eagle 1	BBRAM	48K	896	•	•	•	•			•	•							•			•	•			•	•	•	•		
Eagle 2	BBRAM	48K	896	•	•	•	•		•	•	•							•			•	•			•	•	•	•		
Eagle 3	BBRAM	48K	2048	•	•	•	•		•	•	•							•			•	•			•	•	•	•		
Eptak 7000	BBRAM	48K	2048	•	•	•	•		•	•	•							•			•	•	•		•	•	•	•		
CP 8000	BBRAM	128K	1536	•	•	•	•		•	•	•							•			•	•	•		•	•	•	•		
Eason																														263
1000	BBRAM, EPROM	64K	8			•	•			•				•				•			•	•			•	•				
1100	BBRAM, EPROM	128K	120	•		•	•			•	•			•				•			•	•			•	•				
Eaton Corp.																														264
D100CR14	RAM, EEPROM	1K	34	•	•	•	•				•				•			•		•						•	•	•		
D100CR14	RAM, EEPROM	1K steps	60	•	•	•	•			•	•							•	•											
D100CR20A	RAM, EEPROM	1K steps	40	•	•	•	•				•				•			•		•						•	•	•		
D100CR40A	RAM, EEPROM	1K steps	80	•	•	•	•				•				•			•		•						•	•	•		
D100CR40H	RAM, EEPROM	1K steps	120		•	•	•				•				•			•		•						•	•	•		
D200PR4	RAM, EEPROM	4K steps	240	•	•	•	•	•		•	•				•			•		•	•					•	•	•		
D200PR4C	RAM, EEPROM	4K	240	•	•	•	•	•		•	•				•			•		•	•					•	•	•		
Entertron Industries																														265
Smart-Pak	EAROM, RAM,	8K	24			•	•			•	•							•			•				•	•	•			
SK 1600R-SA	BBRAM, EPROM	8K	56	•		•	•			•	•							•			•				•	•	•			
SK 1600	EPROM	4K	64		•	•					•							•							•	•	•			
SK 1800	EPROM	8K	88		•	•	•	•		•	•							•							•	•	•			
Festo Corp.																														266
FP 101B	RAM, EPROM	12K	117			•	•			•	•		•	•	•			•			•				•	•	•			
FPC 101AF	RAM, EPROM	12K	127	•		•	•	•		•	•		•					•			•			•	•	•	•			
FPC 202	RAM, EPROM	26K	128		•	•				•	•	•	•					•			•				•	•	•			
FPC 405	RAM, EPROM	768K	2272	•	•	•	•	•	•	•	•		•					•		•	•	•			•	•	•	•	•	
Furnas																														267
96/M	E/EPROM	4K	320	•	•	•	•	•		•	•	•	•		•		•	•		•	•	•			•	•	•	•	•	
96/M Plus	RAM, EPROM	8K	704	•	•	•	•	•		•	•	•	•					•		•	•	•			•	•	•	•	•	
96MFMXX	RAM, EPROM	8K	208	•	•	•	•	•		•	•	•	•					•		•	•	•			•	•	•	•	•	
GE Fanuc																														268
Series 90-70 914 CPU	RAM, EEPROM	512K	12288	•	•	•	•	•	•	•	•		•	•				•	•	•	•	•	•	•	•	•	•	•	•	
Series 90-70 781/782/CPU	RAM, EEPROM	512K	12288	•	•	•	•	•	•	•	•		•	•				•	•	•	•	•	•	•	•	•	•	•	•	
Series 90-70 771/772 CPU	RAM, EEPROM	512K	2048	•	•	•	•	•	•	•	•		•	•				•	•	•	•	•	•	•	•	•	•	•	•	
Series 90-70 731/732 CPU	RAM, EEPROM	82K	512	•	•	•	•	•	•	•	•		•	•				•	•	•	•	•	•	•	•	•	•	•	•	
Series 90-30 CPU 341	RAM, E/EPROM	40K	1024	•	•	•	•	•	•	•	•	•			•	•		•		•	•	•			•	•	•	•		
Series 90-30 CPU 331	RAM, E/EPROM	8K	1024	•	•	•	•	•	•	•	•	•			•	•		•		•	•	•			•	•	•	•		
Series 90-30 CPU 313	RAM, E/EPROM	3K	320	•	•	•	•	•	•	•	•	•			•	•		•		•	•	•			•	•	•	•		
Series 90-30 CPU 311	RAM, E/EPROM	3K	520	•	•	•	•	•	•	•	•	•			•	•		•		•	•	•			•	•	•	•		
Series 90-20 CPU 211	RAM, E/EPROM	1K	28	•	•	•	•				•	•			•	•		•		•	•	•			•	•	•	•		
Series Six	Static RAM	64K	8K	•	•	•	•	•	•	•	•					•		•		•	•	•		•	•	•		•	•	
Giddings & Lewis																														269
Pic 90	CMOS, EPROM	512K	128	•	•	•	•	•	•	•	•			•				•			•	•	•	•	•	•	•	•		
Pic 900	CMOS, EPROM	512K	3168	•	•	•	•	•	•	•	•			•				•		•	•	•	•	•	•	•	•	•		
Grayhill																														270
ProMux	EEPROM	8K	24		•	•								•				•		•					•					
MicroDAC	FLASH, BBRAM	128K	32	•	•	•	•							•				•		•					•					
72-MDC-32ADC																														
HMW Enterprises, Inc.																														271
MPCII E	RAM, EPROM	1 3M, 768K	512	•	•	•		•			•				•	•	•	•		•	•		•	•	•	•	•	•		
MPCII C	RAM, EPROM	1 3M, 768K	512	•	•	•		•			•			•	•	•	•	•		•	•		•	•	•	•	•	•		
MPCII T	RAM, EPROM	300K, 256K	256		•	•		•			•				•	•	•	•		•	•		•	•	•	•	•			
MPCII	RAM, EPROM	62KW, 28KW	512	•	•	•		•			•				•	•	•	•	•	•						•	•			
Honeywell																														272
620-12	PROG	2K	256	•	•	•	•	•	•	•	•					•		•		•	•	•	•	•	•	•	•	•	•	
620-18	PROG	8K	1K	•	•	•	•	•	•	•	•					•		•		•	•	•	•	•	•	•	•	•	•	
620-36	PROG	32K	2K	•	•	•	•	•	•	•	•					•		•		•	•	•	•	•	•	•	•	•	•	
S9000	PROG	2K	256	•	•	•	•	•	•	•	•		•			•		•		•	•	•	•	•	•	•	•	•	•	
S9100	PROG	8K	1K	•	•	•	•	•	•	•	•		•			•		•		•	•	•	•	•	•	•	•	•	•	
S9200	PROG	32K	2K	•	•	•	•	•	•	•	•		•			•		•		•	•	•	•	•	•	•	•	•	•	

FIGURE 4–1 (continued)

I&CS Guide to programmable controllers

Manufacturer & Model	Memory		I/O Capability								Languages				Programming Devices					Networking					Other					Reader Service No.
	Type	Size	No.I	Types																										
			Total I/O	Analog	AC	DC	H.S. Counter	Positioning	PID	ASCII	Ladder	Boolean	Grafcet/sfc	Other	Manual	CRT	Tape	Computer	Other	Remote I/O	Host Comp.	PLC-to-PLC	Data Highway	MAP	Math	Diagnostics	Documentation	Color Graphics	Multiple CPUs	
Homer Electric																														273
HE2000	EEPROM	32K	200	•	•	•	•		•	•	•			•				•		•	•	•			•	•	•	•		
Icon Corp.																														274
MC Motion Comptr/PLC			144	•		•	•	•			•				•	•		•			•				•	•			•	
Idec Corp.																														275
Micro 1	EEPROM	6K	28		•	•					•	•	•		•			•			•					•	•			
FA-2 Junior	RAM, E/EEPROM	4K	256	•	•	•	•				•	•	•		•			•		•	•				•	•	•			
FA-3S-CP11	RAM, E/EEPROM	1 or 4K	256	•	•	•	•				•	•	•		•			•			•				•	•	•			
CP11T	EEPROM INTEG	4K	256	•	•	•	•				•	•	•		•			•		•	•	•			•	•	•		•	
CP12	ROM, EEPROM	4K	256	•	•	•	•				•	•	•		•			•		•	•	•			•	•	•		•	
CP13	CMOS, EEPROM	1K, 4K, 8K	256	•	•	•	•				•	•	•		•			•		•	•	•			•	•	•		•	
International Parallel Machines																														276
IP1612	EEPROM	500 steps	28	•	•	•	•				•							•							•	•	•			
IP1612 I	EEPROM	500 steps	28	•	•	•	•				•							•			•				•	•	•			
IP1612 DC	EEPROM	500 steps	28	•		•	•				•				•			•			•				•	•	•			
IP1612A	EEPROM	500 steps	28	•		•	•				•							•			•				•	•	•			
IP1610	EEPROM	500 steps	28	•	•	•	•				•							•			•				•	•	•			
IP1612-200	EEPROM	500 steps	28	•	•	•	•				•							•							•	•	•			
Jumo Process Control																														277
PR-100	EEPROM/RAM	Up to 1Meg	95	•	•	•	•	•	•		•	•	•	•	•	•	•	•	•	•	•	•	•	•	•	•	•	•		
PRf-100	EEPROM/RAM	Up to 1Meg	95	•	•	•	•	•	•		•	•	•	•	•	•	•	•	•	•	•	•	•	•	•	•	•	•		
Klockner-Moeller																														278
PS3	RAM, EEPROM	3 8K	152	•	•	•	•				•	•		•	•	•	•	•		•	•				•	•	•			
PS306	RAM, EEPROM	32KW	333	•		•	•				•	•		•				•		•	•	•			•	•	•			
PS316	RAM, EEPROM	32KW	4624	•	•	•	•	•		•	•	•		•				•		•	•	•			•	•	•		•	
PS32	RAM, EEPROM	32KW	4992	•	•	•	•	•		•	•			•				•		•	•	•			•	•	•			
PC 520-12	CMOS, RAM	2KW	256	•	•	•	•	•	•	•	•			•				•			•	•	•	•	•	•	•	•	•	
PC 520-13	CMOS, RAM	4KW	512	•	•	•	•	•	•	•	•			•				•		•	•	•	•	•	•	•	•	•	•	
PC 520-16	CMOS, RAM	6KW	1024	•	•	•	•	•	•	•	•			•				•		•	•	•	•	•	•	•	•	•	•	
PC 520-26	CMOS, RAM	16KW	2048	•	•	•	•	•	•	•	•			•				•		•	•	•	•	•	•	•	•	•	•	
PC 520-38	CMOS, RAM	32KW	2048	•	•	•	•	•	•	•	•			•				•		•	•	•	•	•	•	•	•	•	•	
Keyence Corp.																														302
KV-10R/T	EEP	4K	10			•	•			•	•				•			•			•					•	•	•		
KV-16R/T	EEP	4K	16			•	•			•	•				•			•			•					•	•	•		
KV-24R/T	EEP	21K	24			•	•			•	•				•			•			•					•	•	•		
KV-40R/T	EEP	21K	40			•	•			•	•				•			•			•					•	•	•		
Mitsubishi																														279
FX	EEPROM	6K	30		•	•	•				•	•	•	•	•	•	•	•	•	•	•	•			•	•	•	•		
FX	RAM, E/PROM	6K	256	•	•	•	•	•	•	•	•	•	•	•	•	•	•	•	•	•	•	•			•	•	•	•		
XTS	RAM, E/PROM	6K	256	•	•	•	•	•	•	•	•	•	•		•	•	•	•		•	•	•	•	•	•	•	•	•		
A2C	RAM, E/PROM	16-448K	2048	•	•	•	•	•	•	•	•	•	•		•	•	•	•		•	•	•	•		•	•	•	•		
A2N	RAM, E/PROM	16-448K	2048	•	•	•	•	•	•	•	•	•	•		•	•	•	•	•	•	•	•	•	•	•	•	•	•	•	
A2N	RAM, E/PROM	16-448K	2048	•	•	•	•	•	•	•	•	•		•	•	•	•		•	•	•	•	•	•	•	•	•			
Modicon																														280
PC-A984-120	CMOS RAM	3 5K	1024	•	•	•	•	•			•		•	•				•	•		•		•							
PC-A984-130	CMOS RAM	6K	1024	•	•	•	•	•			•		•	•		•		•	•		•		•							
PC-A984-131	CMOS RAM	6K	1024	•	•	•	•	•			•		•	•				•	•		•		•		•	•	•	•		
PC-A984-141	CMOS RAM	10K	1024	•	•	•	•	•			•		•	•		•		•	•		•		•		•	•	•	•		
PC-A984-145	CMOS RAM	10K	1024	•	•	•	•	•			•		•	•		•		•	•		•		•		•	•	•	•		
PC-0984-380	CMOS RAM	8K	1024	•	•	•	•	•	•	•	•		•	•		•		•	•		•		•		•	•	•	•		
PC-0984-381	CMOS RAM	8K	1024	•	•	•	•	•	•	•	•		•	•				•	•		•		•		•	•	•	•		
PC-E984-381	FLASH PROM	16K	1024	•	•	•	•	•	•	•	•		•	•				•	•		•		•		•	•	•	•		
PC-0984-385	CMOS RAM	8K	1024	•	•	•	•	•	•	•	•		•	•				•	•				•		•	•	•	•		
PC-E984-385	FLASH RAM	18K	1024	•	•	•	•	•	•	•	•		•	•				•	•				•		•	•	•	•		
PC-0984-480	CMOS RAM	18K	7168	•	•	•	•	•	•	•	•		•	•				•	•				•		•	•	•	•		
PC-E984-480	FLASH PROM	18K	7168	•	•	•	•	•	•	•	•		•	•				•	•				•		•	•	•	•		
PC-0984-485	CMOS RAM	18K	7168	•	•	•	•	•	•	•	•		•	•				•	•				•		•	•	•	•		
PC-E984-485	FLASH PROM	18K	7168	•	•	•	•	•	•	•	•		•	•				•	•				•		•	•	•	•		
PC-E984-680	RAM EPROM	12/20K	32768	•	•	•	•	•	•	•	•		•	•				•	•				•	•	•	•	•	•		
PC-0984-685	CMOS RAM	12/20K	32768	•	•	•	•	•	•	•	•		•	•				•	•	•	•	•	•	•	•	•	•	•	•	
PC-E984-685	FLASH PROM	26K	32768	•	•	•	•	•	•	•	•		•	•				•	•	•	•	•	•	•	•	•	•	•	•	
PC-0984-700	CMOS RAM	26/42K	32768	•	•	•	•	•	•	•	•		•	•				•	•	•	•	•		•	•	•	•	•	•	
PC-L984-785	CMOS RAM	26/80/96K	65536	•	•	•	•	•	•	•	•		•					•		•	•	•	•	•	•	•	•	•	•	
PC-E984-785	FLASH PROM	80/96K	65536	•	•	•	•	•	•	•	•		•	•				•	•	•	•	•	•	•	•	•	•	•	•	
PC-984X-008	CMOS RAM	10K	32768	•	•	•	•	•	•	•	•		•	•				•	•	•	•	•	•	•	•	•	•	•	•	
PC-984A-XXX	CMOS RAM	18/34K	32768	•	•	•	•	•	•	•	•		•	•				•	•	•	•	•	•	•	•	•	•	•	•	
PC-984B-XXX	CMOS RAM	42/74/138K	55536	•	•	•	•	•	•	•	•		•	•				•	•	•	•	•	•	•	•	•	•	•	•	
AM-0984-AT2	CMOS RAM	18K	7168	•	•	•	•	•	•	•	•		•	•				•	•	•	•	•	•		•	•	•	•	•	
AM-0984-MC0	CMOS RAM	18K	7168	•	•	•	•	•	•	•	•		•	•				•	•	•	•		•		•	•	•	•	•	
AM-0984-QX0	CMOS RAM	22K	7168	•	•	•	•	•	•	•	•		•	•				•	•	•	•		•		•	•	•	•	•	
AM-0984-VM0	CMOS RAM	22K	7168	•	•	•	•	•	•	•	•		•	•				•	•	•	•		•		•	•	•	•	•	
Micro CPU311	RAM FLASH	1 4K	165		•	•				•	•		•	•	•			•	•		•	•	•		•	•	•	•		
Micro CPU411	RAM FLASH	1 4K	165		•	•	•			•	•		•	•	•			•	•		•	•	•		•	•	•	•		
Micro CPU512	RAM FLASH	4K	395	•	•	•	•	•	•	•	•		•	•	•			•	•		•	•	•		•	•	•	•		
Micro CPU612	RAM FLASH	4K	395	•	•	•	•	•	•	•	•		•	•	•			•	•		•	•	•		•	•	•	•		

FIGURE 4–1 (continued)

I&CS Guide to programmable controllers

Manufacturer & Model	Memory		I/O Capability								Languages				Programming Devices					Networking					Other					Reader Service No.
	Type	Size	No. I	Types																										
			Total I/O	Analog	AC	DC	H.S. Counter	Positioning	PID	ASCII	Ladder	Boolean	Grafcet/sfc	Other	Manual	CRT	Tape	Computer	Other	Remote I/O	Host Comp.	PLC-to-PLC	Data Highway	MAP	Math	Diagnostics	Documentation	Color Graphics	Multiple CPUs	
Nolatron Inc.																														281
ELC 5	EEPROM	2KX8	48			•								•	•				•							•				
System IV	EEPROM	2KX8	16			•								•	•															
Omega Engineering																														282
OME-STL24	SRAM	512 Bytes	24		•	•	•	•		•	•	•		•				•		•	•		•				•			
Omron																														283
SP-10	RAM, EEPROM SRAM	144 words	10		•						•	•			•				•			•			•	•				
SP-16/20	RAM, EEPROM SRAM	348 words	20		•	•	•				•	•			•							•			•	•				
C20K Series	RAM, ROM EPROM	1194 words	148	•	•	•	•				•	•			•	•	•	•		•	•									
C20H Series	RAM, E/EEPROM	2 8K	240	•	•	•	•				•	•			•	•	•	•		•	•	•			•	•	•			
C200H	RAM, E/EEPROM	7K	720	•	•	•	•	•	•	•	•	•		•	•	•	•	•		•	•	•	•		•	•	•			
C500	RAM, ROM EPROM	8K	512	•	•	•	•	•	•	•	•	•		•	•	•	•	•	•	•	•	•	•		•	•	•			
C1000H	RAM, ROM EPROM	36K	1024	•	•	•	•	•	•	•	•	•		•	•	•	•	•	•	•	•	•	•		•	•	•			
C2000H	RAM, EEPROM EEPROM, SRAM	30K	2048	•	•	•	•	•	•	•	•	•	•	•	•	•	•	•	•	•	•	•	•		•	•	•			
CV1000	RAM, ROM EEPROM, SRAM	62K	4096	•	•	•	•	•	•	•	•	•	•	•	•	•	•	•	•	•	•	•	•		•	•	•			
Opto 22																														284
Mistic 100	RAM ROM	64K 32K	4096	•	•	•	•	•	•	•			•	•		•		•		•	•	•	•		•	•	•	•		
Mistic 200SX	RAM ROM	256K 2M	7500	•	•	•	•	•	•	•			•					•		•	•	•	•		•	•	•	•		
Mistic 200	RAM ROM	4M 4M	18k	•	•	•	•	•	•	•			•					•		•	•	•	•		•	•	•	•		
LC4	RAM, ROM	64K, 32K	4096	•	•	•	•	•	•	•			•	•		•		•		•	•	•	•		•					
LC2	RAM, ROM	32K, 32K	1024	•	•	•	•	•	•	•				•		•		•		•	•	•	•		•					
PEP Modular Computers																														285
IUC 9000	ROM, SRAM EEPROM	1M, 1M 64K	924	•		•	•		•	•	•		•	•				•			•	•			•	•	•	•		
VME 9000	ROM, SRAM EEPROM	1M, 1M 64K	416	•		•	•	•		•	•		•	•				•			•	•			•	•	•	•	•	
VME 9030	ROM, DRAM SRAM, EEPROM	2M, 4M 1M, 256K	416	•		•	•	•		•	•		•	•				•			•	•			•	•	•	•	•	
VME 9040	ROM, DRAM SRAM, EEPROM	2M, 4M 512K, 256K	416	•		•	•	•		•	•		•	•				•			•	•			•	•	•	•	•	
Phoenix Contact																														286
Interbus-R	RAM	Varies	4096	•	•	•				•				•		•		•		•	•					•				
Interbus ROM-DOS	RAM, EPROM FLASH	512K, 256K 128K	128	•	•	•				•	•	•		•		•		•			•	•			•	•	•			
Interbus-S	RAM	Varies	4096	•	•	•	•				•	•	•	•		•		•		•	•	•			•	•	•	•	•	
Pro-Log Corporation																														287
486 ATIPLC	RAM	96K	2048	•	•	•	•	•	•		•					•		•		•	•	•	•		•		•	•	•	
Pyramid Industries																														288
Small PLC	EEPROM	200 bytes	12		•	•					•							•												
Reliance Electric																														289
AutoMax DCS	BBRAM	512K	12K	•	•	•	•	•	•	•	•			•	•	•		•		•	•	•	•		•	•	•	•	•	
AutoMate 15	EEPROM NVRAM	1K	69		•	•					•				•	•		•			•	•	•			•	•	•		
AutoMate 15E	EEPROM, RAM NVRAM	2K	64	•	•	•					•				•	•		•		•	•	•	•		•	•	•	•		
AutoMate 20	EEPROM, RAM NVRAM	2K	256		•	•					•				•	•		•		•	•	•	•		•	•	•	•		
AutoMate 20E	EEPROM, RAM NVRAM	4K	256	•	•	•	•			•	•				•	•		•		•	•	•	•		•	•	•	•		
AutoMate 30	EEPROM, RAM NVRAM	8K	512	•	•	•	•		•	•	•				•	•		•		•	•	•	•		•	•	•	•	•	
AutoMate 40	EEPROM, RAM NVRAM	104K	8192	•	•	•	•		•	•	•				•	•		•		•	•	•	•		•	•	•	•	•	
Shark X	E/EEPROM	1K, 2K	60	•	•	•	•				•	•			•		•	•							•	•	•			
Shark XL	E/EEPROM	1K, 2K	160	•	•	•	•				•	•			•		•	•							•	•	•			
Semix																														290
RC 207	EEPROM	1 7K	20/16			•	•	•					•					•		•										
RC 231	EEPROM	4 14K	8/8			•	•	•						•					•		•									

FIGURE 4–1 (continued)

I&CS Guide to programmable controllers

Manufacturer & Model	Memory		I/O Capability								Languages				Programming Devices					Networking					Other					Reader Service No.
	Type	Size	No.I	Types																										
			Total I/O	Analog	AC	DC	H.S. Counter	Positioning	PID	ASCII	Ladder	Boolean	Grafcet/sfc	Other	Manual	CRT	Tape	Computer	Other	Remote I/O	Host Comp.	PLC-to-PLC	Data Highway	MAP	Math	Diagnostics	Documentation	Color Graphics	Multiple CPUs	
Siemens Industrial Automation																														291
SIMATIC S5-100U	RAM, E/EEPROM	20K	256	•	•	•	•	•	•	•	•	•	•	•		•		•	•		•	•			•	•	•			
SIMATIC S5-115U	RAM, E/EEPROM	96K	2048	•	•	•	•	•	•	•	•	•	•	•		•		•	•	•	•	•	•	•	•	•	•	•		
SIMATIC S5-135U	RAM, EPROM	64K	8192	•	•	•	•	•	•	•	•	•	•	•		•		•	•	•	•	•	•	•	•	•	•	•	•	
SIMATIC S5-155U	RAM, EPROM	896K	1M +	•	•	•	•	•	•	•	•	•	•	•		•		•	•	•	•	•	•	•	•	•	•	•	•	
SIMATIC T1575	CMOS RAM	320K	8192	•	•	•	•	•	•	•	•	•	•			•		•	•	•	•	•	•	•	•	•	•	•	•	
SIMATIC T1305	RAM, E/EEPROM	7.4K	160	•	•	•	•			•	•	•		•		•	•	•	•	•	•		•		•	•	•	•		
SIMATIC T1405	RAM, E/EEPROM	21K	1152	•	•	•	•	•		•	•	•		•			•	•	•	•	•	•	•		•	•	•	•		
SIMATIC T1515	RAM, E/EEPROM	448K	2048	•	•	•	•	•	•	•	•		•			•		•		•	•	•	•	•	•	•	•	•		
SIMATIC T1555	RAM, E/EEPROM	2Mbyte	6192	•	•	•	•	•	•	•	•		•			•		•		•	•	•	•	•	•	•	•	•		
SKH Systems, Inc.																														292
SKH-19	EPROM RAM	24-56K	256	•	•	•	•		•	•	•			•				•			•				•	•	•			
SKH-23	EPROM RAM	23K	36	•	•	•	•		•	•	•			•				•			•				•	•	•			
SKH-29	EPROM RAM	23K	244		•	•	•		•	•	•			•				•			•	•			•	•	•			
Square D																														293
MICRO-1	EPROM	600 steps	28		•	•					•	•			•			•			•					•	•			
SY/MAX Model 50	RAM, E/EEPROM	4K steps	256	•	•	•	•				•	•			•			•					•	•	•		•			
SY/MAX Model 300	RAM, UVPROM	5.1 or 2K	256	•	•	•	•	•	•	•	•				•	•	•	•			•	•	•	•	•	•	•			
SY/MAX Model 400	RAM, UVPROM	4, 8, 16K	4K	•	•	•	•	•	•	•	•				•	•	•	•		•	•	•	•	•	•	•	•	•		
SY/MAX Model 600	RAM	16 or 26K	8K	•	•	•	•	•	•	•	•				•	•	•	•		•	•	•	•	•	•	•	•	•		
SY/MAX Model 650	RAM	16 or 26K	8K	•	•	•	•	•	•	•	•				•	•	•	•		•	•	•	•	•	•	•	•	•	•	
SY/MAX Model 200	RAM	64K	8K	•	•	•	•	•	•	•	•				•	•	•	•		•	•	•	•	•	•	•	•			
TSX 17-10	RAM, EPROM	8K, 8K	120		•	•	•					•	•		•	•		•					•		•	•	•	•		
TSX 17-20	RAM, E/EEPROM	24K	160	•	•	•	•		•	•	•		•			•		•				•			•	•	•			
TSX 47-10	RAM, EPROM	34K	256		•	•					•		•					•	•				•		•	•	•			
TSX 47-20	RAM, EPROM	34K	256	•	•	•	•	•			•		•					•	•		•	•	•		•	•	•			
TSX 47-40	RAM, EPROM	112K	1024	•	•	•	•	•			•		•	•				•		•	•	•	•		•	•	•	•		
TSX 67-40	RAM, EPROM	224K	2048	•	•	•	•	•			•		•	•				•		•	•	•	•	•	•	•	•	•		
TSX 67-40	RAM, EPROM	352K	2048	•	•	•	•	•			•		•	•				•		•	•	•	•	•	•	•	•	•		
TSX 107-40	RAM, EPROM	352K	2048	•	•	•	•	•			•		•	•				•		•	•	•	•	•	•	•	•	•		
Systems Eng'g Assoc.																														294
S3000	RAM, EPROM	52K	256	•	•	•	•	•		•	•			•				•			•	•			•	•	•		•	
M4000	RAM	26K	48			•	•				•			•				•				•			•	•	•			
Toshiba																														295
EX148	RAM, EEPROM	1K	34		•	•	•				•							•	•						•	•	•	•	•	
EX20 Plus	RAM, EEPROM	1K	40		•	•	•				•							•	•						•	•	•	•	•	
EX40 Plus	RAM, EEPROM	1K	80		•	•	•				•							•	•						•	•	•	•	•	
M20	RAM, EEPROM	4K	40	•	•	•	•	•		•	•							•	•	•	•	•	•		•	•	•	•	•	
M40	RAM, EEPROM	4K	80	•	•	•	•	•		•	•							•	•	•	•	•	•		•	•	•	•	•	
EX100	RAM, EEPROM	4K	480	•	•	•	•	•		•	•							•	•	•	•	•	•		•	•	•	•	•	
EX500	RAM, EEPROM	8K	512	•	•	•	•	•	•	•	•							•	•	•	•	•	•		•	•	•	•	•	
T2	SRAM, EEPROM	9.5K	512	•	•	•	•	•	•	•	•							•	•	•	•	•	•		•	•	•	•	•	
Triconex																														296
TRICON V6	BBRAM	1M	1500	•	•	•	•		•	•	•			•				•		•	•		•		•	•	•	•	•	
TRICON V7	BBRAM	512K	2500	•	•	•	•		•	•	•			•				•		•	•		•		•	•	•	•	•	
TRICON TSS6000	BBRAM	512K	128	•	•	•	•		•	•	•			•				•			•				•	•	•	•	•	
Uticor Technology																														297
Director 4001	CMOS	6K	384	•	•	•	•		•		•					•		•							•	•	•	•		
Director 4002	CMOS	6K	64	•	•	•	•		•		•					•		•							•	•	•	•		
Director 6001	CMOS	256K	4096	•	•	•	•		•		•							•		•	•				•	•	•	•	•	
Westinghouse																														298
PC-50	RAM, E/EEPROM	2KW	64	•	•	•	•	•		•	•	•			•			•			•	•				•	•			
PC-55	RAM, E/EEPROM	8KW	256	•	•	•	•	•	•	•	•	•			•			•			•	•			•	•	•			
PC-500	RAM, E/EEPROM	10KW	256	•	•	•	•	•	•	•	•	•			•			•			•	•			•	•	•			
PC-2000	RAM, E/EEPROM	48KW	2048	•	•	•	•	•	•	•	•	•			•			•		•	•	•			•	•	•	•		
Wizdom Systems																														299
386 Coprocessor	RAM	256/512K	4096	•	•	•		•	•	•	•	•		•				•		•	•	•			•		•	•	•	
186 Coprocessor	RAM	128K	4096	•	•	•		•	•	•	•	•		•				•		•	•	•			•		•	•	•	
Local Controller II	RAM	256/512K	4096	•	•	•	•	•	•	•	•	•		•				•		•	•	•			•		•	•	•	
SPLIC	RAM	128K	80	•	•	•	•	•	•	•	•	•		•				•		•	•	•			•		•	•	•	
SPLIC JR	RAM	128K	64	•	•	•		•	•	•	•	•		•				•		•	•	•			•		•	•	•	
Yaskawa																														300
GL40S3	RAM, E/EEPROM	8KW	2056	•	•	•	•	•	•	•	•				•	•		•	•		•	•			•	•	•	•		
GL60S2	SRAM	32KW	12288	•	•	•	•	•	•	•	•		•		•	•		•	•	•	•	•			•	•	•	•		
GL60S3	SRAM	32KW	12288	•	•	•	•	•	•	•	•		•		•	•		•	•	•	•	•			•	•	•	•		
GL60 high speed	RAM, E/EEPROM	64KW	12288	•	•	•	•	•	•	•	•		•		•	•		•	•	•	•	•			•	•	•	•		
GL70H high speed	RAM, E/EEPROM	64KW	12288	•	•	•	•	•	•	•	•		•		•	•		•	•	•	•	•			•	•	•	•		
2-World Engineering																														301
Little PLC	RAM, E/EEPROM	32KB	8		•	•				•				•				•			•	•			•		•			

Editor's Note: Information in this chart came directly from questionnaires returned by the companies listed. Questionanaire mailings were made to all known manufacturers of programmable controllers. To obtain further information about any of the products listed, circle the appropriate number on the reader service card.

FIGURE 4–1 (continued)

time to study any documentation at length before deciding on a particular PLC manufacturer. You will be glad you did.

4–5 PLC MANUFACTURERS

To give you a better idea about what is available in the PLC marketplace, we present, in figure 4–1, the full *I&CS Guide to Programmable Controller Hardware*, listing over 450 PLC models, which lists controller manufacturers and their products. Each April the *I&CS* journal publishes an exhaustive guide. The editors send out a questionnaire to all known manufacturers of PLCs. Information in the chart came directly from questionnaires returned by the companies listed. In the guide shown here, dozens of companies responded.

As you can see in the figure, the chart begins with the manufacturer's name and a list of the PLC models produced. Next, memory, both type and size, is provided. Information on I/O capability, analog availability, and the presence of the PID feature is provided. Languages are tabulated (with ladder logic being the most popular). Programming devices include manual types (hand-held programmers), and CRT, tape, and computer units. In addition, network capabilities are examined. Finally, a miscellaneous category ("Other") is provided that includes math processing, diagnostics, and documentation. All in all, it is a useful guide that should familiarize you with what is available and aid you in selecting an appropriate PLC.

EXERCISES

1. Interview a PLC manufacturer's representative to find out how he or she determines a user's PLC needs.
2. Interview an industrial manager to find out how he or she makes a PLC purchasing decision.
3. Write to a dozen PLC manufacturers asking for sales literature on their products.
4. Investigate the use of microcontrollers, such as the m68HC11, to see what impact they are having on the PLC market.
5. Investigate the use of industrial PCs to see what impact they are having on the PLC market.

Basic PLC Programming

5

Programming On–Off Inputs to Produce On–Off Outputs

OUTLINE

OBJECTIVES

At the end of this chapter, you will be able to

- Describe the contact (input) functions of the PLC.
- Describe the coil (output) function of the PLC.
- Describe the procedure to install a PLC on–off program.
- Convert industrial control problems to PLC logic diagrams.
- Show the advantage of PLCs over relay logic in simpler connection diagrams and wiring.
- Program PLC circuits to include a fail-safe operation.
- Develop a PLC ladder circuit for an industrial problem.

5–1 INTRODUCTION

In this chapter we illustrate how to program a PLC for circuit operations with on–off inputs and outputs. The inputs are labeled in different manners by different manufacturers. The labels commonly used include Examine On/Examine Off, words, letters, numbers, letter–word combinations, and instructions. In this book we use input prefixes and numbers: for example, IN009. Outputs are also labeled in different manners by different manufacturers. Common output designations are OUT, numbers, letters, and CR (control relay). We use the designation CR in this book: for example, CR013.

The first section of the chapter covers inputs of various kinds. The next section covers outputs and their relation to PLCs, then some typical operational procedures for inputting programs for on–off programming. There are many examples of on–off programming in the next section, including latch/unlatch circuits. A discussion of fail-safe circuits in included in the next-to-last section. The final section provides an industrial example of on–off programming.

5–2 PLC INPUT INSTRUCTIONS

As we said in the Introduction, there are a number of types of inputs that turn functions on and off. The inputs are called by different names by various manufacturers. The inputs are called inputs, words, functions, and instructions, such as Examine On and Examine Off. In this book we call them *inputs*. The various types of inputs include:

1. *Normally open contact.* When this contact closes, the function carries out some kind of action.
2. *Normally closed contact.* When this contact opens, the function carries out some kind of action.
3. *Latch/unlatch system.* Actuating the latch input turns the function on or causes it to change state. The function then stays on even if the latch input is turned off. To turn the function off, another input, unlatch, is turned on, which turns the function off. If unlatch is then turned off, the function remains off.
4. *Differentiation up, or rising-edge actuation.* This involves turning the function on for one scan time at the leading edge of an input signal pattern.
5. *Differentiation down, or falling-edge actuation.* This involves turning the function on for one scan time at the trailing edge of a signal pattern.

In this chapter we cover types 1, 2, and 3. For discussions of 4 and 5, see manufacturers' manuals.

In a PLC system, each input is assigned a number on the input module and in the CPU. The number may be a reserved block of numbers or letters. In other PLCs, some prefix is used, such as IN. In a prefix system, the fifth input would correspond to the PLC program number IN0005. A typical input scheme is shown in figure 5–1. The input terminals

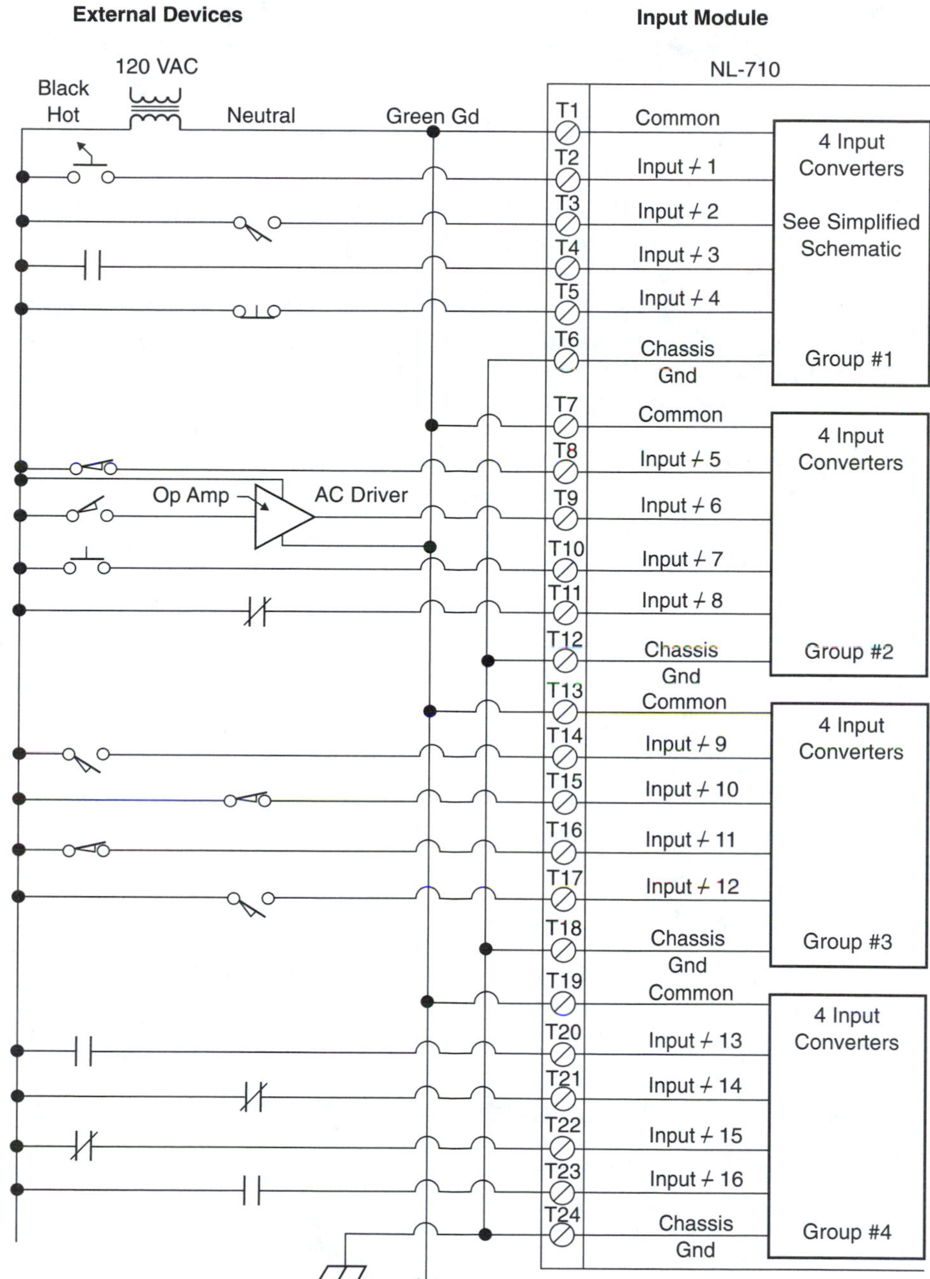

FIGURE 5–1
Typical PLC Input Scheme

correspond to a series of numbers, such as IN0001 through IN0016. The numbers of a module are set by SIP or DIP switches, as described in chapters 2 and 3.

Suppose that we apply power to terminal 5. All contacts programmed in the PLC as IN0005 will change states. The input is examined and the proper action takes place. All normally open IN0005 contacts will go to the closed state in the PLC program. Also, all normally closed IN0005 contacts go to an open state. This opening of normally closed contacts is a key concept in understanding PLC programming.

Figure 5–1 shows 120 VAC as the voltage used; 24 VDC is also a common voltage used. Different voltages can be used with an input module group, provided that the input module is rated for the voltage used.

Some typical input contact devices are shown in figure 5–2. More are shown in appendix C. Note that the value of input supply voltage (for example, 120 VAC in figure 5–1) must correspond to the voltage rating of the input module.

There is one other key point on contacts as related to inputs. Suppose that a contact in the internal program is labeled IN0018. Also suppose that the only inputs connected are IN0001 through IN0016. Would the IN0018 programmed contact ever change state from external signals? No. There is no energizing signal available from an input module to have an effect on the internal CPU status.

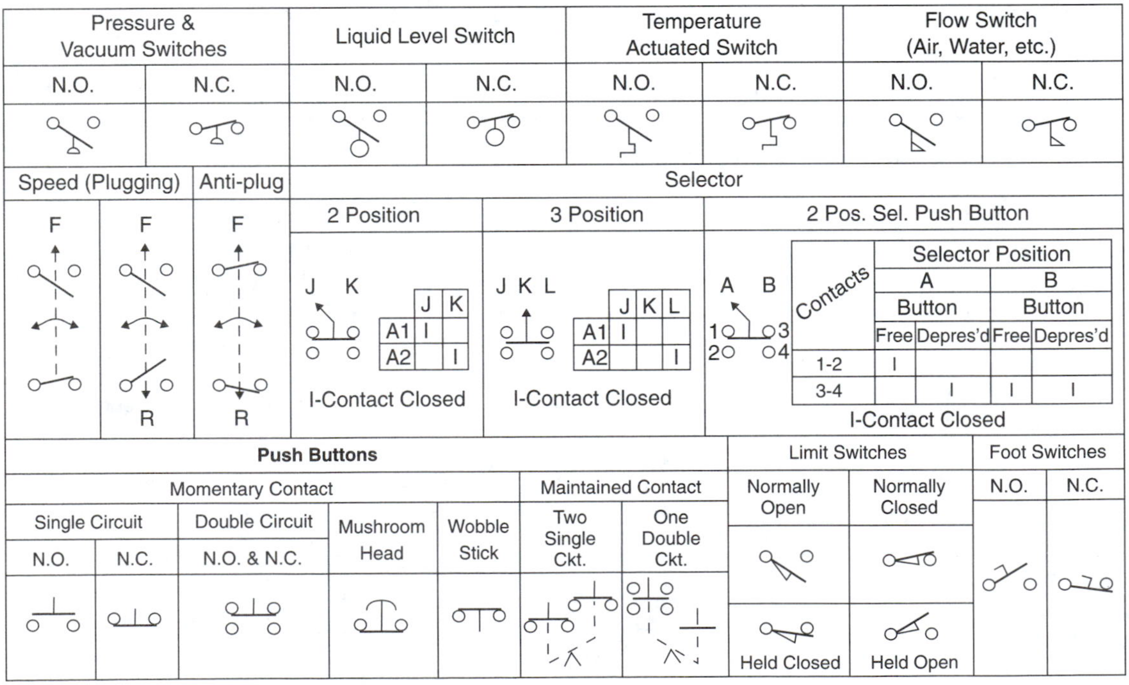

Pressure & Vacuum Switches		Liquid Level Switch		Temperature Actuated Switch		Flow Switch (Air, Water, etc.)	
N.O.	N.C.	N.O.	N.C.	N.O.	N.C.	N.O.	N.C.

Speed (Plugging): F, R

Anti-plug: F, R

Selector

2 Position: J K

	J	K
A1	I	
A2		I

I-Contact Closed

3 Position: J K L

	J	K	L
A1	I		
A2			I

I-Contact Closed

2 Pos. Sel. Push Button: A B, 1 3, 2 4

Contacts	Selector Position A		Selector Position B	
	Button Free	Button Depres'd	Button Free	Button Depres'd
1-2	I			
3-4		I	I	I

I-Contact Closed

Push Buttons

Momentary Contact					Maintained Contact	
Single Circuit N.O.	Single Circuit N.C.	Double Circuit N.O. & N.C.	Mushroom Head	Wobble Stick	Two Single Ckt.	One Double Ckt.

Limit Switches		Foot Switches	
Normally Open	Normally Closed	N.O.	N.C.
Held Closed	Held Open		

FIGURE 5–2
Typical PLC Input Devices

5–3 OUTPUTS: COILS, INDICATORS, AND OTHERS

Coils in an internal PLC program are related to output signals that are sent to external devices. An output is energized through the output module when its corresponding coil number is turned on in the PLC ladder diagram. Note that not all coils in a program have a corresponding output. Many coils are used for internal logic only. A typical output scheme is

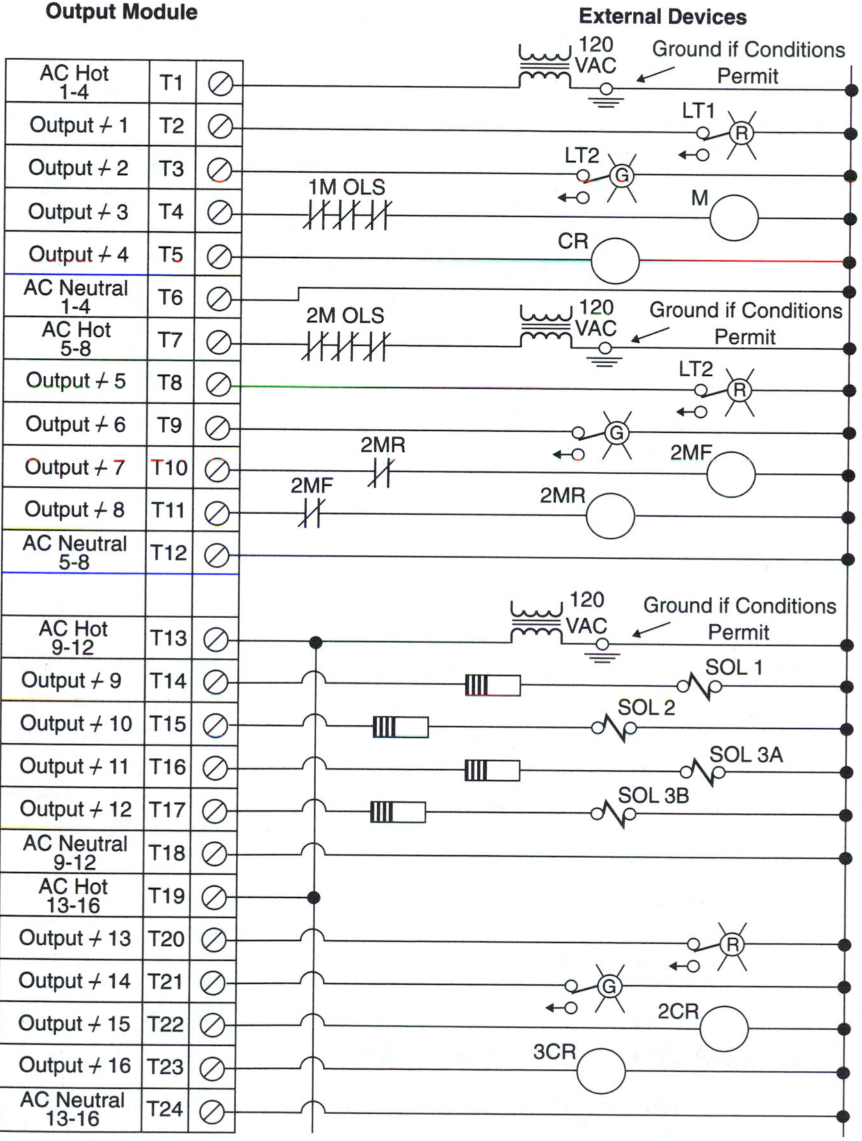

FIGURE 5–3
Typical PLC Output Scheme

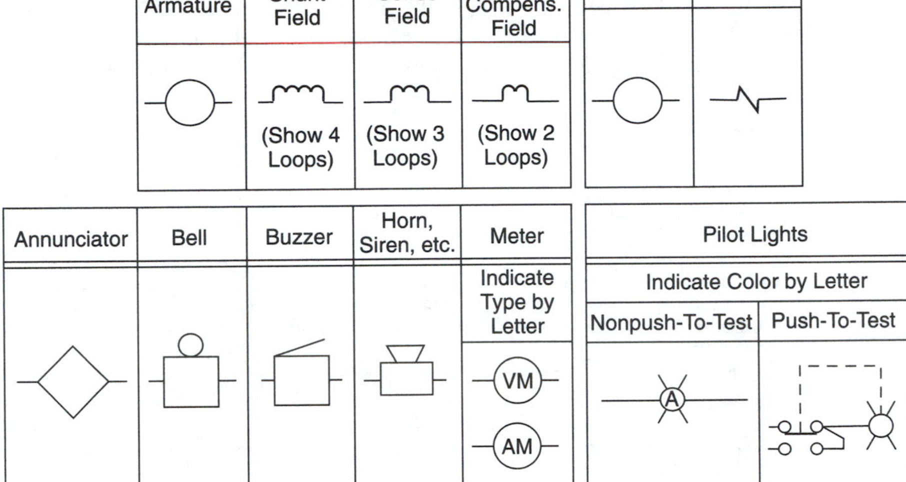

FIGURE 5–4
Typical PLC Output Devices

shown in figure 5–3. The output device's voltages and current requirements must be matched for the output module values. The figure shows 120 VAC used for the outputs. By using different ratings of output modules, other voltages can be used (in groups) as needed.

In a manner similar to inputs, output numbers must correspond. For example, only outputs CR0017 through CR0032 are connected to a CPU through an output module. If program coils have numbers such as CR0014 and CR0034, neither will affect any output. There is no corresponding coil for the output signal to affect. If CR0018 is turned on, output 18 will turn on.

Some typical output devices for coil outputs are shown in figure 5–4. A key point mentioned previously is the presence of a small output module leakage current when the PLC output is off. The leakage current must be considered if the output device is sensitive to a low value of voltage. The output device might not turn off even when the output module is technically in the off state. Figure 5–4 illustrates some of the typical output devices used in processes. More are shown in appendix C.

5–4 OPERATIONAL PROCEDURES

A simple program will indicate how to begin utilizing a PLC. Suppose that you wish to program and connect a PLC to accomplish the following discrete operational procedure: A relay coil is to actuate when two toggle switches and one limit switch are actuated.

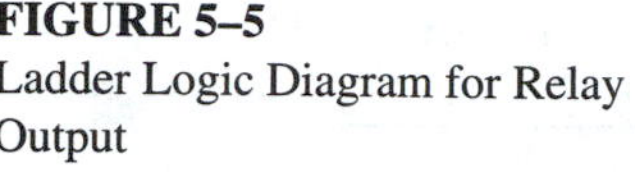

FIGURE 5–5
Ladder Logic Diagram for Relay Output

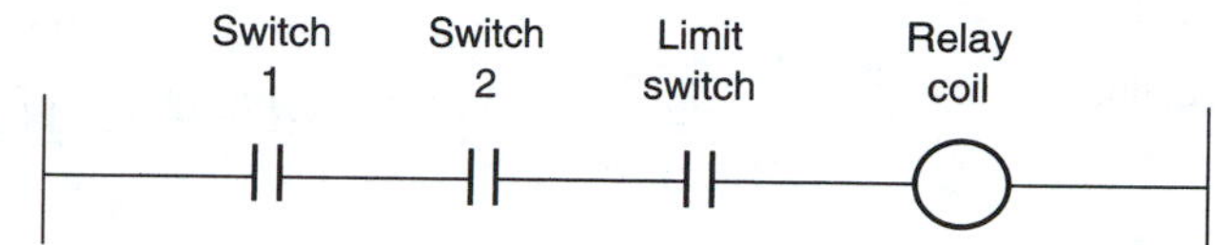

The first step is to assign individual PLC identification numbers to the inputs and outputs. Inputs normally have the prefix I or IN. Outputs normally have the prefix O or CR (control relay). The following numbers could be assigned:

Switch 1 for relay	IN001
Switch 2 for relay	IN002
Limit switch for relay	IN003
Relay output	CR001

Next, sketch a ladder logic diagram to represent the operational circuit. This is shown in figure 5–5.

Next, figure out how the inputs and outputs will be connected to the input and output modules. Assume an eight-terminal input and an eight-terminal output. It is necessary to set the module switches so that the modules recognize signals as inputs 1 through 8 and outputs 1 through 8. The connections from the inputs and outputs then are made according to figure 5–6. Note that each component is connected to one of the modules. No external interconnections are made.

Finally, the ladder program must be entered into the CPU by means of the keyboard. A general procedure for entering the program in ladder format is

1. Clear the PLC program memory with the CPU on Stop. The procedure will be outlined on a screen menu or in the operation manual for the PLC.

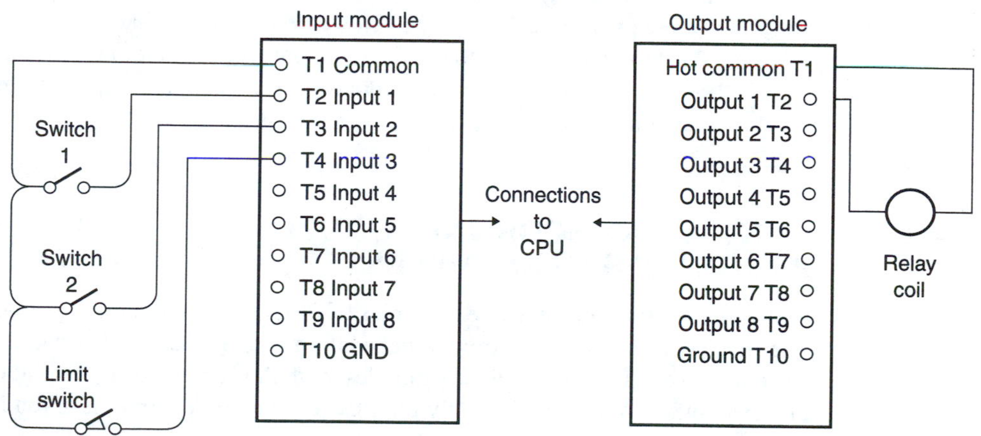

FIGURE 5–6
Connection Diagram for Figure 5–5 Circuit

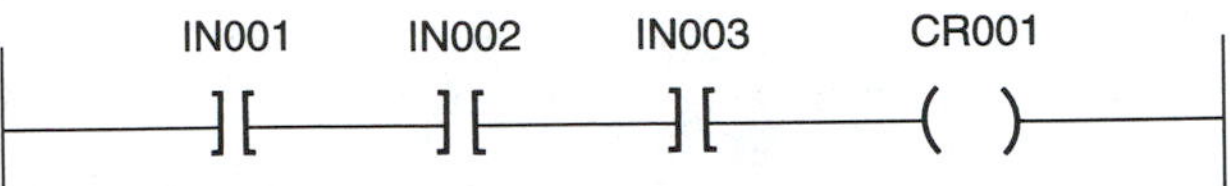

FIGURE 5–7
PLC Screen Ladder for Figure 5–5 Circuit

2. Insert the relay control line as follows, in the EDIT mode:
   a. Push the No contact key.
   b. Push the Input key.
   c. Push 001 numeric keys.
   d. Push the Enter key. The contact should appear on the monitor.
   e. Move the cursor one space to the right.
   f. Repeat steps a and b.
   g. Push the 002 numeric keys.
   h. Push the Enter key. The second contact should appear on the monitor.
   i. Move the cursor one more space to the right, and repeat the process for 003.
   j. Continue the line to the right.
   k. Push the Coil/output key. The coil should appear on the monitor.
   l. Push 001 numeric keys.
   m. Push Enter.
   n. If the line now looks correct (check it), push the Insert ladder key and then Enter.

The resulting PLC diagram should look as shown in figure 5–7. When the PLC switch is set to Run, the circuit will operate as outlined.

What if the programmer were a small hand-held one with no screen? The programming would be similar to steps a through k. Three differences in the sequence would be as follows:

1. Instead of moving the cursor to the right in step e, you would press the And key.
2. Before pressing CR in step k, you would push a Load key.
3. You might not be able to use the same number (001) for both an input and an output. Check how numbers are allocated to inputs and outputs by referring to the operating manual.

5–5 CONTACT AND COIL INPUT/OUTPUT PROGRAMMING EXAMPLES

Following are six representative examples of PLC programming using contacts and coils. The first five examples range from basic to intermediate. The sixth is a more complex alarm system problem. For the first three examples, both PLC and relay logic solutions are shown. For examples 5-4 through 5-6, only the PLC connection diagram is given in the problem solution.

The six examples are:

Example 5–1 Simple one-contact, one-coil circuit.

Example 5–2 Standard start–stop–seal circuit; alternate latch-unlatch circuit.
Example 5–3 Forward–reverse–stop with mutual interlocks.
Example 5–4 Forward–reverse–stop with direct reversal.
Example 5–5 Start–stop–jog.
Example 5–6 Alarm system.

EXAMPLE 5–1 The first example is a simple circuit with one switch as a contact and one output as a coil. As the switch is opened or closed, the output goes on or off. Figure 5–8 shows the ladder diagrams for relay logic and ladder logic. They are identical for this example. The control voltage for M can be 24 VDC, 120 VAC, or another voltage. The current supplied to M is relatively small. The current being controlled on–off through the power contacts can be very large for either relay or PLC control. The power voltage can be any value of DC or AC, depending on the rating of the power contacts.

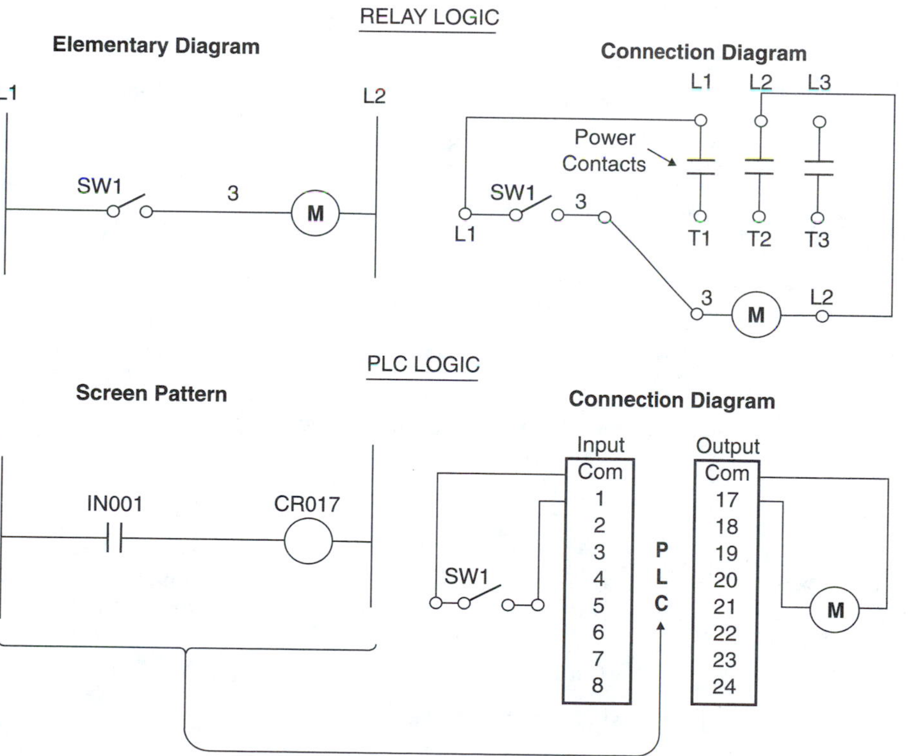

FIGURE 5–8
Example 5–1: Simple One-Switch, One-Coil Control

EXAMPLE 5–2 The second example is a start–stop–seal circuit. When the start button is depressed, the coil energizes. When the button is released, the coil remains on. It is held on by a sealing contact that is in parallel with the start button. The seal contact closes when the output coil goes on. If the stop button is depressed, the coil goes off and stays off. Also, if the control power goes off, the coil goes off. The advantage of this example over example 5-1 is that when failed control power returns, Start must be depressed to reenergize the coil. For example 5–1, the coil would immediately restart, possibly posing a safety hazard to an unsuspecting operator or repair person.

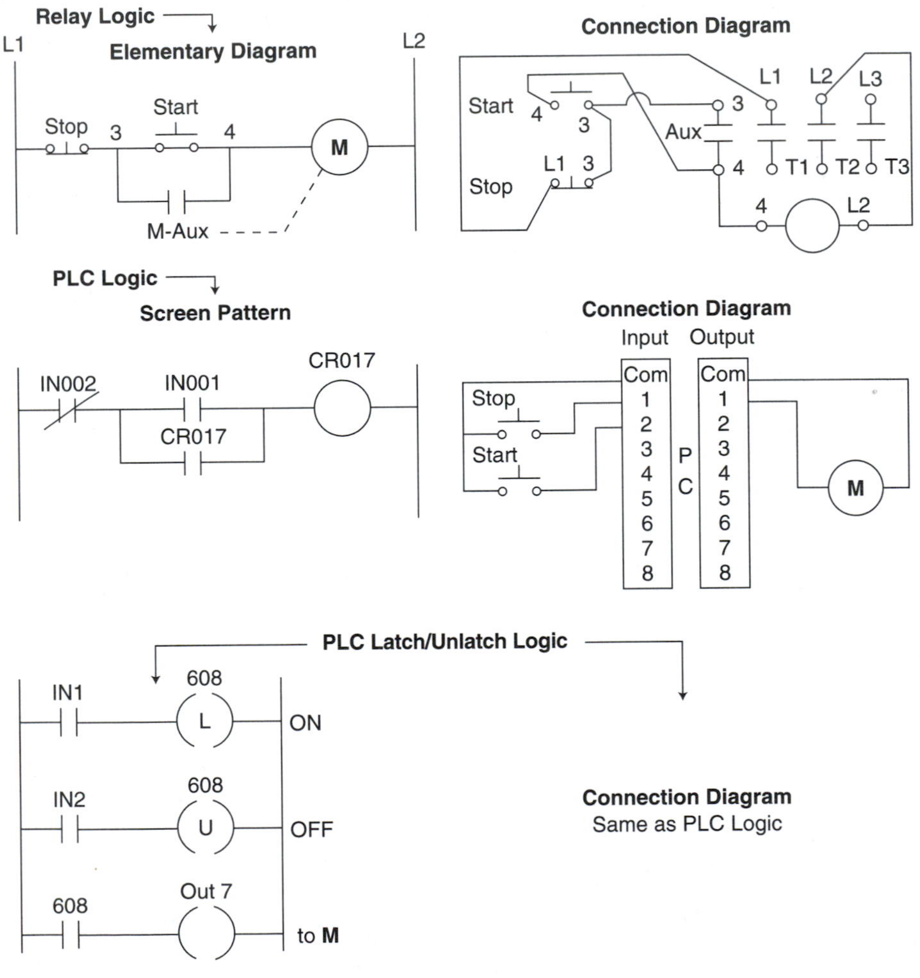

FIGURE 5–9
Example 5–2: Standard Start–Stop–Seal Circuit

The relay logic and PLC logic diagrams for this example are shown in figure 5–9. A major difference between relay and PLC connections is in the physical location of the seal contact. In relay logic, the seal contact is attached physically to, and goes on and off with, the output coil. In the PLC control, the seal is generated internally in the PLC logic. In PLC logic, the seal closes or opens as the output coil goes on or off.

Figure 5–9 also illustrates the LATCH/UNLATCH function used in many types of PLCs. One input switch latches the output on. A different input switch latches the output off. The latch/unlatch PLC coil may be an internal one. A contact from its coil may have to be used to control a coil that corresponds to an output terminal.

Note that if CR017 is associated with a motor starter coil with an overload relay and contact, special precautions are needed. If the overload coil in series with the starter coil opens up, the starter will open up, but it will not unseal CR017 in the PLC program. If the overload then cools off and its contact recloses, the motor will restart unexpectedly. Further circuitry is necessary, as discussed in section 5–6. Also note that for an latch/unlatch system, power must be applied to shut it off. This is another possibly unsafe situation which requires further circuitry.

EXAMPLE 5–3 The third example is a standard forward–reverse circuit. Each direction's coil has its own start button. The single stop button stops either coil's operation. In this circuit you must push Stop before changing direction. Interlocks are provided so that both outputs cannot be energized at the same time. This particular circuit works for other control applications as well. It could be used for low-speed/high-speed or part-up/part-down control systems. Figure 5–10 shows the circuit. IN0001 stops operation in either direction. IN0002 is for forward, CR0017, and IN0003 is for reverse, CR0018.

There are two notes in figure 5–10 that should be explained. The note below the relay logic elementary diagram concerns fail-safe operation. If the motor overloads, the normally closed contact, OL, opens up; when the motor overheats, the circuit is shut down for relay logic. If the motor cools off and the OL contact recloses, the motor won't restart until the start button is activated. In the PLC circuit, the OL contact is in the power circuit. When the OL contact opens, it does not affect the control circuit on the left. The CR017 or CR018 circuit remains sealed. When the motor cools off and the OL closes, the motor will restart automatically, which can be dangerous. To prevent this situation, the OL contact can be wired to the input module and the circuit programmed to run each logic line through an OL normally closed PLC contact.

The other note above the PLC screen pattern refers to the programming of the stop button contacts. For our diagram, the stop button contacts used are normally closed. IN001 is programmed as a normally open contact. When the power circuit is energized, the PLC contact closes as shown. In some programming schemes (not shown), the stop button contact used is normally open, so the PLC contact is programmed as a normally closed contact. In any case, refer to section 5–6 for how to make the circuit truly fail-safe.

At this point we begin to see that the PLC connections are simpler than connections for relay logic. Compare the two connection diagrams; the relay logic connections are quite complicated in comparison to the PLC connections.

Relay Logic

Elementary Diagram

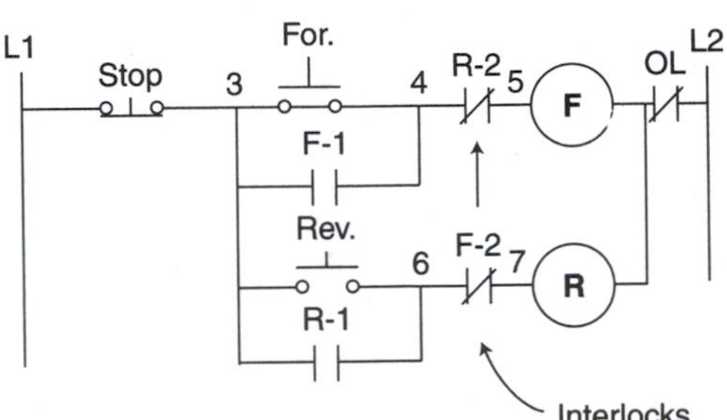

Connection Diagram

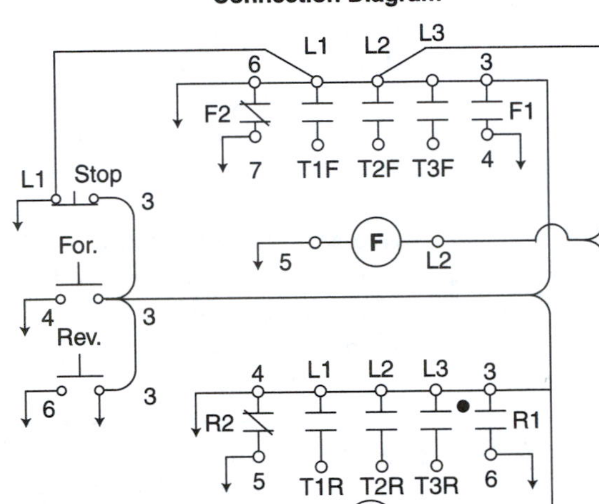

Note: Relay circuit above is "fail safe." It unseals when OL (overload) opens. The PLC circuit below is not "fail safe" since the OL does not unseal the circuit.

To Numbered Wire Arrows Above

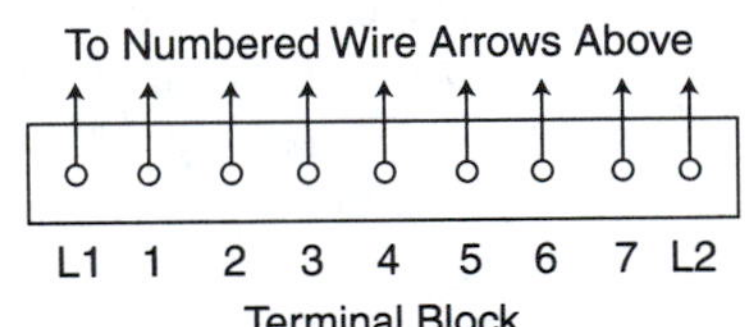

Terminal Block

Note: IN001 is programmed open. Will close when control power is applied.

PLC Logic

Screen Pattern

IN001 IN002 CR18 CR017 F

CR017

Interlocks

IN001 IN003 CR17 CR018 R

CR018

Connection Diagram

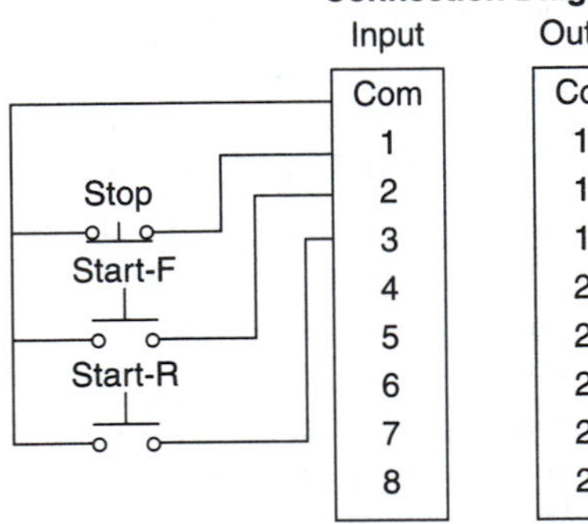

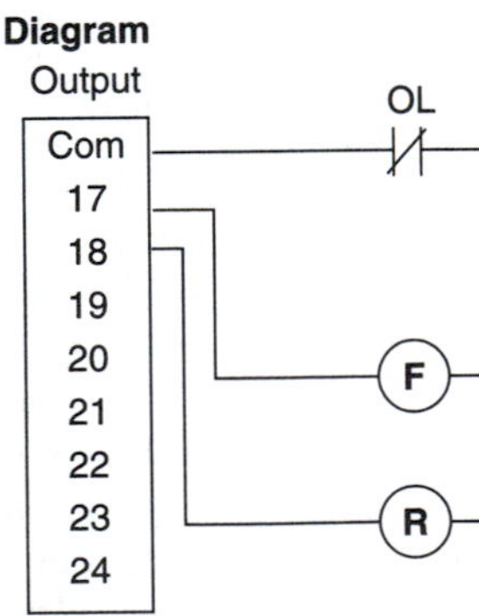

FIGURE 5–10
Example 5–3: Forward–Reverse Control

EXAMPLE 5–4 The fourth example, shown in figure 5–11, is similar to example 5-3. The major difference is that in this example you may go directly from one direction to another without first depressing Stop. Another difference is the added directional pilot light indicators. The key for input identification is shown on the diagram.

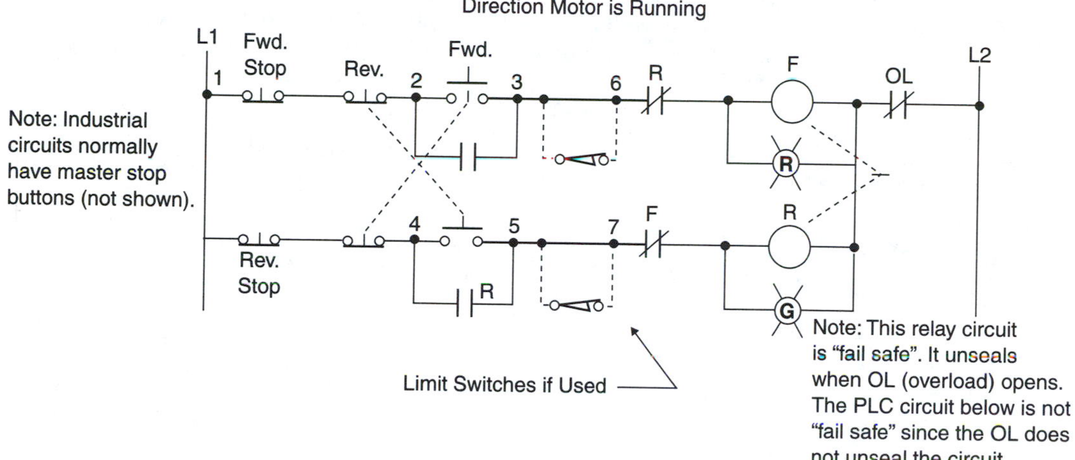

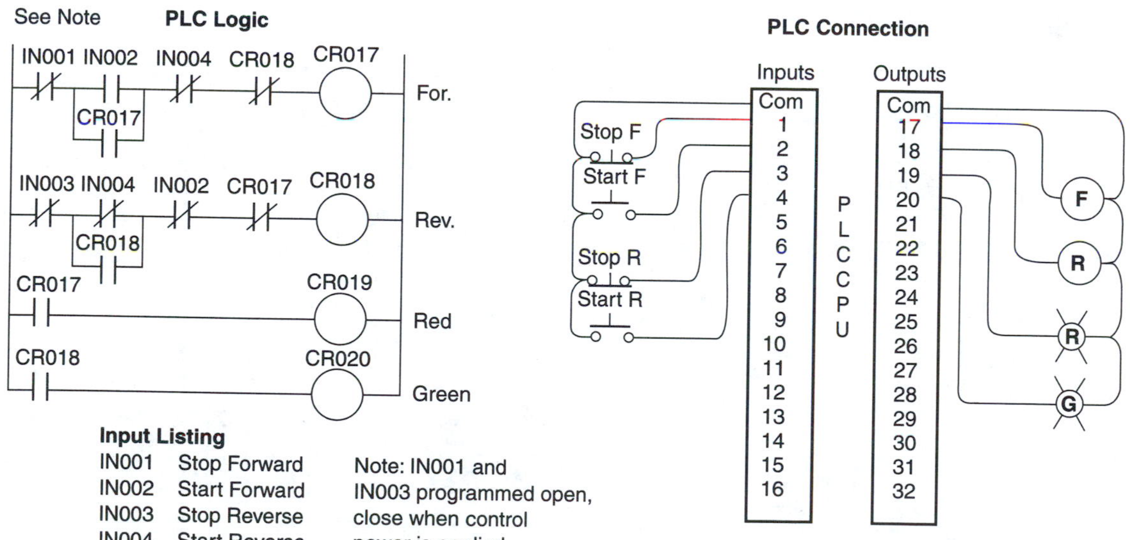

FIGURE 5–11
Example 5–4: Instant Forward–Reverse Change Circuit

The quick reversal may be desirable in some operational applications; however, in some cases it may not be a good system. If, for example, we have a large flywheel as an output device connected to an electric motor, applying instant reversal would cause undue stress on the motor, the power distribution system, and the mechanical parts and mountings. A more advanced circuit with a time delay would be required.

The two notes about fail-safe and how to program the Stop button explained previously also apply to figure 5–11. Additionally, there is a specific listing identifying the inputs, for clarity.

EXAMPLE 5–5 In some cases you may wish to have the output on momentarily only at times. The momentary on is called *jog*. At other times you might like the output on all the time, as in previous examples. Two possible circuits for start–stop–jog are shown in figure 5–12. The

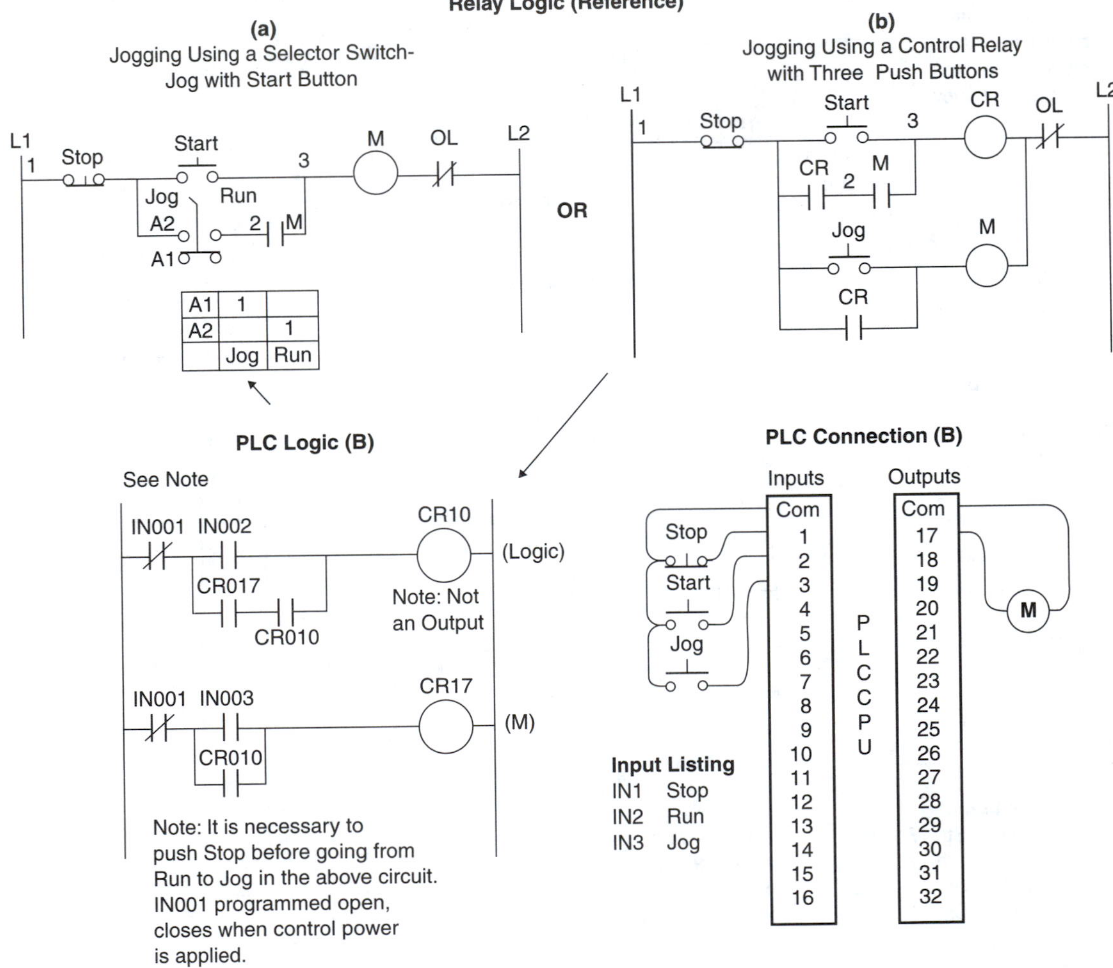

A1	1	
A2		1
	Jog	Run

FIGURE 5–12
Example 5–5: Start–Stop–Jog Circuits

PLC program and the PLC connection diagram are included also. Note that it is necessary to push Stop before going from run to jog in the circuit illustrated here.

EXAMPLE 5–6 The sixth example is an alarm system. There are four hazard inputs to the alarm system that go on as some operational malfunction occurs. We do not define what the hazards are; for PLC operation illustration, we only use the fact that there are four. The system operates as follows:

- If one input is on, nothing happens.
- If any two inputs are on, a red pilot light goes on.
- If any three inputs are on, an alarm siren sounds.
- If all four are on, the fire department is notified.

Since this example is somewhat more involved than the previous ones, let us take time to specify input and output numbers. The PLC program numbers for the inputs and outputs are assigned as follows:

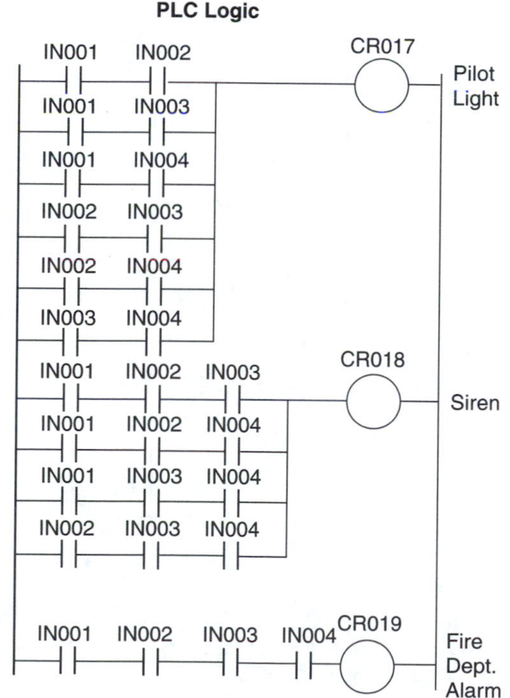

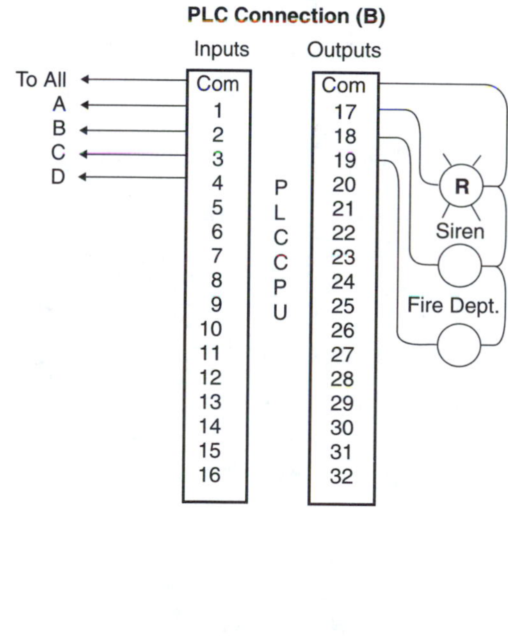

FIGURE 5–13
Example 5–6: Alarm System

	Inputs	Outputs
A IN001	Red Pilot Light	CR017
B IN002	Alarm (Siren)	CR018
C IN003	Fire Department Notify	CR019
D IN004		

A PLC logic diagram to accomplish the circuit requirements is shown in figure 5–13. One final note for this example: Connecting the PLC terminal to the output alarms is very simple; if you had a relay switch system, the connections would be very involved and complicated.

5–6 A LOOK AT FAIL-SAFE CIRCUITS

Some PLC circuits are programmed to be turned off by applying a signal voltage. For example, the LATCH/UNLATCH function requires an unlatch signal to turn the coil or out-

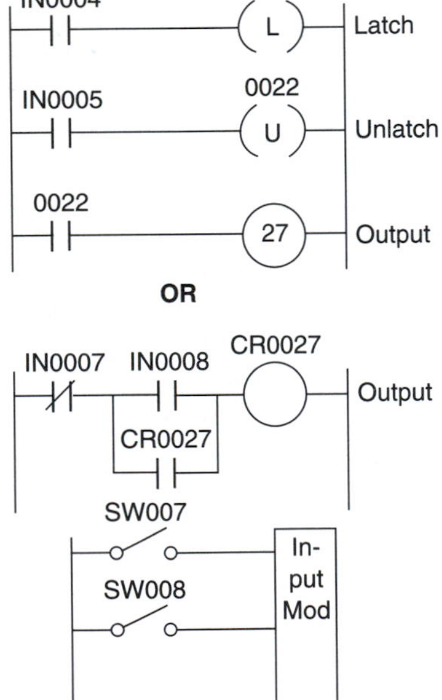

Not Fail Safe
Both require control power available to turn off the output 0027.

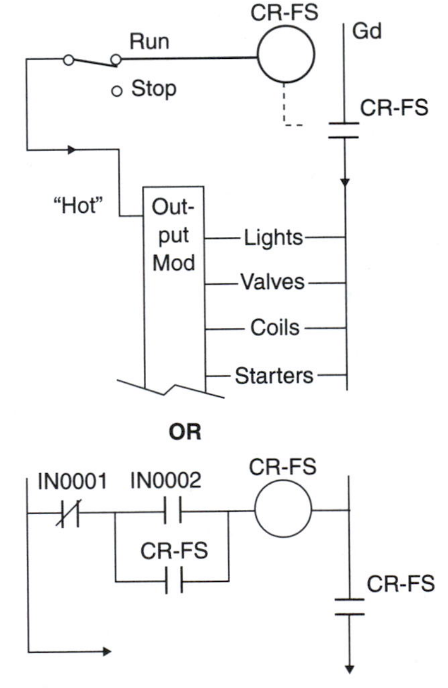

Fail Safe
Circuitry in addition to the PLC circuitry is used. Depressing stop switch *or* loss of control power turns outputs off.

FIGURE 5–14
Safety Fail-Safe Circuit

put off. If you lose control power, pushing the stop button has no effect and the coil remains on, since control power is needed for system turn-off.

Emergency stop switches or pushbuttons that are independent of the PLC on–off circuits should be included. Figure 5–14 shows a circuit that could be used as a true fail-safe system. Turning the master run-stop switch off in the fail-safe circuit on the right deenergizes all coils.

Of course, in the system, fail-safe should be defined. You may not wish to turn off all coils when the emergency switch or pushbutton is pushed. Suppose that a device is spring return. If you expect the emergency switch to stop the machine where it is, it won't; it springs back. For true fail-safe operation, a complete control system analysis is needed.

5–7 INDUSTRIAL PROCESS EXAMPLE

The next example is more involved than the previous ones in this chapter. To formulate a control system, we generally follow the procedural steps of chapter 4. The problem involves a semiautomatic drill press operation as shown in figure 5–15.

The initial position of the drill press spindle is at the top, as shown. A part to be drilled is placed under the spindle. The drill is then to come down after two start buttons are depressed. (Two pushbuttons are recommended to assure that both hands are out of the way.) The drill spindle rotates as it is brought downward. Downward spindle force is furnished by a pneumatic air cylinder pushing against an upward return spring; pneumatic control air is supplied through an electrical solenoid. When the spindle is completely down and the drill bit goes through the part to be drilled, a down sensor is actuated. The solenoid is then deenergized, and the drill returns up by means of the return-up spring. When the spindle is completely back up, the system is to be reset to the off condition. If no part is in place initially, the drill spindle cannot descend.

In addition to the operation described, a safety shield is included. For extra safety, a screen shield comes down before the drill can start down. The shield returns up at the same

FIGURE 5–15
Drill Press Operation Layout

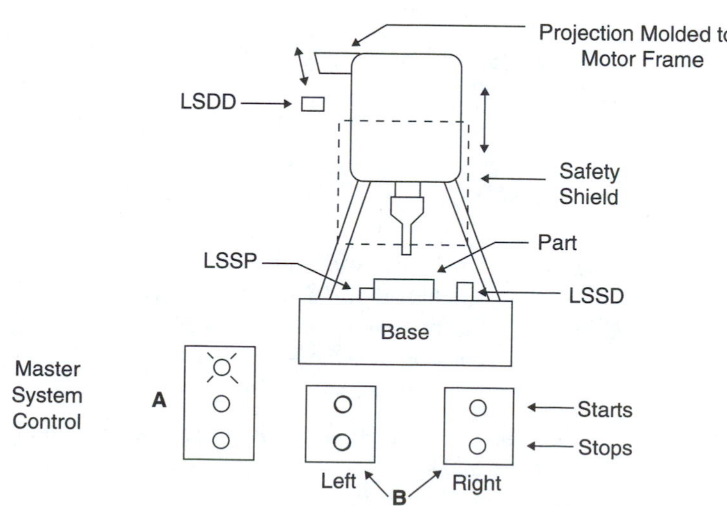

time as the drill by its own spring return. The shield's descent is powered by its own separate pneumatic solenoid.

When the stop button is pushed at any time, the drill and shield return up. Note that this could be a safety hazard. More circuitry would be needed to stop the spindle where it is when the stop button is depressed.

There are a number of procedural steps to go through to arrive at a solution. Previous examples have not been complicated, and we have performed their procedural steps informally. The steps recommended for a problem of this type are:

1. Define the process operation and list the step-by-step sequence of operation.
2. Define and list the input and output devices and sensors required for proper operation.
3. Assign corresponding PLC numbers to the input and output devices.
4. Draw up the PLC scheme. Note that margin notes are helpful.
5. Enter the program into the PLC.
6. Optional step: Check the program sequence by using the FORCE mode. (The FORCE mode is explained in detail in chapter 26.)
7. Wire the PLC system to a simulator and check its operation.
8. Check the actual process operation. Try various out-of-sequence operations to check for hidden safety defects or sequencing problems. For example, what happens if the power fails when the spindle is halfway down?
9. Make modifications as required.

Step one is to list the sequence.

1. Push system start switch.
2. Put part in place to actuate LSPP.
3. Push the two start buttons simultaneously.
4. Safety shield comes down, actuating LSSD.
5. Drill starts rotating and descends.
6. Drill at bottom actuates LSDD.
7. System shuts down. Drill and shield return up by springs.
8. System is reset.

Note that pressing Stop at any time stops the sequence and resets the spindle to the top.

Step two is to list the input and output devices.

- □ System start switch
- □ System stop switch—stops everything
- □ System pilot light
- □ Shield and drill start—left-hand switch
- □ Shield and drill start—right-hand switch
- □ Position indicator—part in place
- □ Position indicator—shield down
- □ Position indicator—drill down

Step three is to assign input and output numbers to all components. This includes switches and sensors.

Inputs		Outputs	
IN001	System start	OUT017	System pilot light
IN002	System stop	OUT018	Shield down solenoid
IN003	LSPP—part in place	OUT019	Motor rotate motor
IN004	Left start	OUT020	Drill down solenoid
IN005	Right start		
IN006	Left stop		
IN007	Right stop		
IN008	LSSD—shield down		
IN009	LSDD—drill down		

Step four is to sketch the PLC system.

Step five is to load the sketch into the CPU. The ladder diagram formulated is shown in figure 5–16.

A sequence by line for the ladder diagram in figure 5–15 is

- Line A. Push Start, IN001. CR17 goes on.
- Line B. CR017 seals on.
- Line C. Put part in place. IN003 closes. IN009 was already closed. Push right and left Starts, IN004 and IN005. CR018 on. Shield starts down.
- Line D. Start buttons IN004 and IN005 must still be pushed. CR018 is closed on. When shield reaches down position, LSSD closes, IN008. CR018 seals on.
- Line E. CR0018 and CR0019 are closed. CR0019 on, motor rotates.
- Line F. CR0019 starts motor down, CR0020.
- Line G. Drill reaches lower position and hits LSDD, IN009. CR0018, CR0019, and CR0020 all go off. Motor goes off and shield and drill spring-return up.

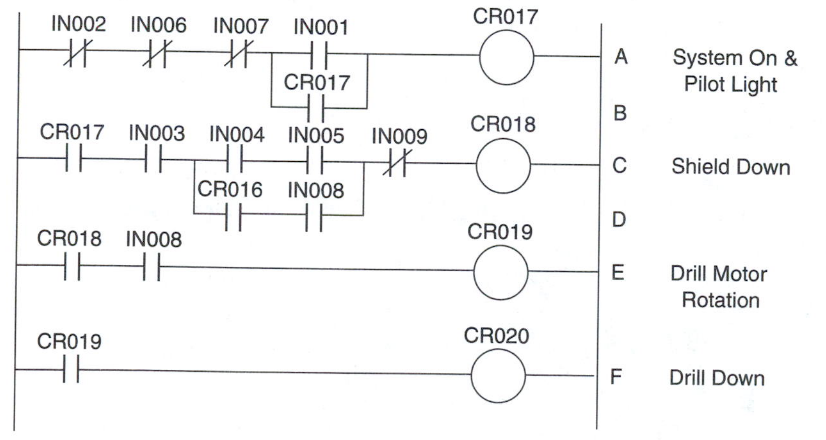

Note: Again, stop buttons are programmed open, but close as soon as control power is applied.

FIGURE 5–16
Drill Press PLC Control Circuit

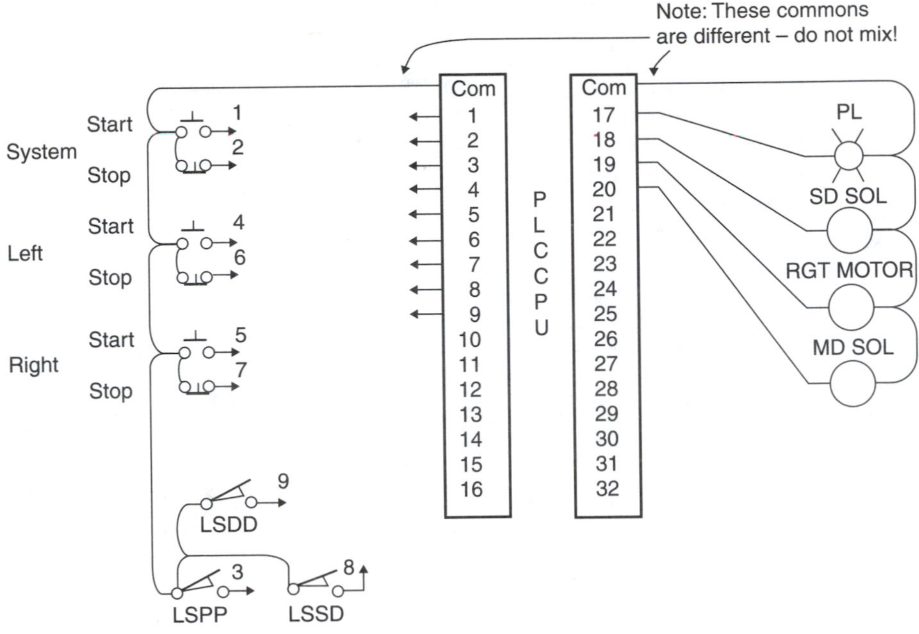

FIGURE 5–17
Input and Output Module Wiring for Drill Press

Step six is an optional FORCE analysis.

Step seven is to wire the system to a simulator. A wiring scheme appears in figure 5–17. Note the connection diagram's simplicity for the PLC—only five output wires and nine input wires.

Step eight, circuit operation, and step nine, modifications, would follow after an analysis of the drill press's actual operation.

TROUBLESHOOTING PROBLEMS

The PLC has been programmed to operate as shown in the figures referred to. However, the circuit has the malfunction noted. What misprogramming or other factor or factors could cause the malfunction?

TS 5–1 Refer to the figure 5–9 (example 5–2) PLC circuit.

1. The output does not go on when IN001 is energized.
2. The output does not seal on when IN001 is released.
3. The circuit cannot be turned off.

TS 5–2 Refer to the figure 5–10 (example 5–3) PLC circuit.

1. When IN002 or IN003 is energized, nothing happens.
2. You can energize the forward output but not the reverse output.
3. When Stop, IN001, is energized, the reverse output does not shut off.

TS 5–3 Refer to figure 5–11 (example 5–4).

1. The output pilot lights remain on even if the coils are off.
2. You can go directly from forward to reverse, but not from reverse to forward.
3. One or both forward and reverse are momentary only.

TS 5–4 Refer to the figure 5–12 (example 5–5) PLC circuit.

1. When Jog is depressed, the coil CR10 seals on.
2. When Start is depressed, the CR10 coil does not seal on.

TS 5–5 Refer to figure 5–13 (example 5–6). State how the system would malfunction if one or more of the inputs were erroneously programmed normally closed instead of normally open.

EXERCISES

Construct PLC ladder diagrams for the problems listed. A sequence could be written first, if necessary. As an option, show the input and output modules along with the device-to-terminal electrical connections.

In the laboratory, you may load the PLC CPU with your program. Connect the PLC to a simulator. Check out the proper operation of the circuit by running the program sequence.

1. Draw figures similar to figures 5–5, 5–6, and 5–7 for the following operation.

a. Switch 11 or switch 12 turns on relay 1.
b. Switch 13 and switch 14 turn on relay 2.
c. Switch 15 or limit switch 31 and relay 1 turn on relay 3.

2. A fan is to be started and stopped from any one of three locations. Each location has a start and a stop button (refer to example 5–2). Note that normally closed stops should be in series and normally open starts in parallel.

3. A two-way hydraulic cylinder has two solenoids controlling it. Energizing one solenoid causes the cylinder to extend and energizing the other solenoid causes it to retract. A limit switch at each end indicates full retraction or full extension. Use two start–stop three-wire controls, one for each direction. Construct a two-directional control system, including interlocks, to control the solenoid. Refer to example 5–3.

4. A milling machine (M) and its lubrication pump (L) both have three-wire start–stop control systems. A three-wire system is shown in figure 5–9 (three control wires from PB station to starter). L must be running before M can be started. Furthermore, if L stops, M must also stop.

5. Two separate start–stop–jog control stations are required for a pump motor. Refer to example 5–5.

6. There are three machines, each with its own start–stop buttons. Only one may run at a time. Construct a circuit with appropriate interlocking.

7. Repeat exercise 6, except that any two may run at one time. Also, any one may run by itself.

8. A temperature control system consists of four thermostats. The system operates three heating units. Thermostats are set at 55, 60, 65, and 70°F. Below 55°F, three heaters are to be on. A temperature between 55 and 60°F causes two heaters to be on. For 60 to 65°F, one heater is to be on. Above 70°F, there is a safety shutoff for all three heaters in case one stays on by mistake. A master switch turns the system on and off.

9. Create a PLC system in a manner similar to the example in section 5–7 for the problem in figure 5–18. When a part is placed on the conveyor at position 1, it automatically moves to position 2.

FIGURE 5–18
Diagram for Exercise 9

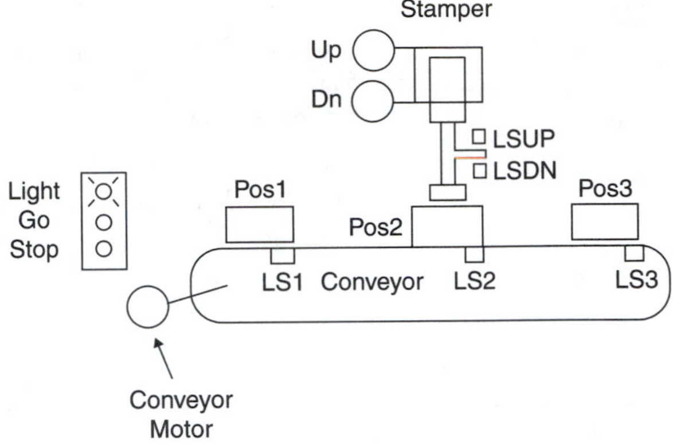

Upon reaching position 2, it stops and is stamped. After stamping, it automatically moves to position 3. It stops at 3, where the part is removed manually from the conveyor. Assume that only one part is on the conveyor at a time. Add limit switches, interlocks, pushbuttons, and other devices as required. If you become stuck at the middle station, you may add a manual restart switch for this point on the conveyor.

Extra credit: Remove the middle restart and operate automatically.

6

Relation of Digital Gate Logic to Contact/Coil Logic

OUTLINE

OBJECTIVES

At the end of this chapter, you will be able to

- □ List the seven basic digital gate types, draw their symbols, and describe their function.
- □ Show the relation of switch contact logic (relay/PLC ladder logic) to digital logic for each gate type.
- □ Create digital systems and PLC/relay logic diagrams from process word descriptions.
- □ Convert from any one of the three programming systems to any other for (1) process operation word description, (2) relay logic diagrams/PLC logic diagrams, and (3) digital gate diagrams.
- □ Write Boolean expressions.

6–1 INTRODUCTION

Large PLC programming systems of the screen/monitor type do not require the use of digital gate logic principles. The programming is normally done by typing in lines, connection nodes, contacts, and coils or functions. However, most smaller programmers with smaller LED displays have keyboard keys with digital logic notations.

These smaller programmers can have digital gate logic keys such as AND, OR, NOT, and others. This chapter shows how to relate these logic terms to the relay and large-screen PLC logic. Once the logic terms are understood, PLC programming using them can be easily accomplished.

Other symbols that appear on some PLC keyboards are a dot, +, −, 0, and =. These are Boolean algebra symbols. Boolean algebra is a shorthand way of writing digital gate diagrams. Since this type of programming format is not found often, Boolean principles are discussed only briefly.

There is another reason for studying digital programming. Some computer-trained persons understand PLC programming best using digital logic. This chapter should help these people to program PLCs properly.

This chapter compares word descriptions, relay ladder diagrams/PLC ladder diagrams, and digital gate diagrams. The relay ladder diagrams and PLC ladder diagrams involve coil/contact programming. More important, in this chapter we show how to translate from one of the three systems to another. The use of a Boolean system and its digital relation is also reviewed briefly. A person who knows one system but must program in another will be able to do so by mastering this chapter's principles. You may be interested in delving deeper into digital gates and Boolean equivalents. Many digital logic texts cover them in detail.

By necessity, the chapter has a high ratio of illustrations to text to fully explain the principles involved.

6–2 DIGITAL LOGIC GATES

We discuss digital logic gates from a PLC logic standpoint, but we do not cover the details of their electronic internal workings or their electrical operation. Figure 6–1 shows the seven basic types of logic gates.

All gates have one output. They are either on (1) or off (0), depending on the logic status of their inputs, on (1) or off (0). A gate-on condition is typical when +5 volts DC comes from the output terminal. Off is typically 0 volts output.

The NOT gate always has one input. The EXCLUSIVE OR and EXCLUSIVE NOR gates usually have two inputs but can have more. The other four types can have two to eight inputs and sometimes more. An input on is typically when +5 volts DC is applied to an input terminal. Off is typically 0 volts applied to an input terminal.

There is internal electronic circuitry that causes the gates to function properly. There are usually four gates of one kind in the digital integrated circuit. In addition to input and output terminals, each chip needs two terminals for power, one for +5 volts, the other for ground.

FIGURE 6–1
Six Basic Digital Logic Gates

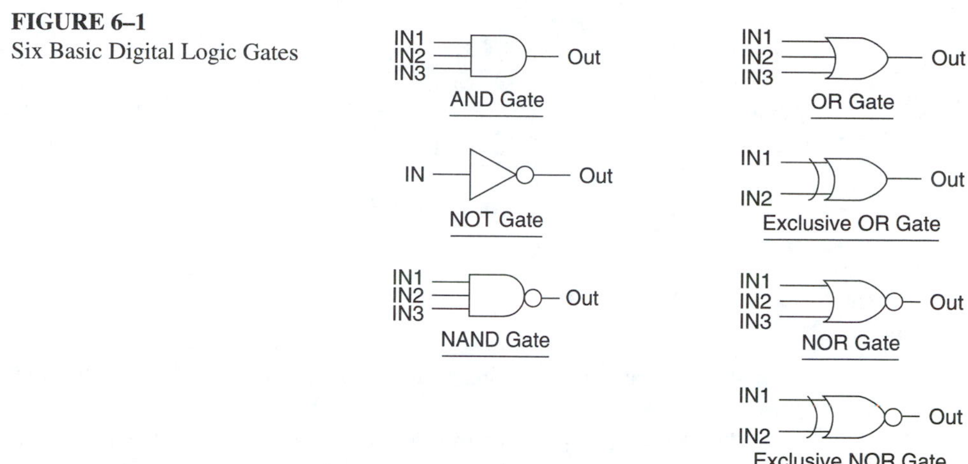

The AND gate and its programming equivalents are shown in figure 6–2. For the AND gate output to be on (1), all inputs must be on (1). The relay programming and PLC programming equivalents are also shown in the figure. For the four-input situation, input 1, input 2, input 3, and input 4 must be on for output 12 to be on. Otherwise, output 12 is off.

For AND gate programming with a digital PLC keyboard, the sequence of key operation is 1, and, 2, and, 3, and, 4, =, 12.

The OR gate operation is shown in figure 6–3. For an OR gate output to be on (1), any one or more of the inputs must be on (1). For the output to be off (0), all inputs must be off (0). The same operational voltages and principles as for AND gates apply to OR gates.

Again, the equivalent relay and PLC programming diagrams are shown in the figure. The word description of this operation is, "For output 17 to be on (1), any one or more of inputs 1, 2, and 3 must be on (1); otherwise, output 17 is off (0)."

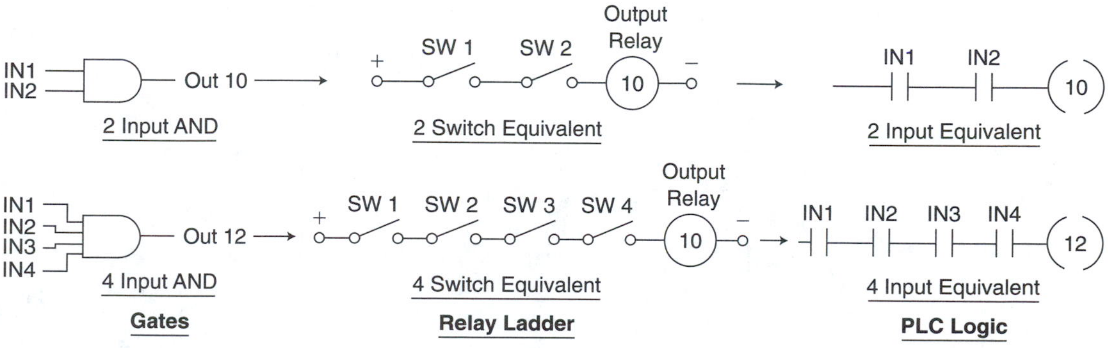

FIGURE 6–2
AND Gate and Relay and PLC Equivalents

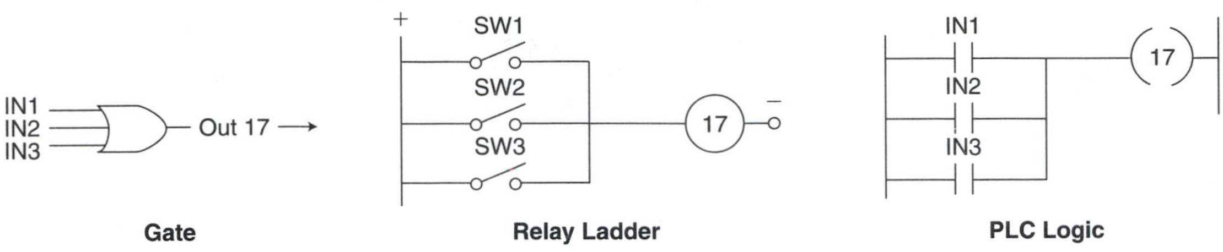

FIGURE 6–3
OR Gate and Relay and PLC Equivalents

To input an OR keyboard program for figure 6–3 we would use the sequence 1, or, 2, or, 3, or, =, 17.

The NOT gate is shown in figure 6–4 along with relay and PLC equivalents. It reverses the input logic status, on or off, from the input to the CPU. The output will then be the reverse of the input, off or on. A NOT key input, inserted at the proper point in the program sequence, carries out the reversal.

The EXCLUSIVE OR gate symbol is included in figure 6–1. Its output is on (1) when one, and only one, of its two inputs is on (1). If both inputs are on, the output is off. Thus this gate's output is on if the inputs are *different*. The EXCLUSIVE OR gate is seldom used in PLCs, so we do not discuss it beyond showing its basic symbol.

The EXCLUSIVE NOR gate symbol is also included in figure 6–1. Its output is on (1) when both of its inputs are on or off (0). If any of its inputs are on, its output is off. In other words, this is gate's output is on if the inputs are the *same*. Like the EXCLUSIVE OR gate, the EXCLUSIVE NOR gate is seldom used in PLCs. Only its basic symbol is shown.

The NAND and NOR gates are the two final basic gates. Both are a combination of two other basic gates. The NAND gate, shown in figure 6–5, is the combination of an AND and a NOT gate. The relay and PLC logic for the NAND gate require a logic relay, as shown. The NAND keyboard program is 1, nand, 2, nand, 3, nand, 4, =, 27.

The NOR gate, shown in figure 6–6, is made up of an OR gate and a NOT gate. Again, a logic relay is required for relay or PLC logic. The NOR keyboard program would be 1, nor, 2, nor, 3, nor, 4, =, 62.

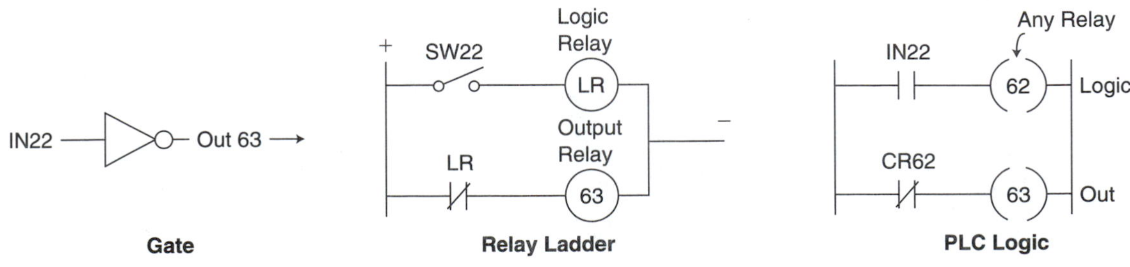

FIGURE 6–4
NOT Gate and Relay and PLC Equivalents

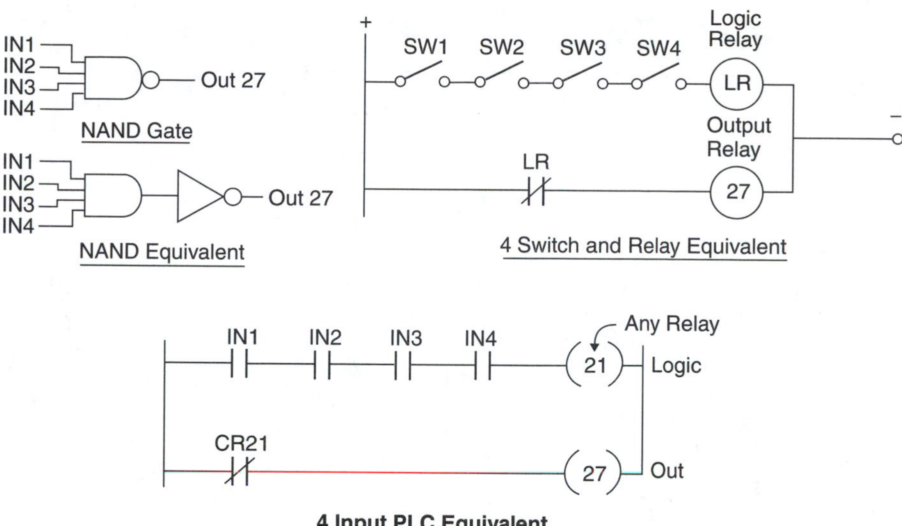

FIGURE 6–5
NAND Gate and Relay and PLC Equivalents

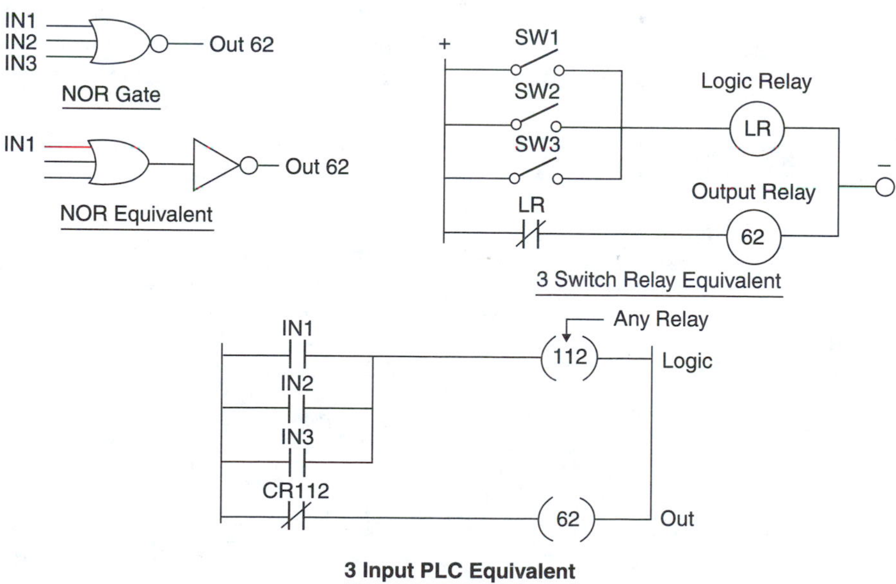

FIGURE 6–6
NOR Gate and Relay and PLC Equivalents

Symbol	Definition	Example of Usage	Meaning- Word Description
•	and	C • D • E	C and D and E
+	or	11 + 12	11 or 12
–	not	$\bar{M}$	Not M
o	invert	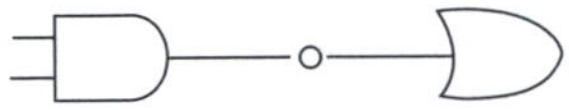	Change
=	results in	F • G = L	L is true (on) if both F and G are true(on)

FIGURE 6–7
Boolean Algebra Symbol Notation

6–3 BOOLEAN ALGEBRA PLC PROGRAMMING

Sometimes you may have to program a PLC in the Boolean algebra system, which is a shorthand method of writing gate diagrams. Complex gate diagrams can be analyzed easily when they are written in Boolean form. The analysis is covered in digital logic texts. This section covers only the PLC programming aspects of Boolean algebra.

The symbols used in the Boolean algebra system are illustrated in figure 6–7. Examples of usage and the meaning of the Boolean expression in words are also given. Figure 6–8 shows some typical gates and how they would be represented in Boolean form.

FIGURE 6–8
Boolean Algebra Equivalents for Digital Gates

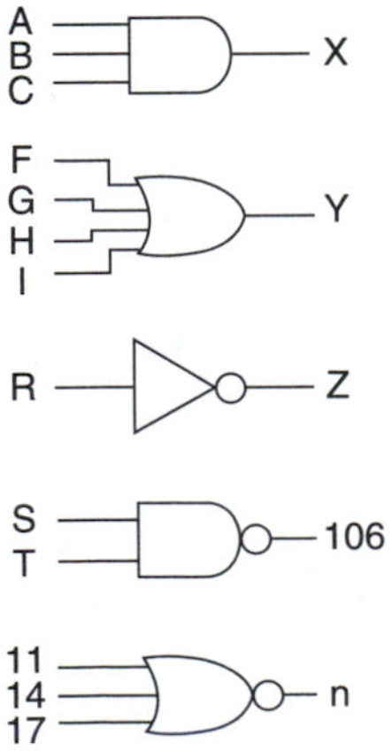

Inputs	Output	Gate	Boolean Form
A, B, C	X	AND	A • B • C = X
F, G, H, I	Y	OR	F + G + H + I = Y
R	Z	NOT	$R = \bar{Z}$
S, T	106	NAND	$S \cdot T = \overline{106}$
11, 14, 17	n	NOR	$11 + 14 + 17 = \bar{N}$

6–4 CONVERSION EXAMPLES

The ten chapter examples that follow illustrate conversion from one PLC programming system to another. The Boolean expressions are included for reference in each example as optional information. The ten examples are divided into three groups as follows:

1. Four examples 6–1 through 6–4, of how to convert a word description into ladder and gate diagrams.
2. Three examples, 6–5 through 6–7, of conversion from a ladder diagram to a gate diagram.
3. Three examples, 6–8 through 6–10, of conversion from a gate diagram to a ladder diagram.

Ladder diagrams will be included for relay logic and PLC logic.

EXAMPLE 6–1 The first example is shown in figure 6–9. Output 122 is to be on only when either inputs 7 and 8 are on or if inputs 17 and 18 are on. Output 122 can be on when all four inputs are on.

EXAMPLE 6–2 The second example is a conveyor control problem. Conveyor C is to run when any one of four inputs is on. It is to stop when any one of four other inputs is on. The ladder relay and PLC control diagrams are shown.

An explanation of the gate diagram is in order. The four starts are inputted to an OR gate, OR 1. When any one of the four is depressed, the OR gate output goes on. When the start is released, the output goes off. The stop buttons are connected to another OR gate, OR 2. OR 2 goes on when any one or more stop buttons are depressed. The OR 2 output is inverted by the NOT gate and sent to the AND gate.

With no stop button depressed, the stop OR 2 gate is off (or low). Due to the NOT inversion, the stop input to the AND gate is on (or high) when no stop buttons are depressed. With no stops depressed, output C can go on if OR 1 is also on. Any time that both OR 1 and OR 2 outputs are high, the inputs to AND 4 are high and output C is on. This example is illustrated in figure 6–10.

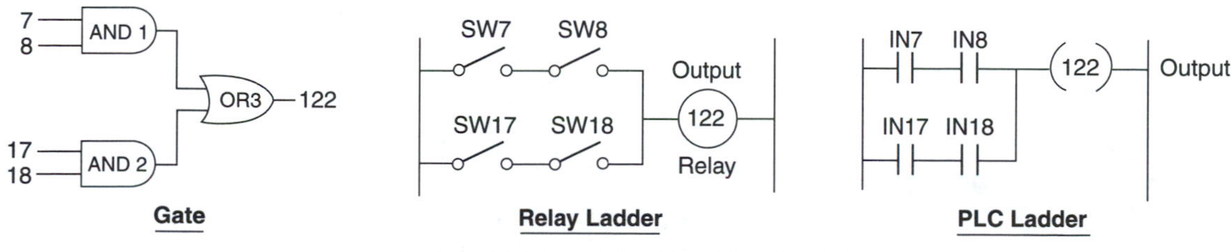

FIGURE 6–9
Example 6–1: Word Description Conversion

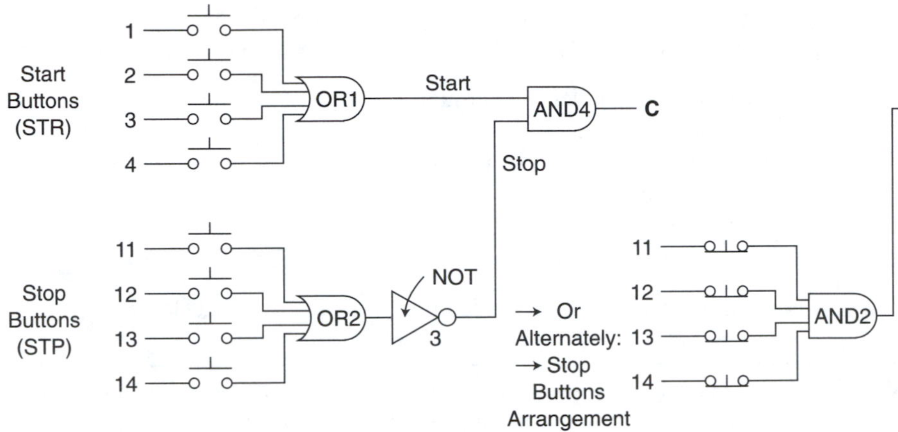

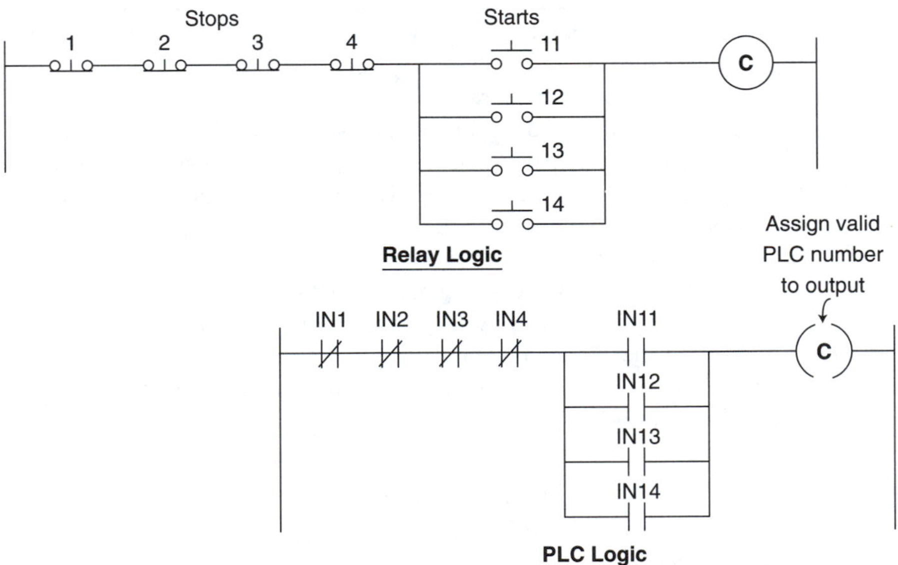

$$(STR1 + STR2 + STR3 + STR4) \bullet \overline{(STP1 + STP2 + STP3 + STP4)} = C$$

Boolean (Ref.)

FIGURE 6–10
Example 6–2: Word Description Conversion

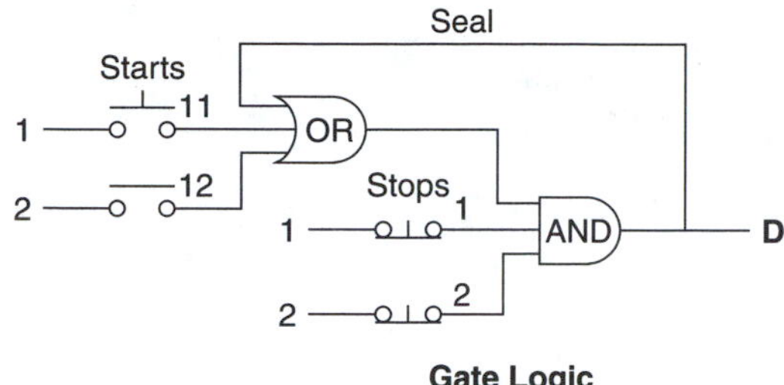

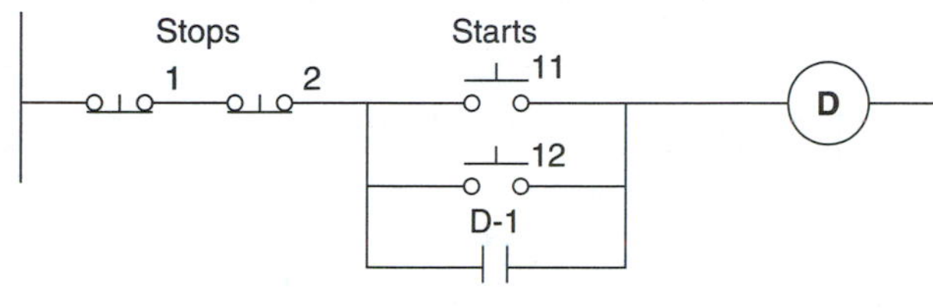

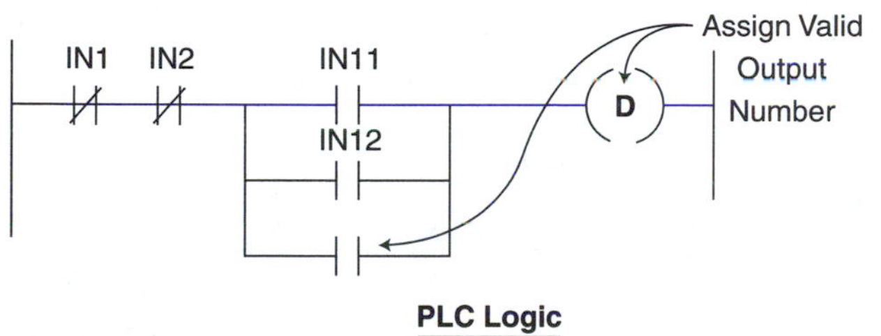

$$(\text{STR1} + \text{STR2} + \text{D}) \bullet \overline{\text{STP1}} \bullet \overline{\text{STP2}} = \text{D}$$

Boolean (Ref.)

FIGURE 6–11
Example 6–3: Word Description Conversion

EXAMPLE 6–3 The third example is a motor control circuit with two start and two stop buttons. When the start button is depressed, the motor runs. By sealing, it continues to run when the start button is released. The stop buttons stop the motor when depressed. This example differs from the previous one in that the system seals on when the start is released (see figure 6–11).

EXAMPLE 6–4 This example is a more complex system. Its solutions are shown in figure 6–12. A process fan is to run only when all of the following conditions are met.

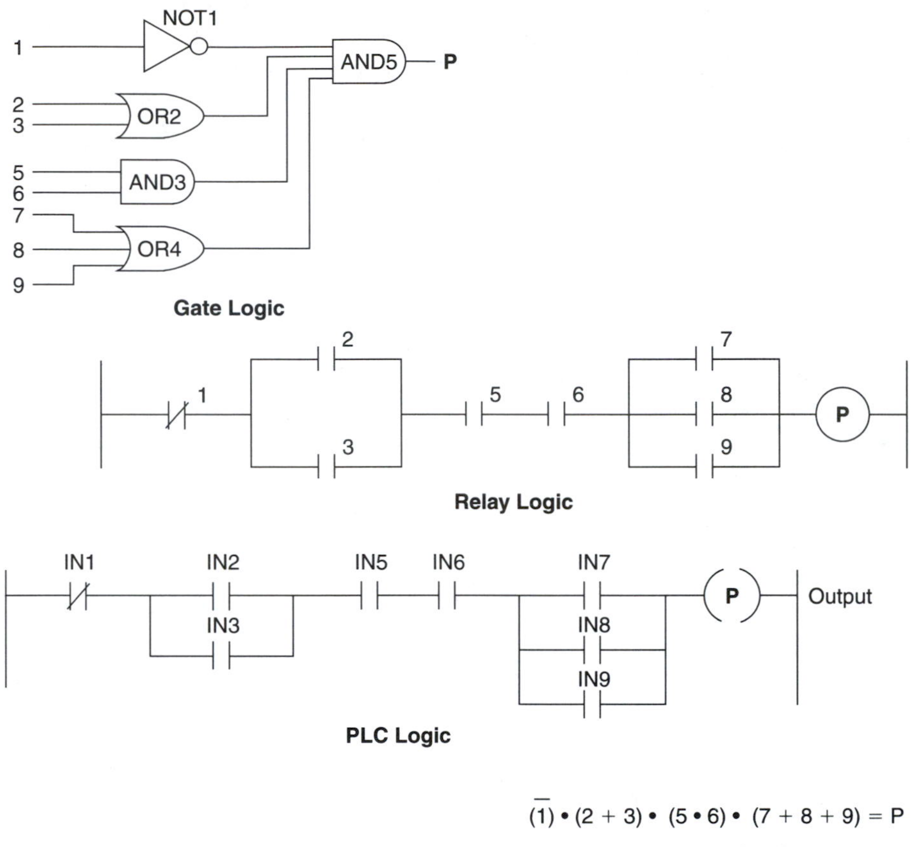

$\overline{(1)} \bullet (2 + 3) \bullet (5 \bullet 6) \bullet (7 + 8 + 9) = P$

Boolean (Ref.)

FIGURE 6–12
Example 6–4: Word Description Conversion

1. Input 1 is off.
2. Input 2 is on or input 3 is on, or both 2 and 3 are on.
3. Inputs 5 and 6 are both on.
4. One or more of inputs 7, 8 or 9 is on.

EXAMPLE 6–5 Examples 6–5 through 6–7 involve converting PLC ladder diagrams to gate diagrams. Conversion to Boolean is an added option. This example, shown in figure 6–13, is a fundamental conversion. Series contacts are converted to AND gates. Parallel contacts are converted to OR gates. Then, the combinations are treated in the same manner—series combinations to AND and parallel combinations to OR.

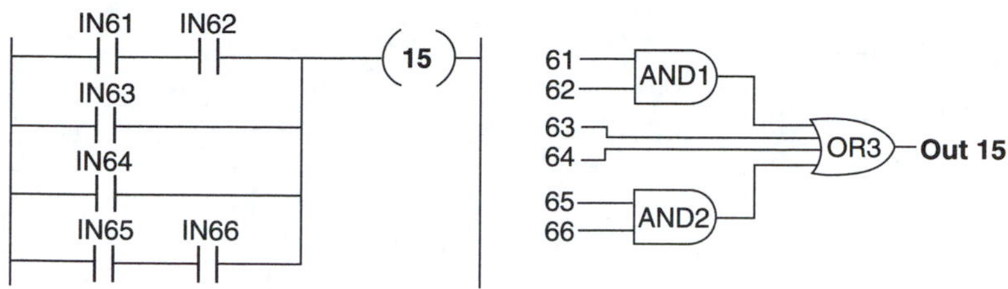

FIGURE 6–13
Example 6–5: Ladder Diagram-to-Gate Conversion

EXAMPLE 6–6 Figure 6–14 illustrates the sixth example. It is more involved than example 6–5 and requires more gates for the gate diagram. It also requires an input inversion. Since contact 10 is normally closed in the PLC ladder diagram, its state must be inverted. The inversion is accomplished at the input of logic gate OR 22. By convention, the bubble performs the same function as a NOT gate.

EXAMPLE 6–7 The seventh example is a more advanced circuit requiring many gates and two inversions. The construction of the equivalent gate diagram follows the principles of the previous two examples. It is shown in figure 6–15.

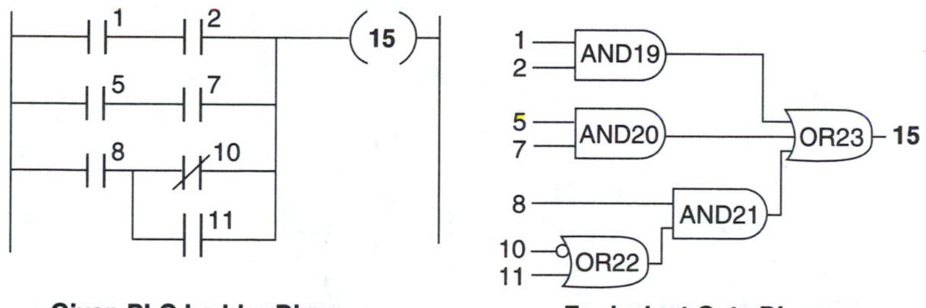

FIGURE 6–14
Example 6–6: Ladder Diagram-to-Gate Conversion

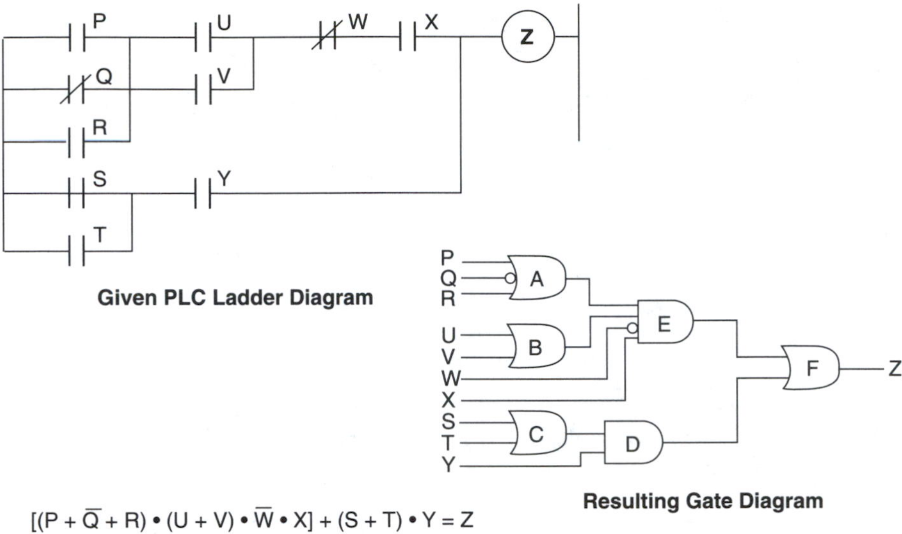

FIGURE 6–15
Example 6–7: Ladder Diagram-to-Gate Conversion

EXAMPLE 6–8 Examples 6–8 through 6–10 show conversions of given digital gate diagrams into ladder diagrams. The ladder diagrams are drawn for PLC logic only. As in previous examples, the Boolean equivalent is given for reference.

The example in figure 6–16 is a fundamental gate-to-ladder diagram conversion. Two different AND gates feed one OR gate. If either or both of the feeder gates are on, the OR gate is on and output R will be on.

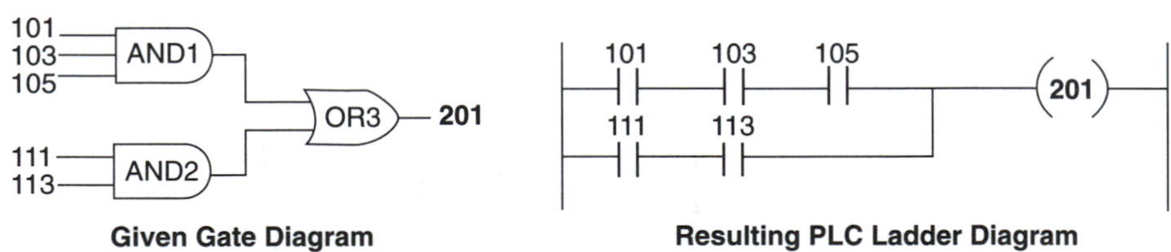

(101 • 103 • 105) + (111 • 113) = 201

Equivalent Boolean Expression (Ref.)

FIGURE 6–16
Example 6–8: Gate-to-Ladder Diagram Conversion

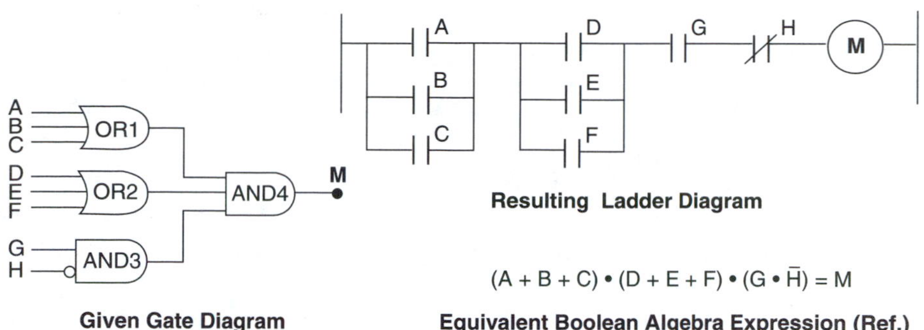

FIGURE 6–17
Example 6–9: Gate-to-Ladder Diagram Conversion

EXAMPLE 6–9 This example has two OR respective gates and one AND gate all feeding an AND gate. Each OR gate is converted to parallel contacts in the ladder diagram. The AND gate is converted to series contacts or to series groups of contacts. The only new concept introduced is the bubble on the H input. The dot means the input is inverted. It is the same as if a NOT gate were in the line between the H and the input point of AND gate 3. Figure 6–17 shows the conversion.

EXAMPLE 6–10 The example in figure 6–18 is another gate-to-ladder diagram conversion. A new feature introduced is an input being fed to two different places, which is shown in the resulting ladder diagram. Two different contacts are needed in the ladder diagram for it to be equivalent to the one gate input number.

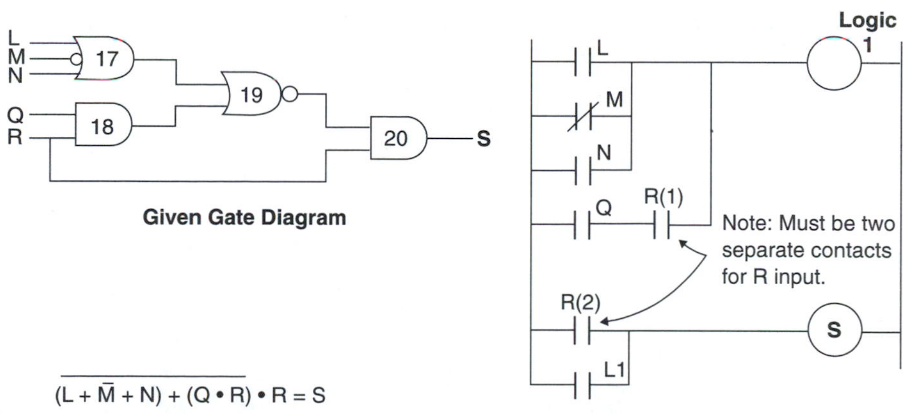

FIGURE 6–18
Example 6–10: Gate-to-Ladder Diagram Conversion

EXERCISES

For exercises 1 through 4, convert the word description to:

a. Gate symbols.
b. PLC/Relay logic ladder diagrams.

1. Switch 8 and switch 11, plus either switch 22 or switch 34, must be on for output 67 to be on.
2. For output 7 to be on, input 6 must be off and either input 8 or input 9 must be on. In addition, one of inputs 1, 2, or 3 must be on.
3. For output H to be on, input A must be on and both inputs C and D must be off. In addition, one or more of inputs E, F, and G must be off.
4. Four pushbutton stations control a fan. Each station has a start and stop button. Two door interlocks must be closed before the fan may run. Pushing any start button will make the fan run, and the fan is sealed on when the start button is released. Pushing any stop button turns the fan off and also prevents the fan from starting or running.

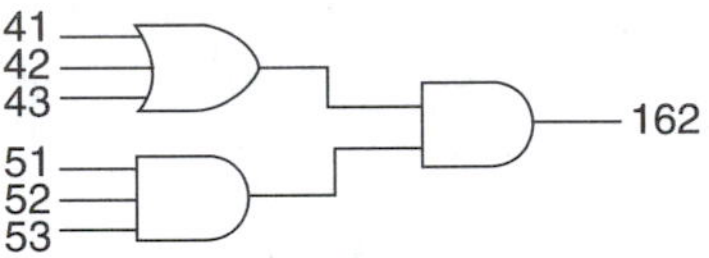

FIGURE 6–19
Diagram for Exercises 5–7

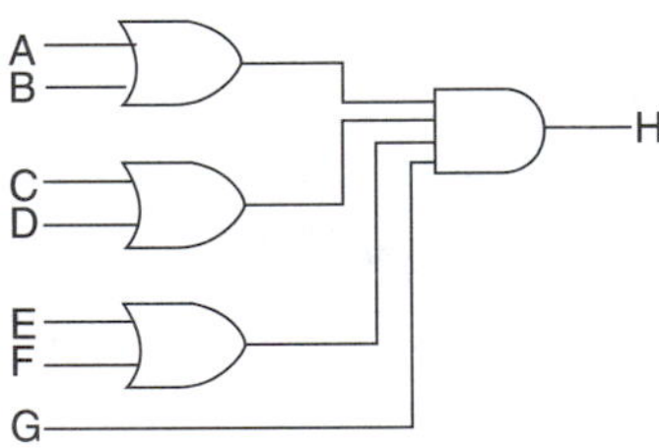

FIGURE 6–20
Diagram for Exercises 5–7

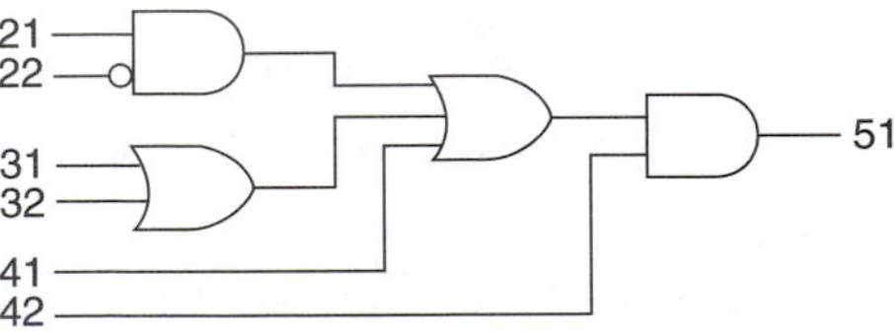

FIGURE 6–21
Diagram for Exercises 5–7

FIGURE 6–22
Diagram for Exercises 8–10

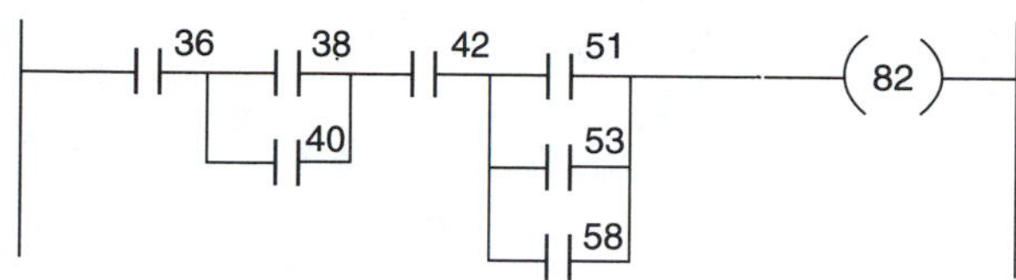

FIGURE 6–23
Diagram for Exercises 8–10

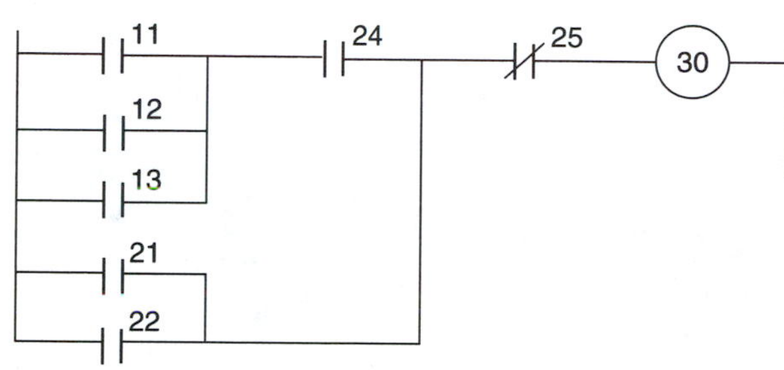

FIGURE 6–24
Diagram for Exercises 8–10

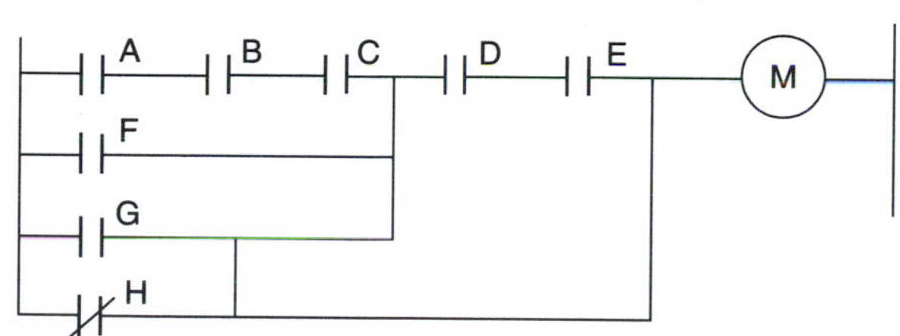

5–7. For exercises 5 through 7, convert the gate diagrams given in figure 6–19, 6–20, and 6–21 to PLC ladder diagrams.

8–10. For exercises 8 through 10, convert the PLC ladder diagrams given in figures 6–22, 6–23, and 6–24 to gate diagrams.

11. (*Optional*) Convert the diagrams of exercises 1 through 10 to Boolean algebra expressions.

7

Creating Ladder Diagrams from Process Control Descriptions

OUTLINE

OBJECTIVES

At the end of this chapter, you will be able to

- □ Create basic ladder diagrams from a sequence of operational steps.
- □ List the major steps in creating a PLC program for an industrial situation.
- □ Describe the content of each of these steps.
- □ Flowchart a process.

7–1 INTRODUCTION

Planning without action is a waste of time and money, and action without planning creates chaos. The purpose of this chapter is to outline some of the planning needed to create good, workable, safe PLC programs—without chaos.

You may want to omit this chapter for now if you work with preprogrammed PLC programs. You should include this chapter if you have to create your own programs, modify programs, or if you doubt the validity of the program you have.

This chapter is written in relay logic. The principles are readily converted to PLC programs as described in chapter 5 for coils and contacts.

7–2 LADDER DIAGRAMS AND SEQUENCE LISTINGS

Ladder diagrams are the most commonly used diagrams for nonelectronic control circuits. They are sometimes called *elementary diagrams* or *line diagrams*. Sometimes they are considered a subtype of schematic diagrams. The term *ladder diagrams* is used in this book. Why are these diagrams called ladder diagrams? They look like a ladder in a way. You start at the top of the ladder and generally work your way down.

Two types of ladder diagrams are used in control systems: the *control ladder diagram* and the *power ladder diagram*. This section concentrates on control ladder diagrams, with only a fundamental explanation of the power ladder diagram.

Figure 7–1 shows two basic control ladder diagrams. The first one, (a), is for a single switch that turns a relay output, CR_5, on and off. The second, (b), is a single-function diagram with parallel lines for control and parallel lines for output. Either or both of two switches turn the output and a pilot light on.

FIGURE 7–1
Basic Control Ladder Diagrams

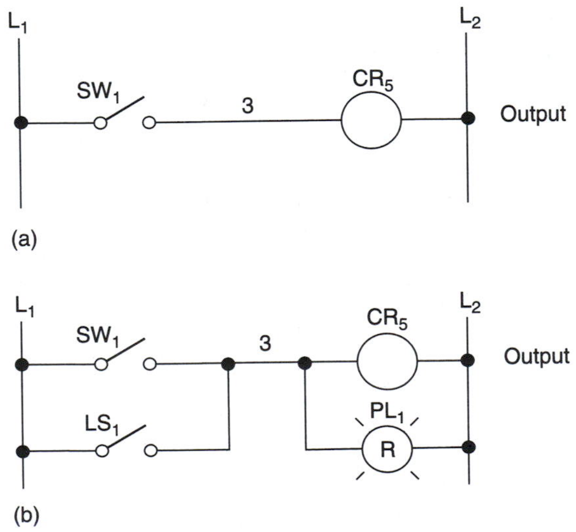

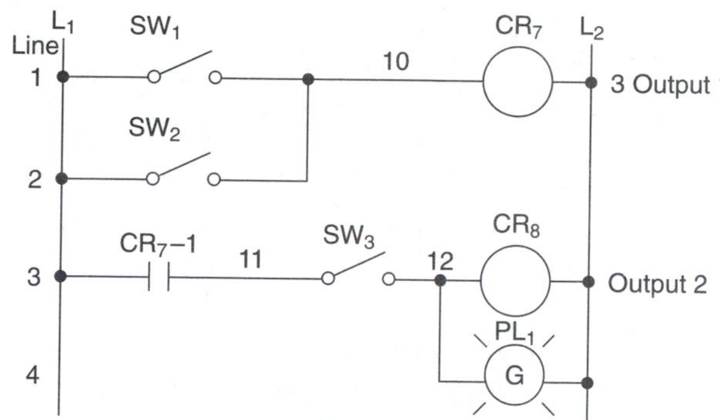

FIGURE 7–2
Two-Function Control Ladder Diagram

The control ladder diagram of figure 7–2 has two active functional lines. Some of the common practices for the format of control ladder diagrams are illustrated by this figure. Those practices are as follows:

- □ All coils, pilot lights, and other outputs are on the right.
- □ An input line can feed more than one output. If it does, the outputs are connected in parallel.
- □ Switches, contacts, and other devices are inserted in the ladder line starting on the left.
- □ Switches, contacts, and other devices may be multiple contacts in series, parallel, or series–parallel.
- □ Lines are numbered consecutively downward on the left.
- □ Every connection node is given a unique identification number.
- □ Outputs can be identified by function on the right, in notes.
- □ A cross-identification system may be included on the right. The contacts associated with the line's coil or output are identified by line location. In figure 7–2, the *3* to the right of line 1 indicates that a normally open contact of relay CR_7 (the coil on line 1) is located on line 3. For a normally closed contact, the number would have an asterisk (*) next to it or a bar over it. Figure 7–5 uses the same system on two different lines.
- □ Relay contacts are identified by the relay coil number plus a consecutive sequence number. For example, we have included contact CR_7-1. If other relay CR_7 contacts were used, the next would be CR_7-2, and so on.

The control ladder diagram in figure 7–2 has an operating sequence as follows:

Straight-Through Sequence

All switches are open to start; both coils are off.
Close SW_1, SW_2, or both; CR_7 is energized.
On line 3, CR_7-1 closes, enabling line 3 (CR_8 is still off).
Closing SW_3 energizes CR_8 and pilot light PL_1.
Opening both SW_1 and SW_2 turns everything off.

FIGURE 7–3
Incorrect Control Ladder Diagram for Figure 7–2

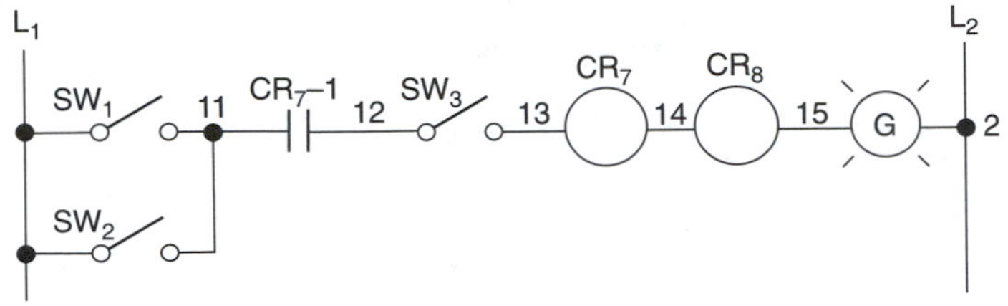

Alternative Possible Sequence

Initially turning on SW_3 causes nothing to energize.
Opening SW_3 when everything is on would turn off CR_8 and PL_1 only.
(Other sequence possibilities exist.)

Figure 7–3 is an incorrect ladder diagram that contains the same components used in figure 7–2. Will this circuit work? No. First of all, if power could get to point 13, the outputs would not work. Each would have 1/3 control voltage across it. Relays would not pull in, and the light would glow dimly or not at all. But the outputs will never go on anyway. If all switches are closed, no power gets through contact CR_7-1. It cannot close until CR_7 is energized, which is impossible.

The operation of the power ladder diagram in figure 7–4 is straightforward. When the power contactor coil is energized, the power contacts close and power is applied to the motor or the load device. Note that the power ladder diagram wiring is shown by thicker lines, to differentiate its wire from control circuit lines.

Additional sequence requirements may call for the construction of additional control ladder lines. The following functional modifications can be added to the ladder diagram of figure 7–2.

SW_4 must be on for CR_7 to go on.
CR_7 must be on for CR_8 to go on.
CR_9 is turned on by CR_7, CR_8, and SW_3.

The extended ladder diagram is shown in figure 7–5. Note that there is a dotted line between the two SW_3 contacts. The dotted line indicates a common single switch with two contacts. (If SW_3 were on the left, only one contact would be needed to run lines 3, 4, and 5.)

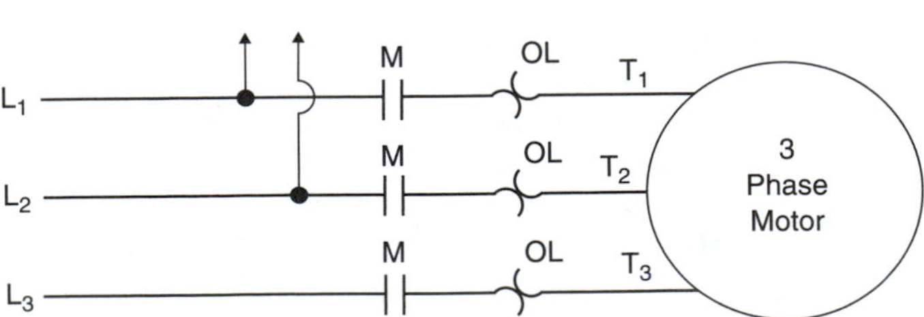

FIGURE 7–4
Power Ladder Diagram

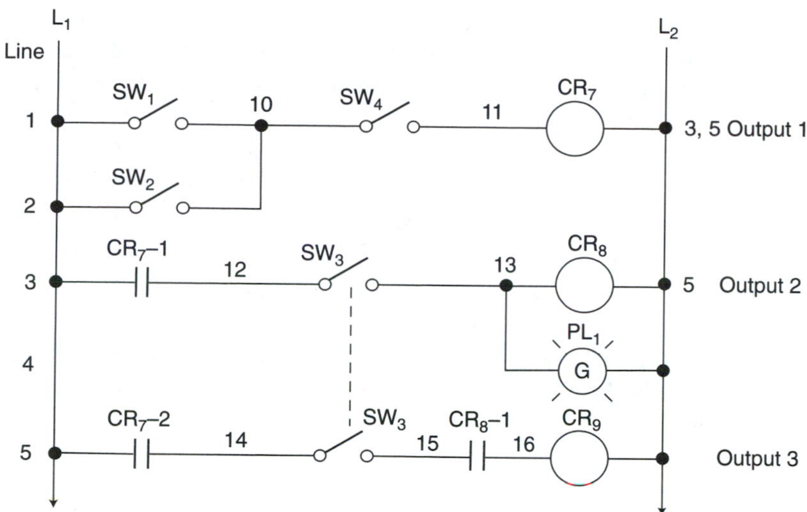

FIGURE 7–5
Extended Control Ladder Diagram for Figure 7–2

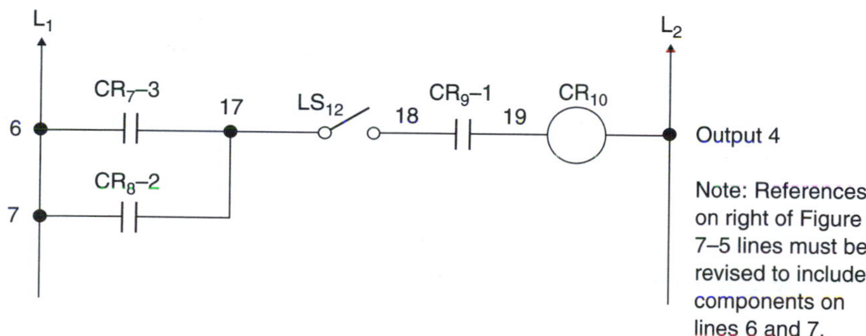

FIGURE 7–6
Added Line for Control Ladder Diagram of Figure 7–5

An added sequence of operation can be determined from an added ladder line. Such an added ladder line is shown in figure 7–6. The added sequence based on this additional line would be as follows: CR_7 or CR_8 or both, plus LS_{12} and CR_9, turn on relay output CR_{10}.

7–3 LARGE-PROCESS LADDER DIAGRAM CONSTRUCTION

Some of the steps in planning a program for a large process are:

1. Define the process to be controlled.
2. Make a sketch of the process operation.
3. Create a written step sequence listing for the process.

4. Add sensors on the sketch as needed to carry out the control sequence.
5. Add manual controls as needed for process setup or operational checking.
6. Consider the safety of the operating personnel and make additions and adjustments as needed.
7. Add master stop switches as required for safe shutdown.
8. Create the ladder logic diagram that will be used as a basis for the PLC program.
9. Consider the "what if's" where the process sequence may go astray.

Some other steps needed in program planning that we will not cover are troubleshooting of process malfunctions, parts list of sensors, relays, and so on, and wiring diagrams, including terminals, conduit runs, and so on.

To illustrate the nine steps of the planning sequence, we use a fundamental industrial control problem. We then go through the creative process to illustrate each of the steps of the planning process.

Step 1

Define the problem.

We wish to set up a system for spray-painting parts. A part is to be placed on a mandrel. (A mandrel is a shaft or bar whose end is inserted into a workpiece to hold it during an operation.) When the part is in place, two pushbuttons are pressed and the mandrel rises. After the part rises to the top and is in the hood, it is to have spray paint applied for a period of 6 seconds. At the end of the 6 seconds, the mandrel returns to the original position. The painted part is then removed from the mandrel by hand. (We assume for our illustration that the part dries very quickly.)

Step 2

Make a sketch of the process (figure 7–7).

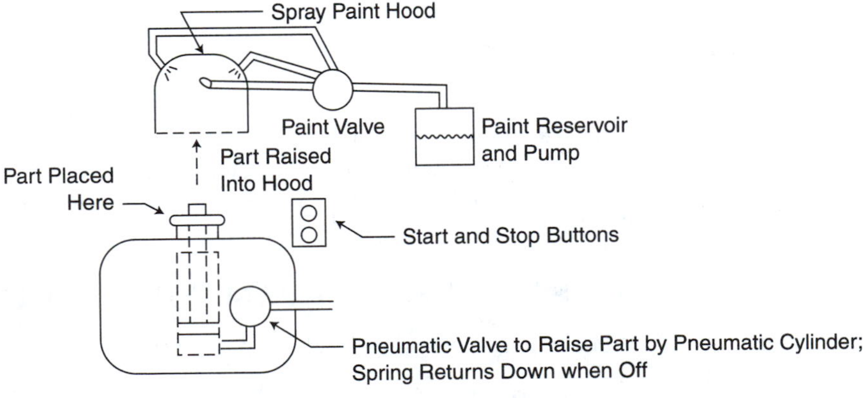

FIGURE 7–7
Sketch of the Spray Process System

Step 3

List the sequence of operational steps in as much detail as possible. The sequence steps should be double or triple spaced so that any omitted steps discovered later may be added. The following is a step sequence for this process.

1. Turn on the paint pump and pneumatic air supply.
2. Turn the system on. This requires pushbuttons other than the system buttons.
3. Put the part on the mandrel. A sensor indicates that the part is in place.
4. Push the Master Start button and the two system start buttons. Having to push two system start buttons (with both hands) reduces the possibility of the operator's hands being injured by the rising mandrel.
5. The mandrel is raised by a pneumatic cylinder energized by the opening of an electrically actuated air valve when the system start buttons are pressed. (The mandrel will return by gravity and downward spring action when the valve is reopened.) When the mandrel rises, the part-in-place sensor at the bottom becomes deenergized. (*Note:* The part-in-place sensor does not rise with the mandrel.)
6. When the part reaches the top and is under the hood, it is held against a stop by air pressure. A sensor has indicated that the part has reached the top.
7. A timer starts and runs for 6 seconds.
8. During the timing period of 6 seconds, paint is applied by the sprayer.
9. At the end of 6 seconds, painting stops and the mandrel, with the part on it, lowers.
10. The up sensor is deenergized when the mandrel with the part on it descends.
11. The part arrives at the bottom, reenergizing the part-in-place sensor.
12. The part is removed from the mandrel.
13. The system resets so that we may start at step 3 again.

Step 4

Add sensors as required. Once we list the sequence, we find that sensors are needed in the machine to indicate process status. We need a sensor (a limit switch placement) to show that the part has been placed on the mandrel initially. We also need a sensor (limit switch up) to indicate when the mandrel is fully extended upward. Among other possible sensors that a process such as this might need is one to make sure the paint sprayer has paint and one to make sure the inserter's hand is out of the way. Depending on the process and the detail of control, other sensors could be required as well. Figure 7–8 includes the two basic sensors, LSP and LSU, and their locations. The figure also shows the enclosures needed, along with the locations of start and stop buttons.

Step 5

Add manual controls as needed. We may need a manual pushbutton to raise the mandrel to the top for setup purposes. The manual up position is needed when we set the spray-gun pressure for optimum paint coverage. We include pushbutton up (PBU) on our ladder diagram to accomplish this manual control.

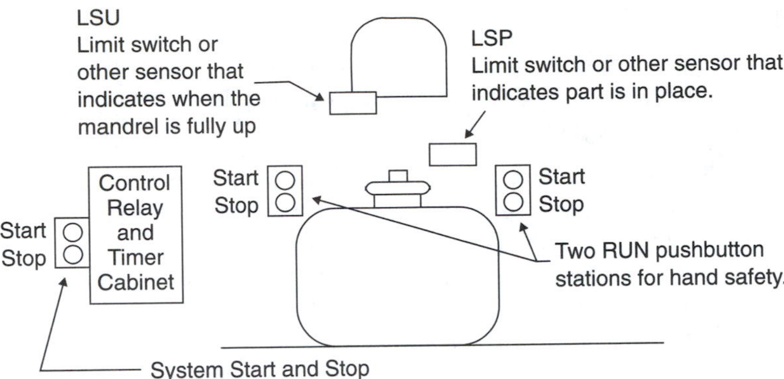

FIGURE 7–8
Sensor, Enclosures, and Pushbutton Locations

Step 6

Consider the safety of the machine operator. One basic way to keep hands out of a process is to have two start buttons. Then both hands must be away from the work to depress both buttons (which works until the operator figures out how to use one knee and one hand). Other considerations, which we do not cover in detail here, might be operating a fan to disperse fumes during spraying, or perhaps a photocell proximity-personnel-system-stop device.

Step 7

Add emergency and master stop switches as needed for operator safety. This may seem to be part of step 6 because both steps deal with operator safety. It is a continuation of the safety issue, but emergency stop switches are so important that they need special consideration as an additional step.

Step 8

Create the ladder logic diagram. The diagram created is to include the steps and considerations of the first seven steps. This is shown in figure 7–9 for our spraying example.

Step 9

Determine the "what if's," or potential problem areas. After the ladder diagram is completed, all possible situations and emergencies should be listed. In this example, some of them might be:

- What if no part is in place when the start buttons are pushed?
- What if the power fails during the cycle when the part is rising, during painting, or at any other time?

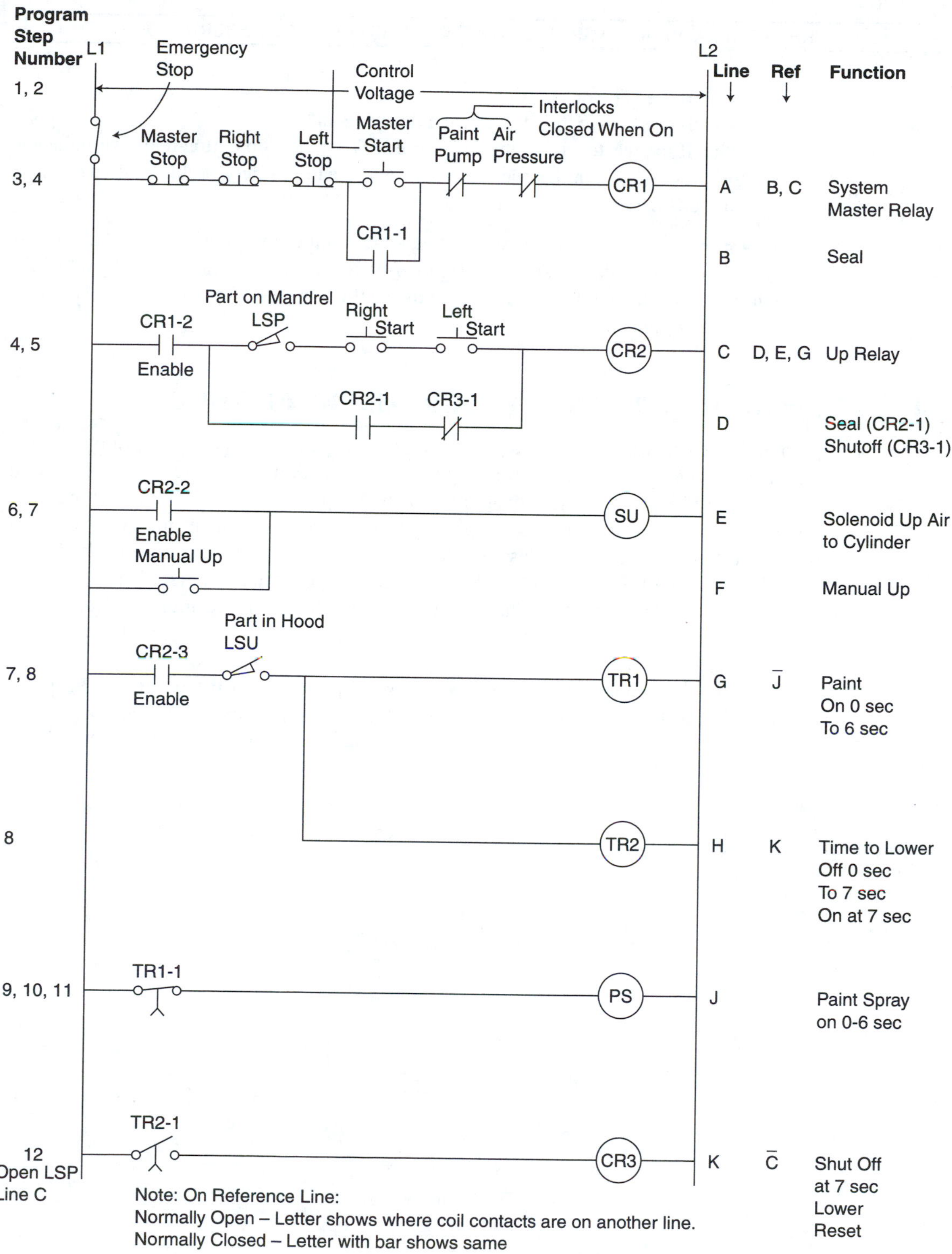

FIGURE 7–9
Ladder Logic Diagram

- □ What if the sprayer runs out of paint?
- □ What if the same part is left in for a double coat?
- □ What if the master stop button is pushed? Does the stop button really stop the entire process, or can the mandrel move and create a safety problem after the stop button is depressed (it can)?

All of these types of questions should be considered in the final sequence and ladder diagram. Review of the ladder diagram in figure 7–9 covers some of these contingent situations but not all of them. Further modifications would be needed for a more complete consideration of contingencies.

7–4 FLOWCHARTING AS A PROGRAMMING METHOD

As mentioned earlier, one of the steps in planning a large-process ladder diagram is to create a written step-sequence listing for the process. Prior to doing so, however, you may wish to develop a flowchart, or flow diagram, which is a pictorial representation of program logic. Such a flowchart can make application program development much easier. The flowchart shows the points of decision, relevant operations, and the sequence in which they should take place to solve a problem. Such a chart lets you think in a graphic manner.

Most flowcharts use four basic symbols: oval, diamond, rectangle, and parallelogram. In addition, connection arrows are used to connect the various symbols (see figure 7–10).

The *oval* symbol indicates either the beginning or end of the program. We place "start" in the oval to begin a program, "end" to finish it. The *diamond* indicates a point of decision. With it, we ask a question, one that can usually be answered by "yes" or "no." The

FIGURE 7–10
Flowchart Symbols

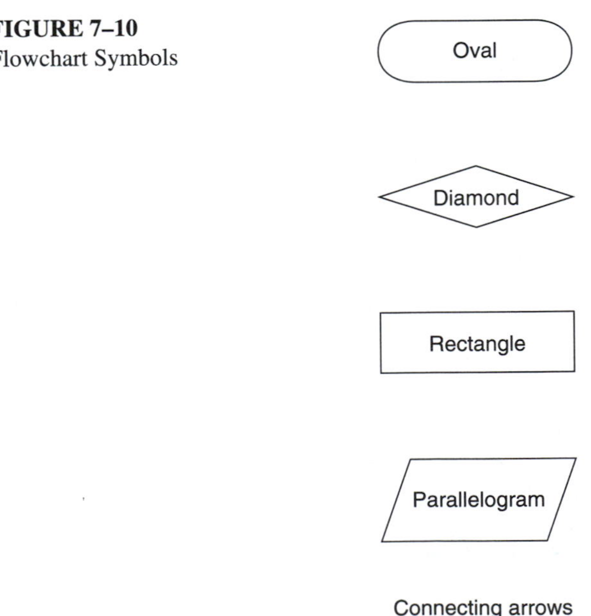

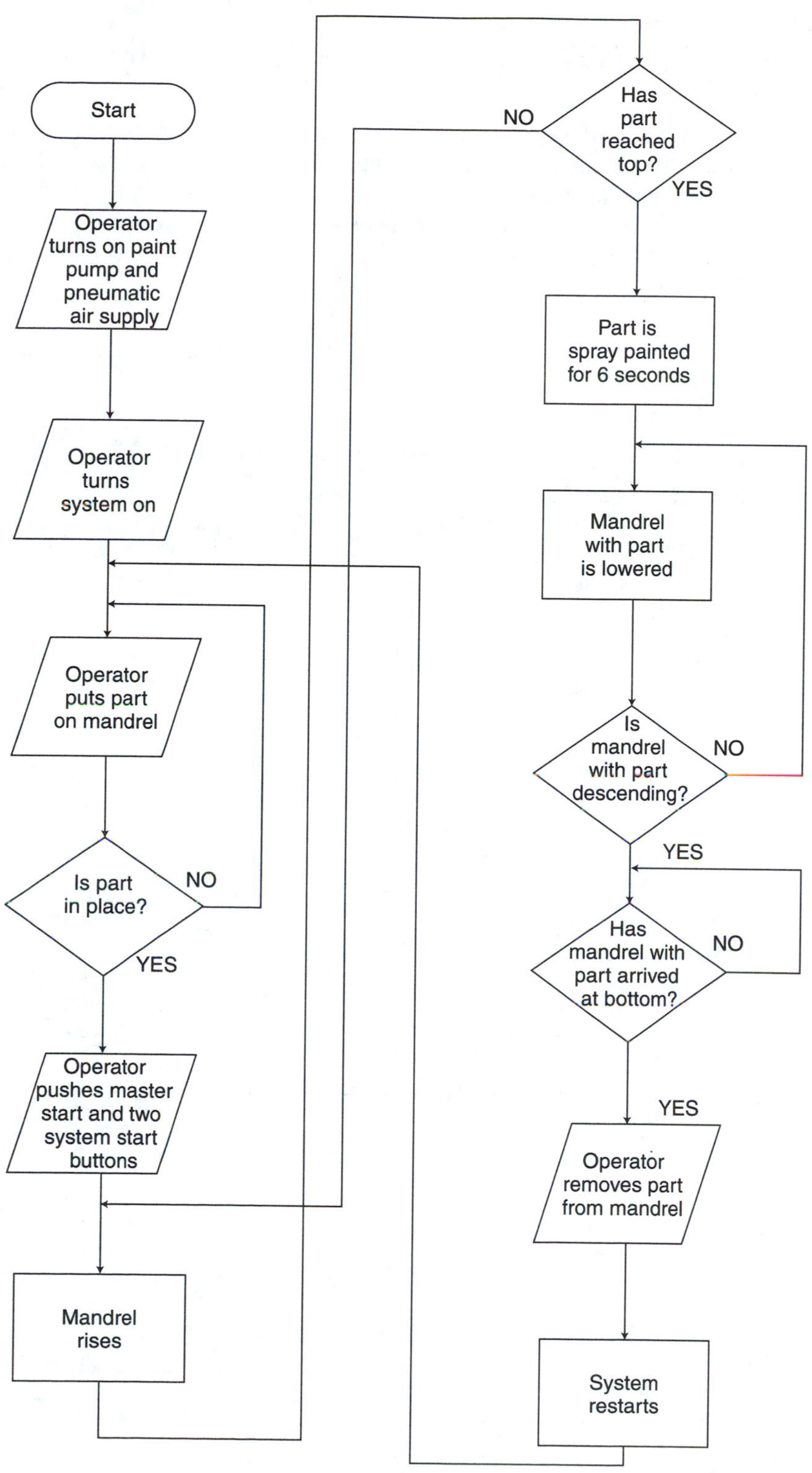

FIGURE 7–11
Flowchart for Spray Process System

rectangle is the process block; some type of computational processing or operation is taking place. The *parallelogram* indicates input/output. It is used to input information into the system or to take it out.

A possible flowchart for the spray process system is shown in figure 7–11. Once the flowchart is created, the written step-sequence listing of the process should go more smoothly. Why not give flowcharting a try?

EXERCISES

Solve the following problems using the nine-step planning sequence:

1. Make a ladder diagram for the following sequence:
   When SW_1 is closed, CR_1 goes on.
   After CR_1 goes on, SW_2 can turn CR_2 on.
   When CR_2 goes on, PL_1 goes off.
2. Make up a sequence listing for the ladder diagram in figure 7–12.
3. Make a flowchart for the ladder diagram in figure 7–12.

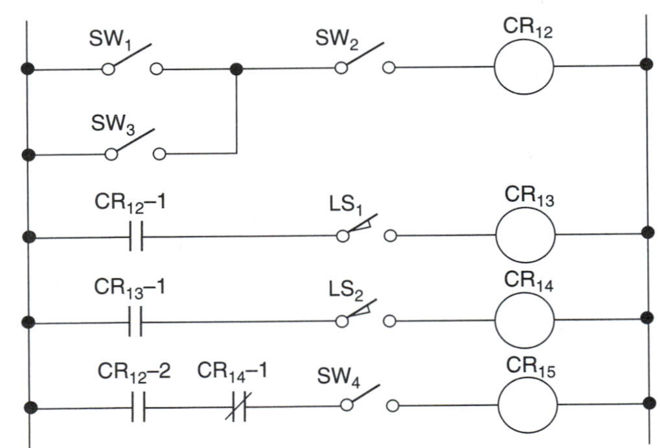

FIGURE 7–12
Ladder Diagram from Which to Make a Sequence

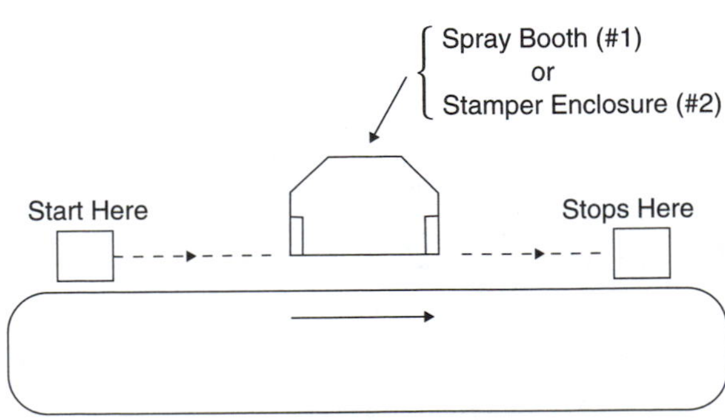

FIGURE 7–13
Diagram for Exercises 4 and 5

4. A part is placed on a conveyor. The part automatically moves down the conveyor. In the middle of the conveyor, the part goes through a 2-foot-long painting section. The sprayer paints for the time the part is under the booth, during which time the conveyor does not stop. When the part reaches the end of the conveyor, the conveyor stops and the part is removed. Assume that only one part can be on the conveyor at one time. (*Hint*: Use one limit switch at the front of the booth and another at the end.) See figure 7–13.
5. Same as exercise 4 except that the part stops in the middle of the conveyor and is stamped, not painted, and then continues to the end of the conveyor.

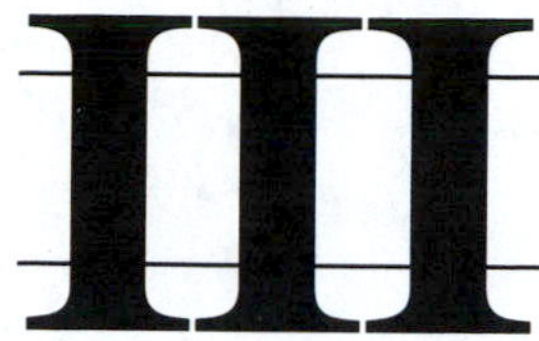

Basic PLC Functions

8

Register Basics

OUTLINE

OBJECTIVES

At the end of this chapter, you will be able to

- □ List the five common types of PLC registers.
- □ Describe the function of each of the five register types.
- □ Describe how each of the five common types of PLC registers is used in PLC operations.

8–1 INTRODUCTION

So far, we have examined the complete PLC system and peeked inside to see how it operates. We have explored general programming procedures and, more specifically, the programming of on–off inputs and on–off outputs. Furthermore, we have looked at auxiliary commands and functions, and then have created a ladder diagram for a process problem. It is now time to move on to timers and counters, two core functions available with all PLCs. But before we do so, it is best to preface our study with an introduction to registers: memory locations that provide temporary storage of data, instructions, information, and, what is particularly relevant, numerical values associated with timers and counters. Keep in mind, though, that this chapter is meant to provide only an introduction to the subject. Further analysis will take place as the book progresses, especially when we encounter data movement systems, the utilizing of digital bits, and the SEQUENCER function in chapters 16, 18, and 19, respectively.

8–2 GENERAL CHARACTERISTICS OF REGISTERS

Within the PLC CPU, registers are found in two locations. The microprocessor has *internal* registers, most of which are not directly accessible by the user. These registers (4, 8, 16, or 32 bits wide, depending on the microprocessor) help the control and arithmetic and logic units within the processor to carry out their tasks. Accumulator registers, data registers, index registers, condition code registers, scratch pad registers, and instruction registers—all work to temporarily store data, which in turn is used to facilitate the carrying out of programmed functions.

In addition to these internal registers, the CPU's RAM also contains slots that are designated to hold variable information. These locations, or addresses, become *external* registers. Throughout this chapter and text we assume these registers are 16 bits wide. There can be a mere handful of such registers or hundreds, depending on the size of the CPU and complexity of the user program.

Each bit location in a register contains, of course, either a 1 or a 0. You can observe register contents on a VDT by calling up the register on a keyboard. In addition, on many models you can print the register contents on a typed printout. Various numbering systems are possible for reading register contents or printing them out. In Chapter 13 we describe a variety of PLC numbering systems. Depending on your PLC capabilities, you may choose to print register values based on one or more different numbering systems. For example, one model allows you to choose between 1—Decimal, 2—Binary, 3—Hex, or 4—ASCII. Other possibilities are Octal and special codes unique to the system being used. Still other PLCs are confined to displaying or printing in only one numbering system, usually the decimal system.

Registers are usually designated using prefixes followed by numbers, as is the case in this chapter. HR256 represents holding register 256; OG2 represents output group register number 2. In other systems, a certain numerical series of addresses may be assigned to a specific task or function. One model of PLC has register addresses 901 through 930 assigned to timers and counters only. It is important to determine the functions for the addresses of your PLC registers from its operational manual.

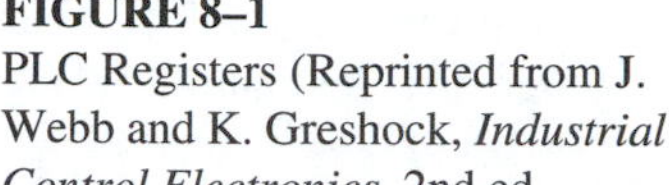

FIGURE 8–1
PLC Registers (Reprinted from J. Webb and K. Greshock, *Industrial Control Electronics*, 2nd ed., Macmillan, Indianapolis, IN)

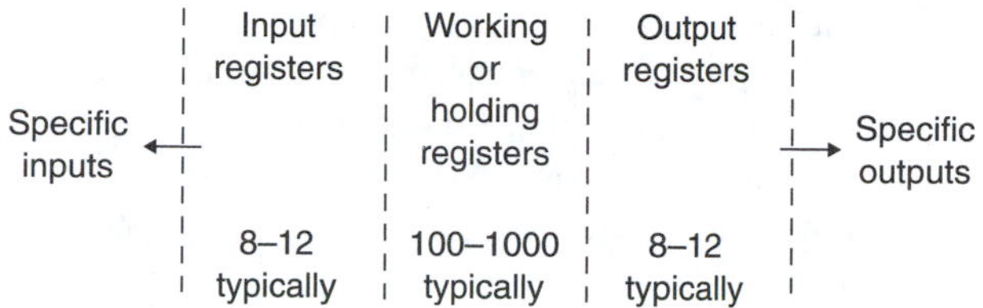

In this chapter we look at the function of five key registers. We see how the holding register, input registers (single and group), and output registers (single and group) operate to temporarily store data for microprocessor manipulation.

8–3 HOLDING REGISTERS

A holding, or working, register (HR) "holds" the contents of a calculation, arithmetic or logic. Conceptually, it is in the "middle" of the CPU, as shown in figure 8–1.

Because in many PLCs, particularly smaller ones, the holding register is not directly accessible to inputs or outputs, input and output registers (single or group) interface the holding register contents to the outside world. Signal data from a specific input device is first "deposited," in the form of 0's and 1's, in an input register. It may then be manipulated by the microprocessor and the results sent to a holding register. Conversely, before the contents of the holding register can affect the output device, they are transferred to an output register. The output register's 0's and 1's "drive" output interface devices such as optoisolators.

To illustrate holding register use, we can look briefly at their function in arithmetic, timer, and counter operations. Keep in mind, though, that not only are these operations covered in greater detail in later chapters, but holding registers are used in numerous other functions covered in future chapters.

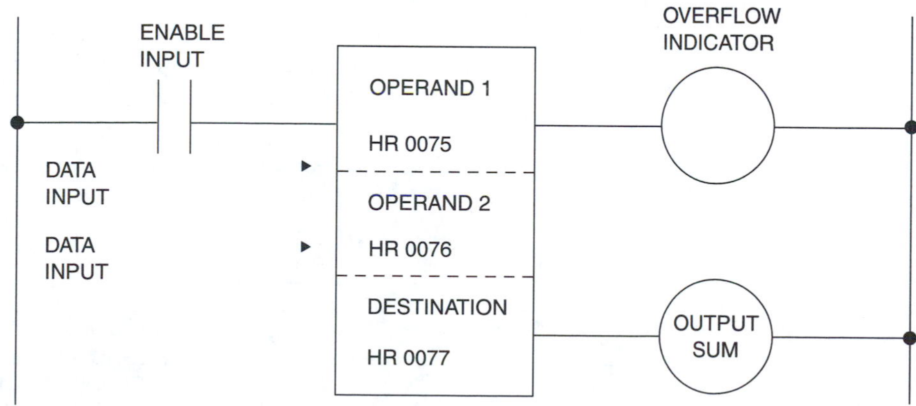

FIGURE 8–2
Registers in Arithmetic Operations

In arithmetic operations, a holding register might contain the first operand; another holding register, the second operand; and a final holding register, the destination of the mathematical manipulation (see figure 8–2).

In the timer function, the preset time value would be placed in a constant or designated register. The holding register is the register in which the count takes place (see figure 8–3a).

The counter function is similar (see figure 8–3b). The preset count value is also placed in a constant or designated register. The holding register, of course, is the register in which the count takes place.

How many holding registers are there? In small PLCs there may be as few as 16. In large machines there are hundreds of holding registers, all accessible for programming use, manipulation, and visual analysis.

FIGURE 8–3
Holding Registers in Timers and Counters

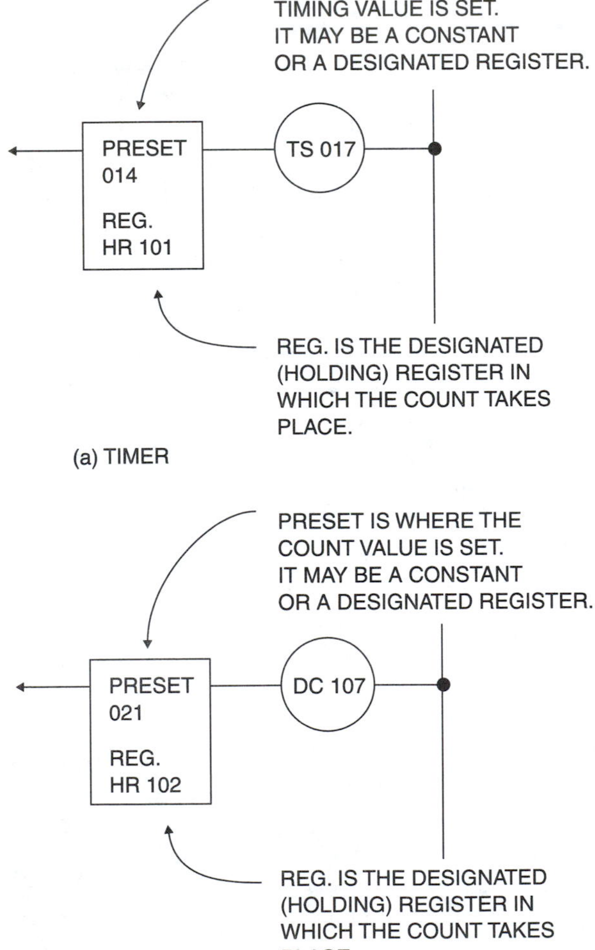

8–4 INPUT REGISTERS: SINGLE AND GROUP

The input register has basically the same characteristics as the holding register, except that it is readily accessible to the input module's terminals or ports. The number of input registers in a PLC is normally one-tenth that of holding registers.

The input group register (IG) is somewhat like the input register. It differs in that each one of the individual 16 bits is directly accessible from one input port. One input group register receives data from 16 consecutive input ports (terminals). Figure 8–4 illustrates how this IG system works. The advantage of the IG system is that only one register is required to service 16 inputs. Without the IG system, you would need 16 registers to service 16 in-

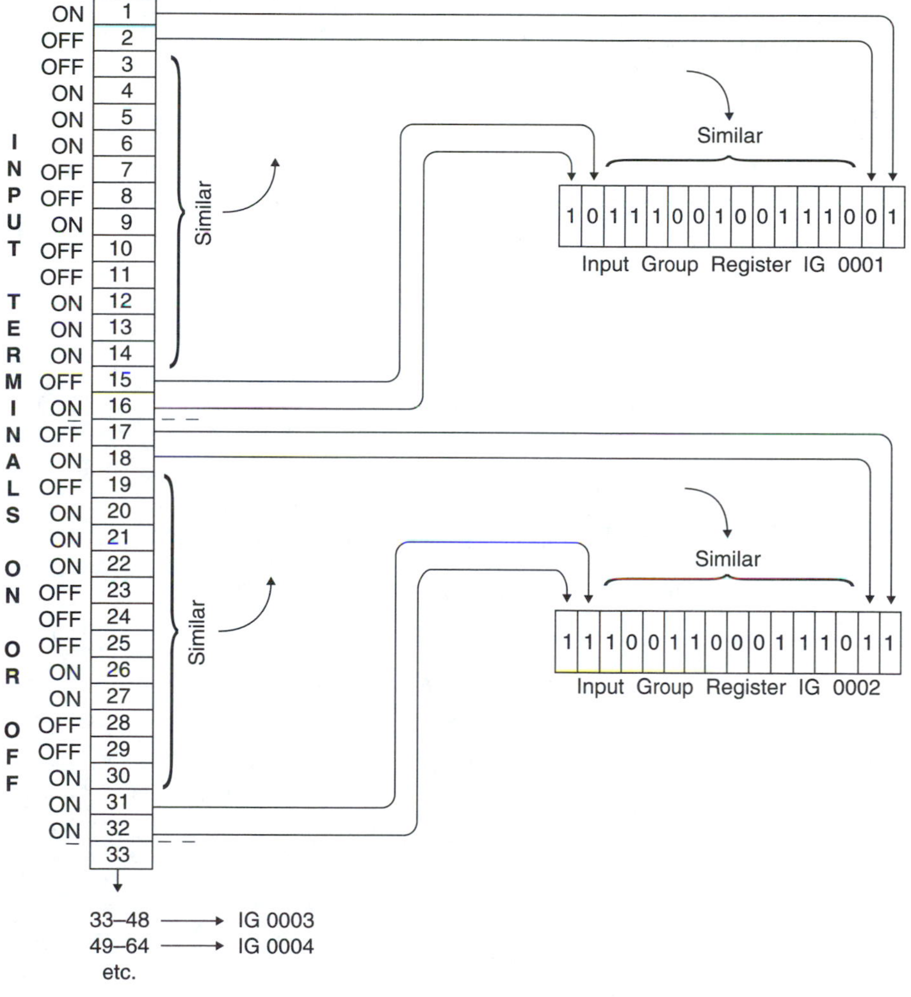

FIGURE 8–4
Input Group Register Scheme

Input Group Register Number	8 Bit System–Inputs Controlled	16 Bit System–Inputs Controlled
1	1–8	1–16
2	9–16	17–32
3	17–40	33–48
4	41–48	49–64
5	49–56	65–80
6	57–64	81–96
7	65–72	97–102
etc.	etc.	etc.

FIGURE 8–5
Input Group/Input Port Numbering Scheme

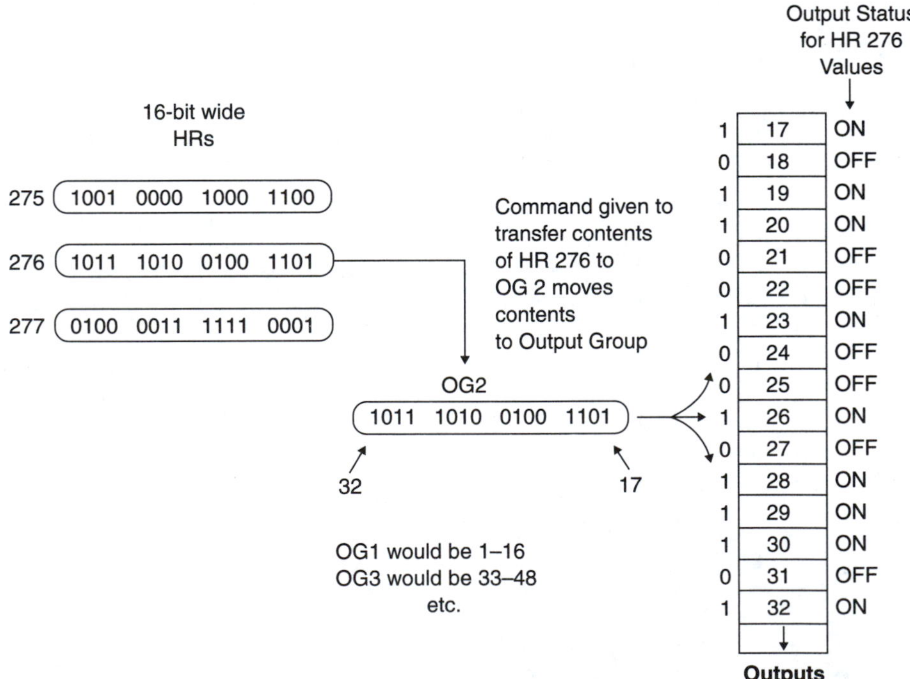

FIGURE 8–6
Output Group Register Scheme

puts. Without the input group system, you would use up more computer memory space to run your programs.

The input module port (terminal) corresponds to a single input group register bit. Each IG register status controls one bit's status. When a port is enabled, or on, it creates a 1 in the corresponding bit slot. If the port is off, it produces a 0 in the corresponding bit slot.

It is necessary to know how your PLC groups the input numbers that correspond to each input group register. A typical scheme is shown in figure 8–5.

8–5 OUTPUT REGISTERS: SINGLE AND GROUP

Like the input register, the output register has the same basic characteristics as the holding register. The output register differs from the holding register, however, in that it is readily accessible to the output module's terminals and ports. The number of output registers is normally equal to the number of input registers.

The output group register (OG) is organized in a manner similar to the input group register. It differs from the IG in a manner similar to the difference between input registers and output registers. Figure 8–6 shows how the OG register functions. One OG register can control 16 outputs. If a 1 is in a bit position, it will turn that bit's corresponding output on. A 0 will turn its corresponding output off. The grouping scheme for output group registers is similar to the input group register system. The grouping scheme is shown in figure 8–6. The output group register is particularly useful in sequencer operation, as discussed in chapter 19.

EXERCISES

1. List the five major types of registers. Use a block diagram to show where each type fits into the PLC scheme of operation.

FIGURE 8–7
Diagram for Exercise 2

A

Input No.	Status
49	ON
50	ON
51	OFF
52	ON
53	OFF
54	OFF
55	ON
56	ON
57	ON
58	OFF
59	ON
60	OFF
61	ON
62	ON
63	OFF
64	ON

B

Input No.	Status
105	ON
106	OFF
107	ON
108	OFF
109	ON
110	OFF
111	OFF
112	ON

(8 Bit PLC)

C

Input No.	Status
209	OFF
210	ON
211	OFF
212	ON
213	ON
214	OFF
215	ON
216	OFF
217	ON
218	ON
219	OFF
220	OFF
221	OFF
222	ON
223	ON
224	OFF

FIGURE 8–8
Diagram for Exercise 3

	Register	Label
A	0011 1100 1010 0111	OG 0007
B	1010 0110	OG 0006
C	0101 1111 0000 0110	IG 0011
D	1100 1011 1011 1000	OG 0021

2. What would the input group register look like for the three input module status arrangements shown in figure 8–7? What would the number of each IG register be? What would the register contents be in binary?
3. What would be the status of the corresponding outputs for the four output and input group registers shown in figure 8–8? What are the corresponding output numbers for the four OG and IG registers shown?

9

PLC Timer Functions

OUTLINE

9–1 Introduction □ **9–2** PLC Timer Functions □ **9–3** Examples of Timer Function Industrial Applications □ **9–4** Industrial Process Timing Application

OBJECTIVES

At the end of this chapter, you will be able to

- □ Describe PLC retentive and delay timer functions.
- □ List and describe eight major timing functions that are commonly used in circuits and processes.
- □ Apply PLC functions and PLC circuitry to process control for each of these eight major timing functions.
- □ Apply PLC timers in multiple timing problems that combine two or more of the basic timing functions.
- □ Apply PLC timers for the control of industrial processes.

9–1 INTRODUCTION

The most commonly used process control device after coils and contacts is the timer. The most common timing function is TIME DELAY-ON, which is the basic function. There are also many other timing configurations, all of which can be derived from one or more of the basic TIME DELAY-ON functions. PLCs have the one basic function timer capability in multiples. This chapter illustrates the basic PLC TIME DELAY-ON function and seven other derived timing functions. Typical of the derived functions are TIME DELAY-OFF, interval pulse timing, and multiple pulse timing of more than one process operation.

Most PLCs have one timer function, the retentive TIME DELAY-ON function. Some have two additional timer functions, TIME ON-DELAY, and TIME OFF-DELAY, whose operation will be explained.

Normally, only one of two types of the basic PLC timing functional blocks is in a PLC. The timing block functions are used with various contact arrangements and in multiples to accomplish various timing tasks. Typical industrial timing tasks include timing of the intervals for welding, painting, and heat treating. Timers can also predetermine the interval between two operations. With a PLC you can utilize as many timer blocks as you need, within the PLC memory limitations.

What does the PLC timer function replace? Detailed descriptions of traditional industrial timers may be found in controls texts, including those listed in the bibliography. Symbols for conventional timers are given in appendix C. PLC timer functions can replace any of these industrial timers. Whether the industrial timer is motor driven, RC time constant, or dash-pot, it can be easily simulated by a PLC.

The digital solid-state electronic timer is one technological step above the three types of industrial timers just listed. These digital timing devices are also discussed in various controls texts. The PLC timing function is more versatile and flexible than either the industrial or the digital electronic timers.

One major advantage of the PLC timer is that its time may be a programmable variable time as well as a fixed time. The variable time interval may be in accordance with a changing register value. Another advantage of the PLC timer is that its timer accuracy, repeatability, and reliability are extremely high because it is based on solid-state technology.

9–2 PLC TIMER FUNCTIONS

A single-input timer called a *nonretentive timer* is used in some PLCs. An example is shown in figure 9–1. Energizing IN001 causes the timer to run for 4 seconds. At the end of 4 seconds the output goes on. When the input is deenergized, the output goes off and the timer resets to 0. If the input IN001 is turned off during the timing interval (for example, after 2.7 seconds), the timer resets to 0.

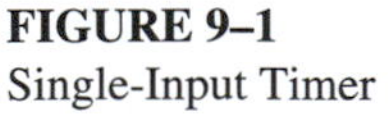
FIGURE 9–1
Single-Input Timer

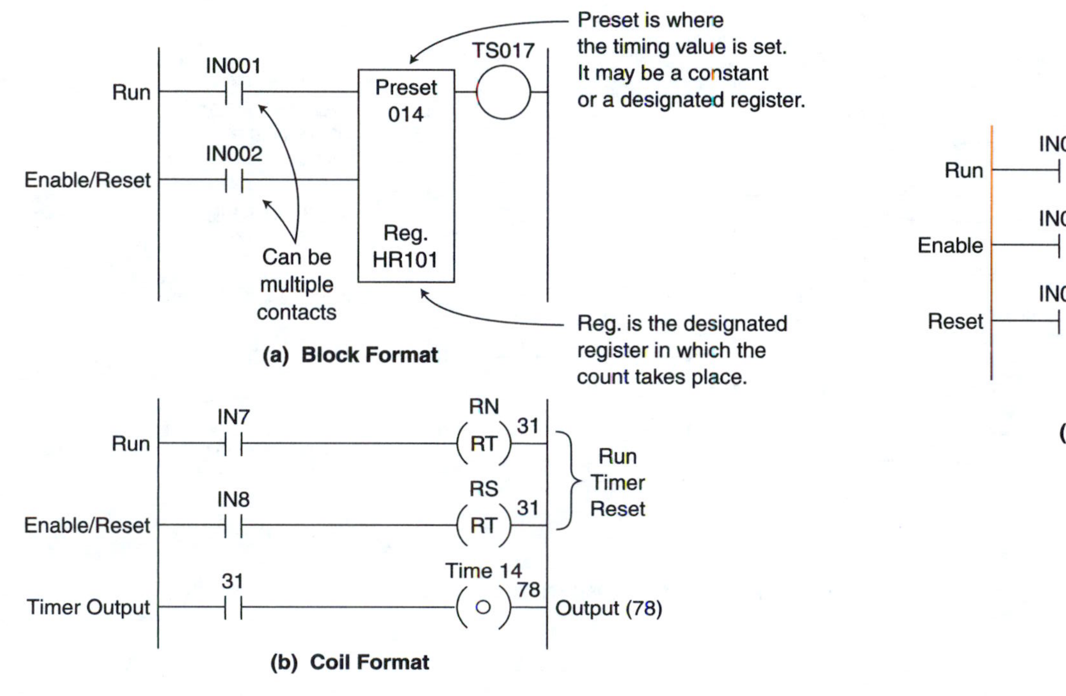

FIGURE 9–2
Double-Input Timer

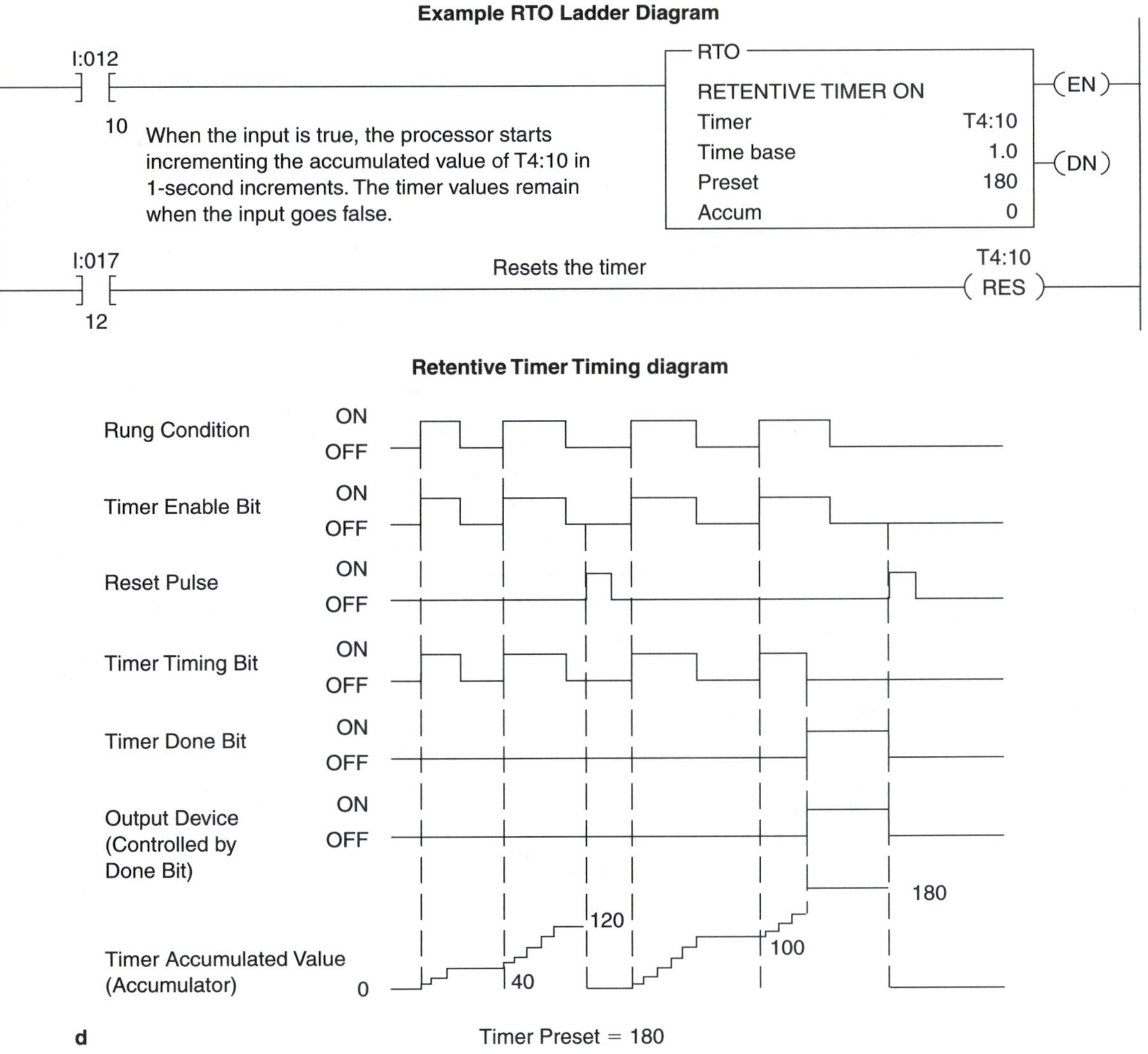

FIGURE 9–2 (continued)

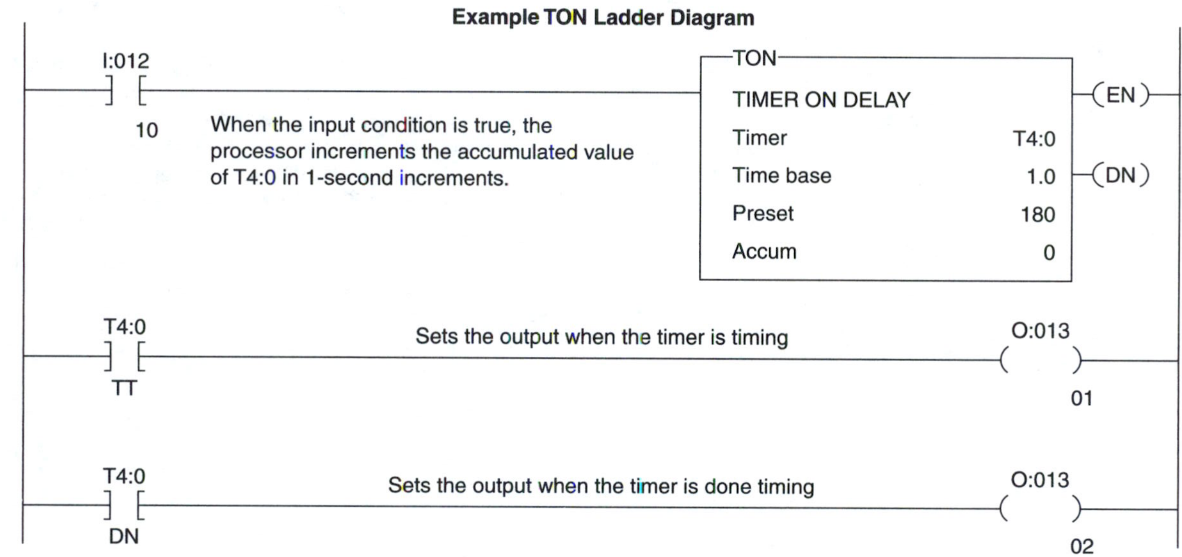

e When bit I:012/10 is set, the processor starts T4:0. The accumulated value increments in 1-second intervals. T4:0.TT is set and output bit O:013/01 is set (the associated output device is energized) while the timer is timing. When the timer is finished (.ACC = .PRE) T4:0.TT is reset (so O:013/01 and the associated output device is deenergized) and T4:0.DN is set (so O:013/02 is set and the associated output device is energized). When the accumulated value reaches 180 or when the rung conditions go false, the timer is reset.

FIGURE 9–2 (continued)

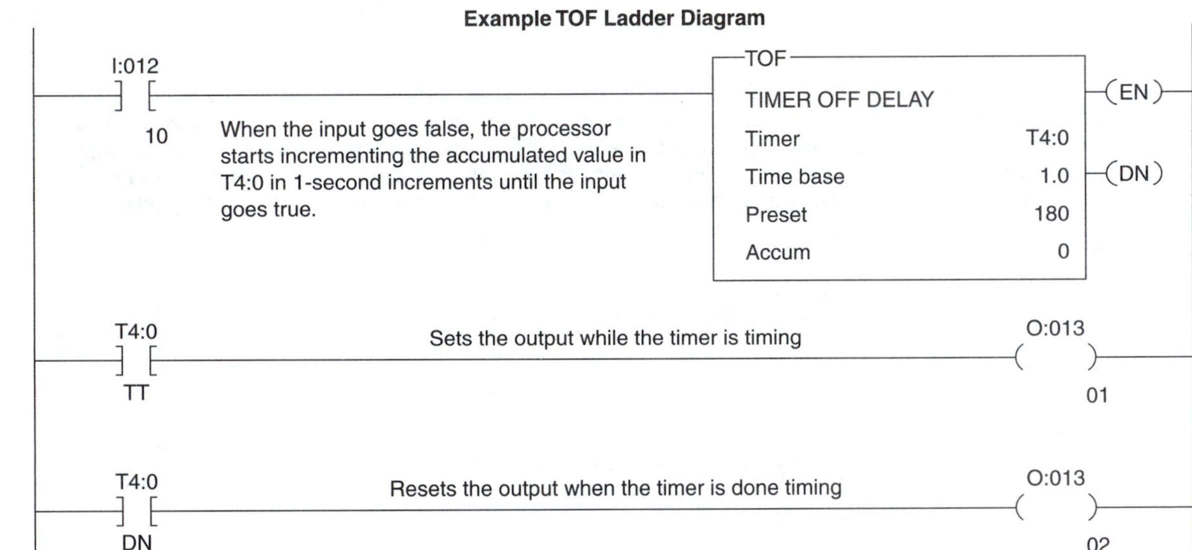

When bit I:012/10 is reset, the processor starts timer T4:0. The accumulated value increments by 1-second intervals as long as the rung remains false. T4:0.TT is set and output bit O:013/01 is set (the associated output device is energized) while the timer is timing. When the timer is finished (.ACC = .PRE), T4:0.TT is reset (so O:013/01 is reset and the associated output device is deenergized) and T4:0.DN is reset (so O:013/02 is reset and the associated output device is deenergized). When the accumulated value reaches 180 or when the rung conditions go true, the timer stops.

f

FIGURE 9–2 (continued)

There are operational disadvantages of the single-input type that are overcome by the multiple-input timer. Figure 9–2 shows three types of formats for PLC timers. Figure 9–2a and b illustrate two common types. Figure 9–2c shows an occasionally used type where the enable and reset are different inputs for use in some special cases.

The block format in figure 9–2a includes the Enable/Reset line, which allows the timer to run when energized. When deenergized, the timer is kept at 0 or reset to 0. The upper line causes the timer to run when the timer is enabled. When enabled, the timer runs as long as the run input is energized. If Run is deenergized while the timer is running, the timing stops where it is and does not reset to 0. Note that for figure 9–2a if both IN001 (Run) and IN002, (Enable/Reset) open and close at the same time, the timer functions in the same manner as the timer in figure 9–1.

For format (a) in figure 9–2, suppose that IN002 is closed and IN001 is turned on. After 6 seconds, IN001 is opened. The timer retains a count of 6. Timing has not reached the preset value of 14 seconds, and the timer output is still off. The timer does not reset unless IN002 is opened. Suppose that sometime later IN001 is reclosed. After 8 more seconds of IN001 being closed, the timer coil will energize, since 6 + 8 = 14.

Format (b) in figure 9–2 is an alternate. IN7 is for timing RT31 = RN. IN8 enables RT31 = RS. When the timer goes on, its output 31 (internal) turns output 78 on. In some PLC formats, the register count is not displayable. Also, in some PLC formats, the preset time value is fixed for the function chosen—for example, 5-second timer, 10-second timer, and so on.

Figure 9–2c illustrates a special case when the enable and reset are two separate inputs rather than a common single input. This configuration can be used for special program requirements when the arrangement is available. The configuration and timing diagram for an Allen-Bradley PLC is illustrated in figure 9–2d.

Two other timer functions are included in some PLCs: TIME DELAY-ON and TIME DELAY-OFF. Their functions are shown in figure 9–2e and f.

An added help in defining timer contact status is shown in figure 9–3. There are three states in a timing cycle: (1) the initial or reset state, (2) the state during timing, and (3) the state after timing is complete. A system of X for on and O for off is normally used. The ex-

FIGURE 9–3
Sequencing Chart.

KEY: O — Off X — On

Reset Status	During Timing	Timed Out		Convention
Open	Open	Closed	⟶	O O X
Open	Closed	Open	⟶	O X O
Closed	Open	Closed	⟶	X O X
	Etc. ...			

X — Contact Closed

O — Contact Open

amples illustrate this convention. In some systems a 1 is used instead of an x to indicate contact closed.

9–3 EXAMPLES OF TIMER FUNCTION INDUSTRIAL APPLICATIONS

Some commonly used timer functions are:

Example 9–1 On delay. Output B comes on at a specific set time after output A is turned on. When A is turned off, B also goes off.

Example 9–2 Off delay. Both A and B have been turned on at the same time. Both are in operation. When A is turned off, B remains on for a specific set time period before going off.

Example 9–3 Limited on time. A and B go on at the same time. B goes off after specific set time period, but A remains on.

Example 9–4 Repeat cycling. An output pulses on and quickly off at a constant preset time interval.

Example 9–5 One-shot operation. Output B goes on for a specified time after output A is turned on. Output B will run for its specified time interval even if A is turned off during the B timing interval.

Example 9–6 Alternate on and off of two outputs. An example of this timing application is two alternately flashing signal lights. The time on for each of the two lights may be the same, or the two times could be set to different intervals.

Example 9–7 Multiple on delay. Two different events start at different time intervals after an initial starting time reference point.

Example 9–8 Multiple off delay. Two different functions remain on for two different time intervals after a process is turned off.

Example 9-9 Interval time within a cycle. We may require that an output come on 7.5 seconds after system startup, remain on for 4.5 seconds, and then go off and stay off. The interval would then be repeated only after the system is shut off and then turned back on.

There are other timing examples we could illustrate as well; however, these nine examples are representative of PLC timing capabilities.

Now we give examples of each of the nine listed timing systems. In most cases, each example is illustrated in terms of an industrial problem. Each example includes a diagram showing time versus output on–off status. The time-status diagrams include the X and O designations for reference.

EXAMPLE 9-1 The first example is the simplest form of time delay. When the circuit is turned on, one action takes place. A specified time later, another action occurs. Both relay logic and PLC logic diagrams are shown in this example only for comparison. Subsequent examples have PLC diagrams only. Figure 9–4 shows the program for this example.

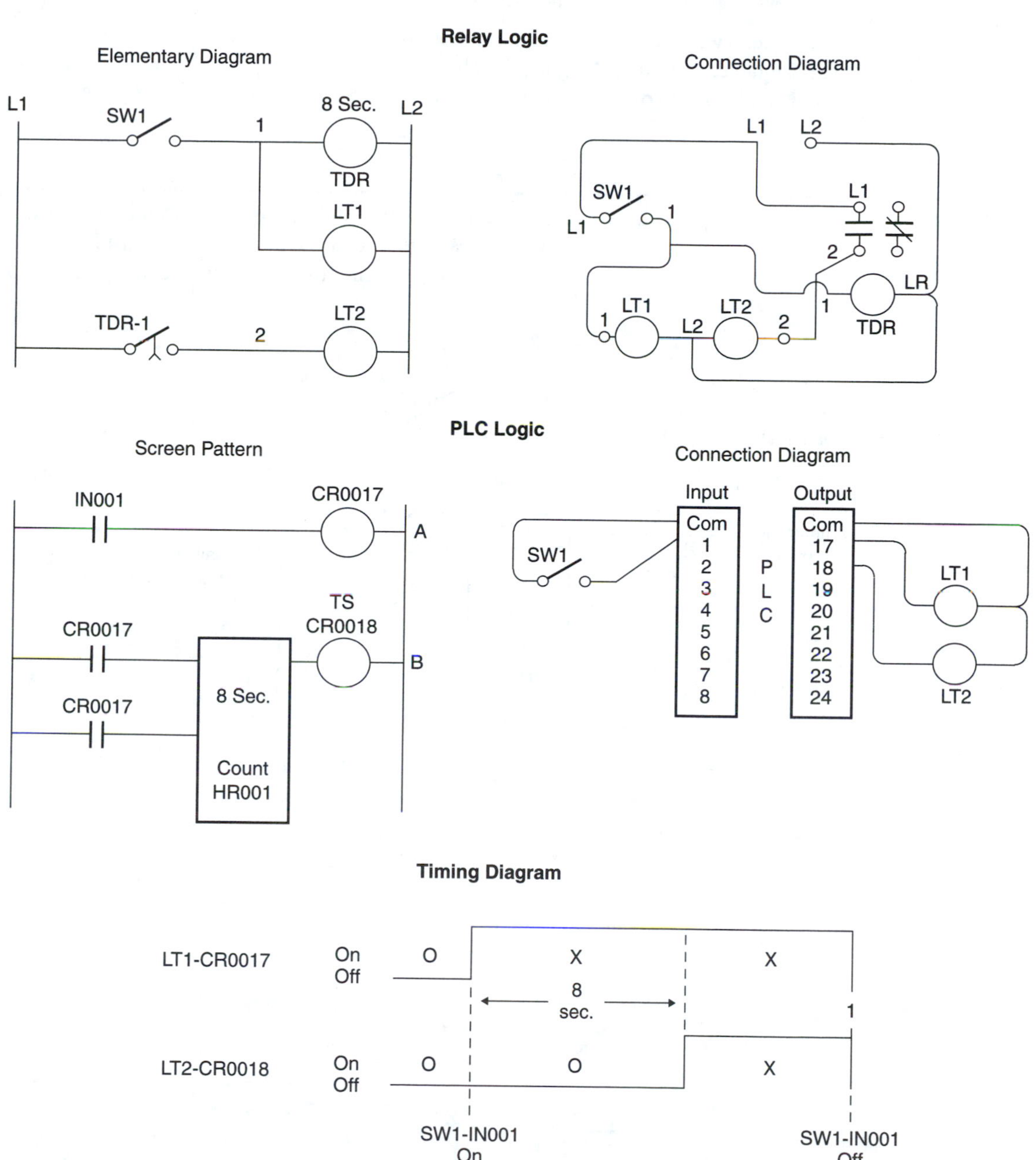

FIGURE 9–4
Example 9–1: TIME DELAY-ON.

The sequence for this example is as follows:

1. When switch 1 is turned on, light A lights.
2. Eight seconds after A lights, B lights.
3. Both lights go off or stay off whenever switch 1 is opened.

EXAMPLE 9-2 The second example is an off-delay circuit. A motor and its lubrication pump motor are both running. Lubrication for main motor bearings is required during motor coastdown. After the main motor is shut off, the lubricating pump remains on for a time corresponding to

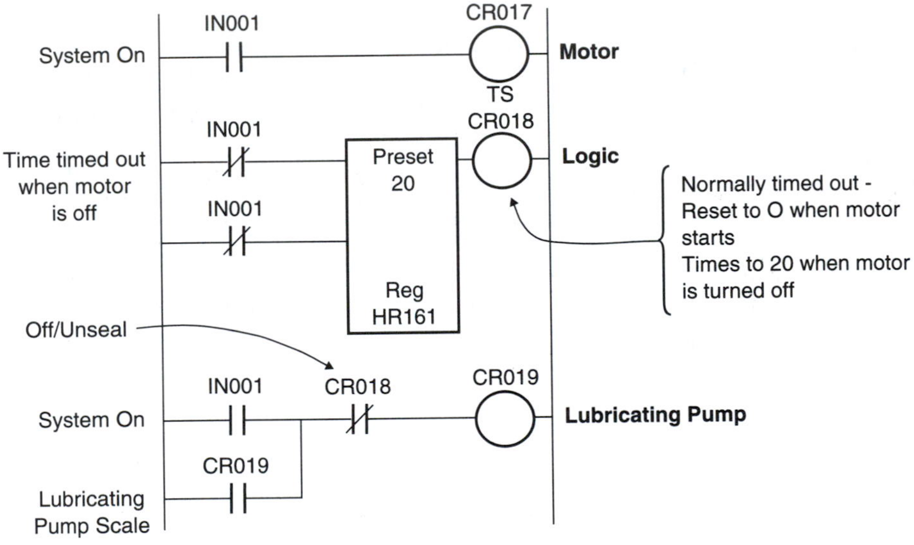

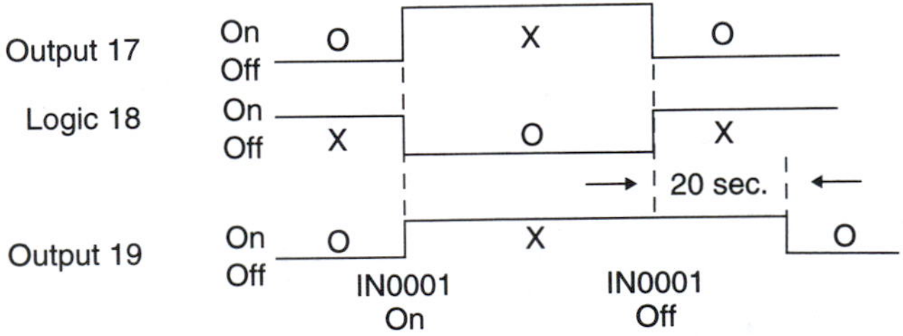

FIGURE 9–5
Example 9–2: TIME DELAY-OFF.

coast-down time. In this example, a lubricating pump remains on for 20 seconds after the main system is shut down. Figure 9–5 shows the required program.

EXAMPLE 9-3 The third example involves two parts. The first part is for a single time interval; the second is a multiple time interval example.

Example 9–3–1, shown in figure 9–6, illustrates a situation in which two inputs go on at the same time. Then, one of them is to go off after a preset period of time. One output, A, stays on; the other output, B, turns off at the end of the timing interval. Resetting is accomplished by turning IN001 and IN002 off.

Example 9-3-2, shown in figure 9–7, is a multiple application timing system. Three outputs turn on at the same time. One stays on. Another, M, shuts off after 8 seconds. The third output, N, shuts off after 14 seconds.

FIGURE 9–6
Example 9–3–1: One Output, Time Interval On at Start.

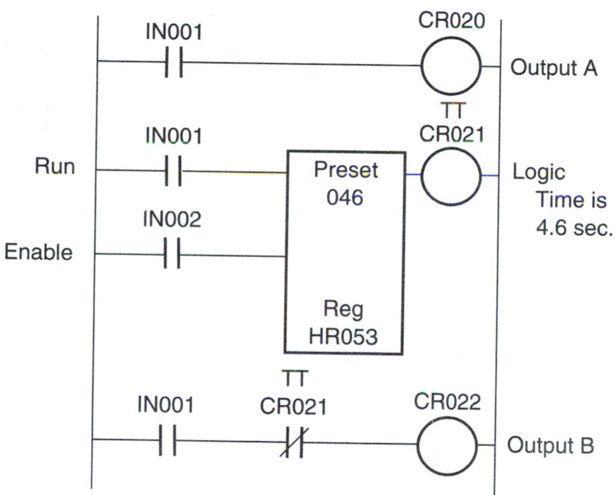

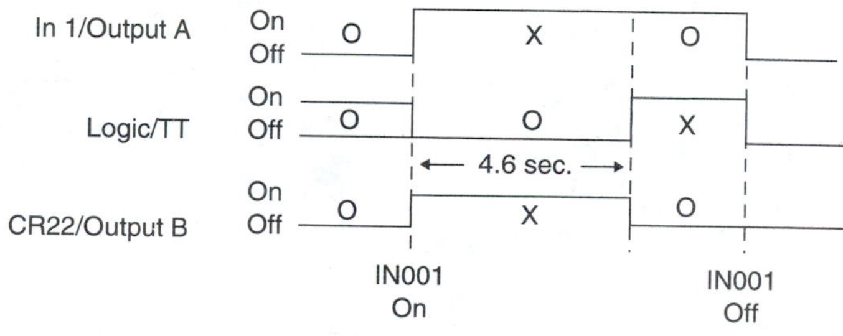

FIGURE 9–7
Example 9–3–2: Two Outputs, Time Intervals On at Start.

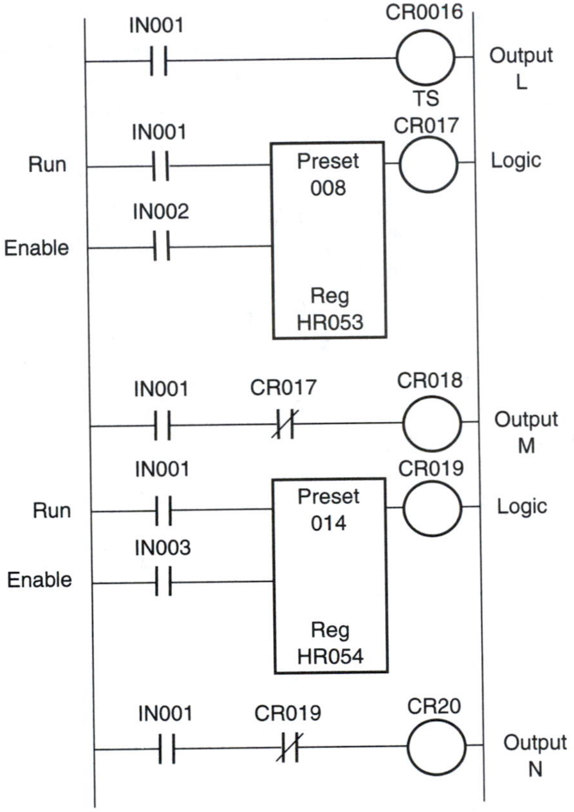

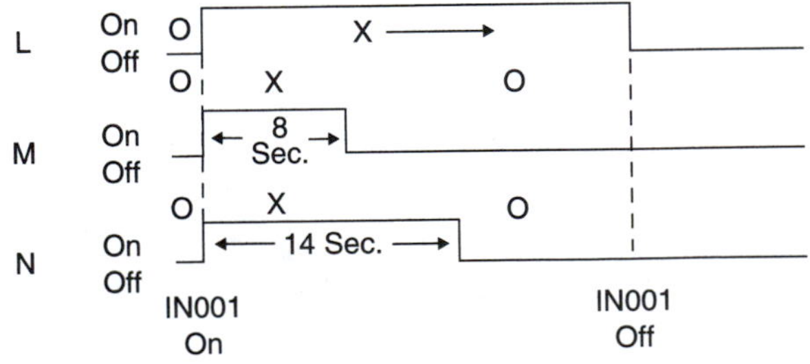

EXAMPLE 9-4 The fourth example is a pulsed timer. A short voltage pulse is produced every 12 seconds. The PLC circuit shown in figure 9–8 will produce the required pulses. The timer is initially turned on by its "time" input. After the timing interval, the timer turns the output on. When the output goes on, one of its contacts, TT013, immediately opens and resets the timer to 0. When the timer is reset to 0, the output is turned off. Then, since the timer is also off, TT013 recloses, restarting the cycle. The pulsed on time is a very short, one-scan cycle time. The process repeats itself continuously.

EXAMPLE 9-5 The fifth example, shown in figure 9–9, is the one-shot system. The output comes on after its specified time period even if the input is turned off during the timing period. IN0011 must be opened and reclosed to reset the system.

EXAMPLE 9-6 The sixth example is an alternating two-output system. Figure 9–10 is an example of this application. It is a two-light, alternately flashing PLC program. Outputs 11 and 12 control the two lights. The outputs alternate on–off and off–on every 5/10 of a second. The on times for each light can be varied by resetting the times in the functional block. The programmed times may be the same value, or each could be set at different time values. IN0001 is the system on–off control.

FIGURE 9–8
Example 9–4: Pulse Repetitive Timer.

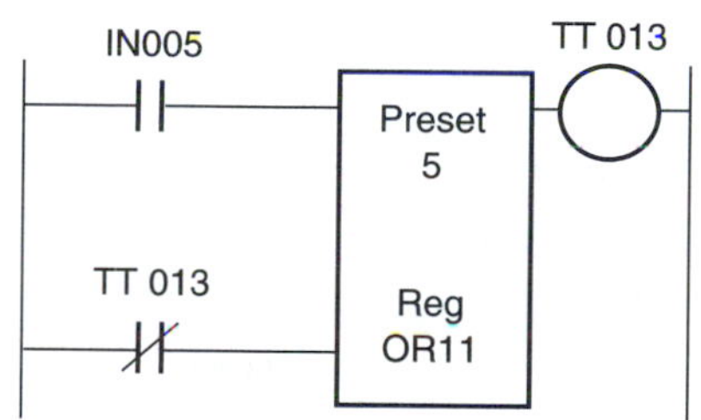

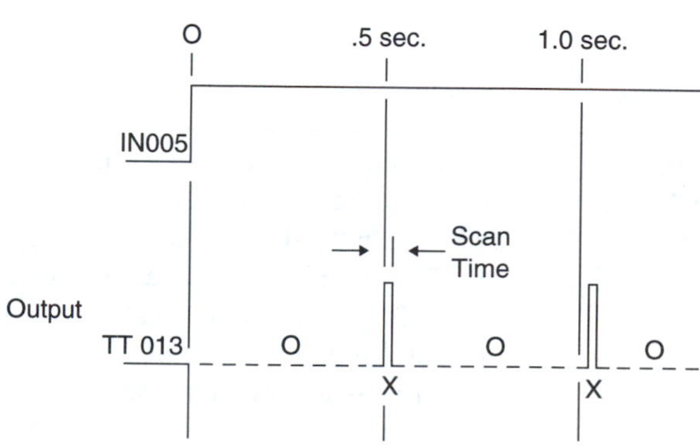

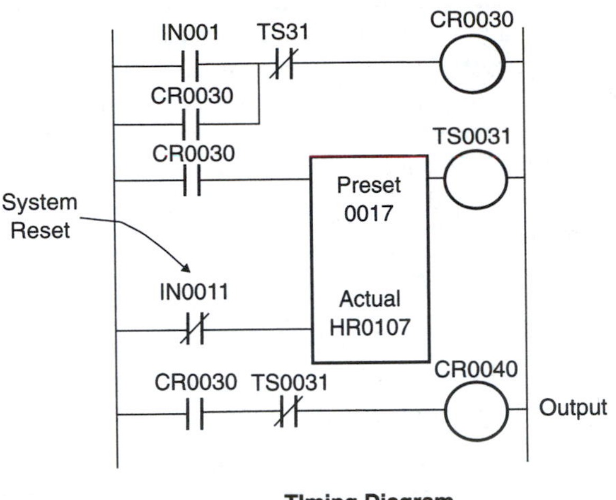

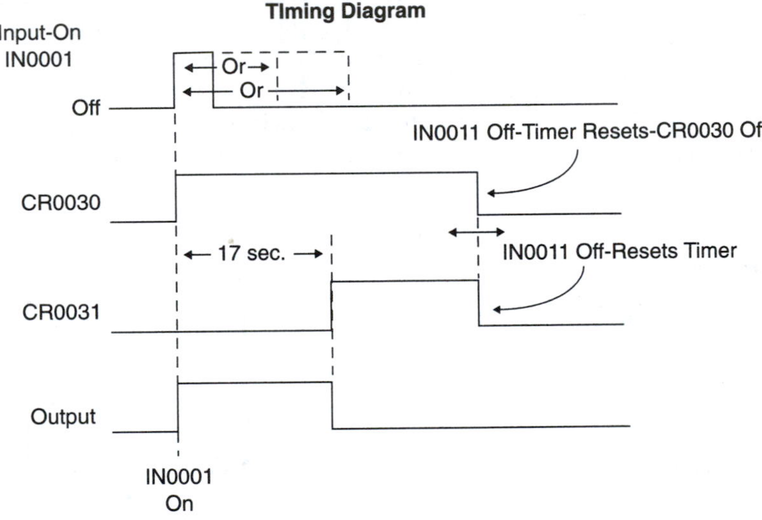

FIGURE 9–9
Example 9–5: One-Shot Timer Operation.

EXAMPLE 9-7 The seventh example is for a time delay on for two outputs with respect to the start of a sequence. There are two ways to accomplish a time delay on, both of which are shown in figure 9–11. In the diagram on the left, both time intervals start at the same time and delay 7 and 12 seconds. The diagram on the right accomplishes the same thing, but chains the timers. The second time delay is accomplished by adding 5 seconds to the first 7 seconds, for a resulting time delay of 12 seconds. Multiple time delay on can be accomplished by appropriately adding more timers.

FIGURE 9–10
Example 9–6: Alternate Flasher System.

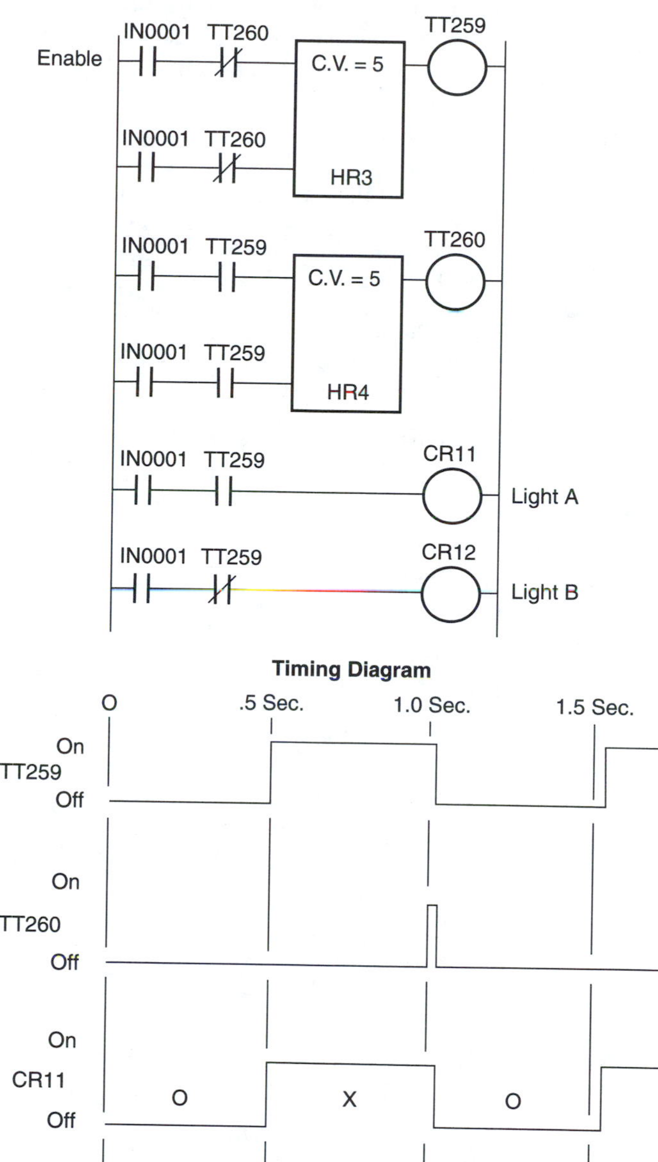

EXAMPLE 9-8 Dual time delay off is illustrated in figure 9–12. We use an actual application for illustration. When the lights are turned off in a building, an exit door light is to remain on for an additional 42 seconds. In addition, the parking lot lights are to remain on for an additional 3 minutes after the door light goes out. The three outputs are identified in the figure. For more delay off outputs, an additional repeated two lines of logic are added as required.

Output 018 energizes after 7 sec. and 019 after 12 sec. (5 more) after IN001 is energized.

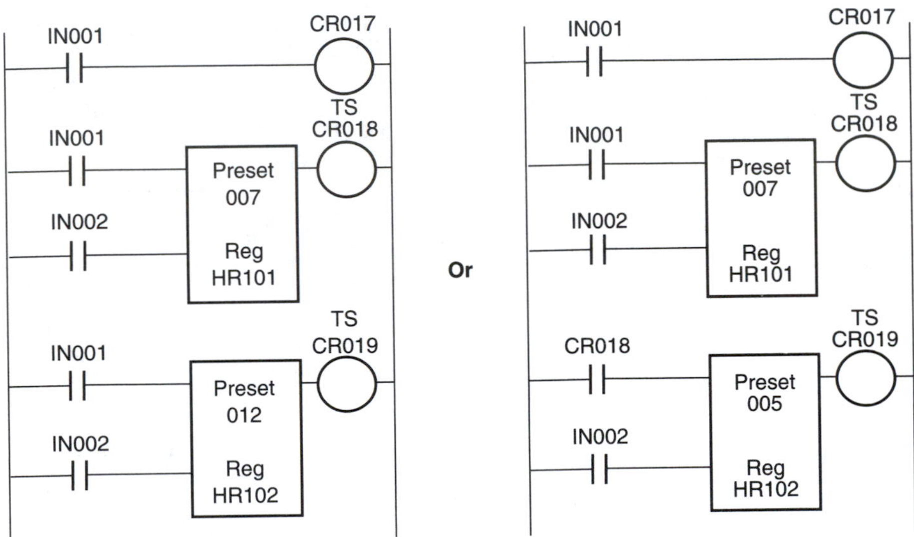

Timing Diagram
IN002 is the enable/reset input.

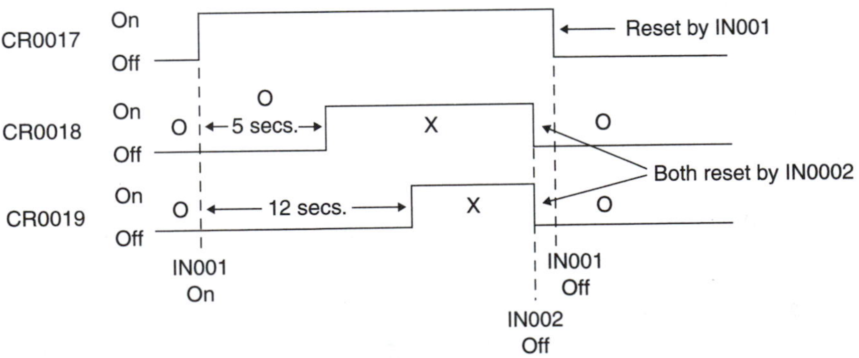

FIGURE 9–11
Example 9–7: Dual On Delay—Two Schemes.

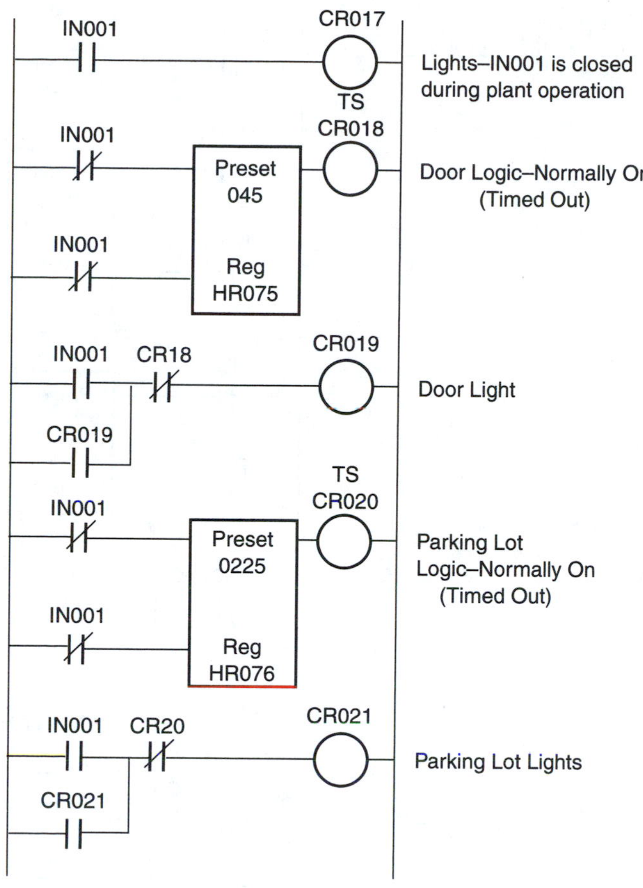

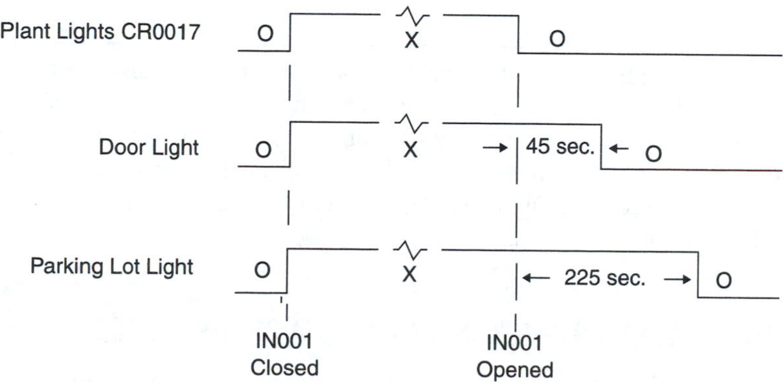

FIGURE 9–12
Example 9–8: Dual Off Delay.

FIGURE 9–13
Example 9–9: Embedded Interval Timing.

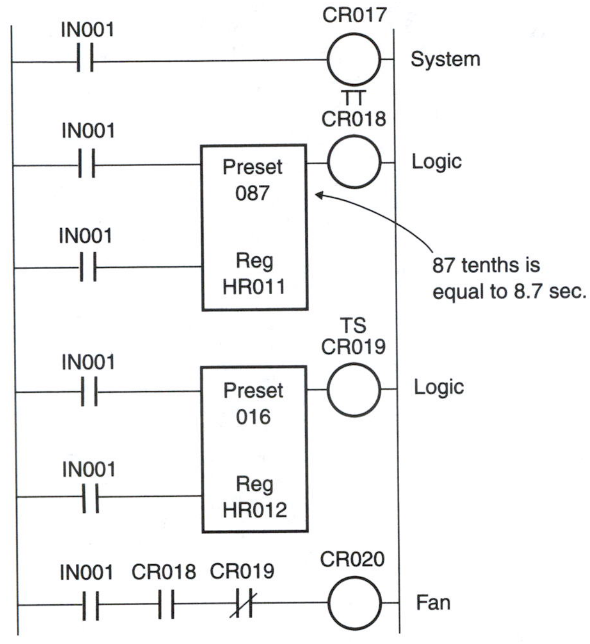

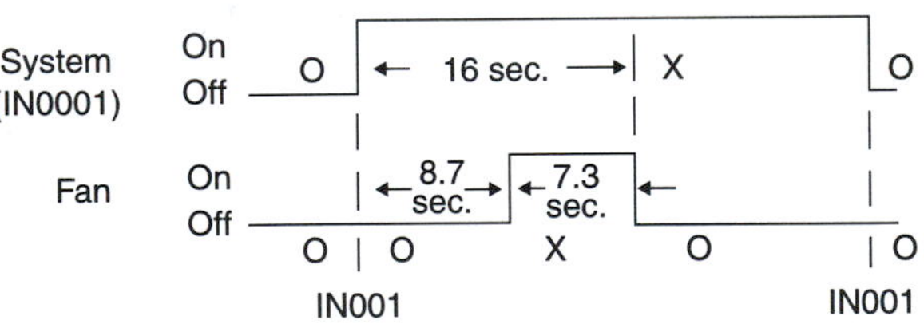

EXAMPLE 9-9 The ninth example is for a timed interval of a number of seconds after the start of a process operation. This type of time interval is sometimes called an *embedded time interval*. This operation uses the special operation of a fan. The fan is to come on 8.7 seconds after a system is turned on. It is then to run until 16 seconds after the system is turned on, which is a net time of 7.3 seconds. Figure 9–13 shows a PLC program that accomplishes this time interval requirement.

9–4 INDUSTRIAL PROCESS TIMING APPLICATION

The following problem requires the use of multiple timing programming as well as contact/coil logic. It consists of a single, operational, heat-treating machine. The station carries out a surface-hardening process on a steel ring. Hardening is accomplished by heating the steel ring to a high temperature, then immediately quenching it (cooling it very rapidly). The metallurgical result is a relatively hard surface on the steel ring.

The heating is done by a noncontact induction heating process. A high current in the circular coil around the outside of the part induces high circulating currents in the part. The part therefore heats up very rapidly. The coil has cooling water circulated through its outer half to keep the heating ring from overheating or even melting. The quench is then done by spraying cold water on the part. The quench water is pumped into the inner half of the induction coil. Spraying the part with cold water through the many holes in the inside of the coil results in fast cooling, which produces a case-hardened surface.

A mechanical layout of the process station is shown in figure 9–14. The processing sequence of this operation is as follows:

1. The master pushbutton is depressed, turning the system on.
2. The part is placed on the mandrel.
3. Both left and right start-up buttons are depressed.
4. At this point, or at any other time, pushing any stop button stops all processing action.

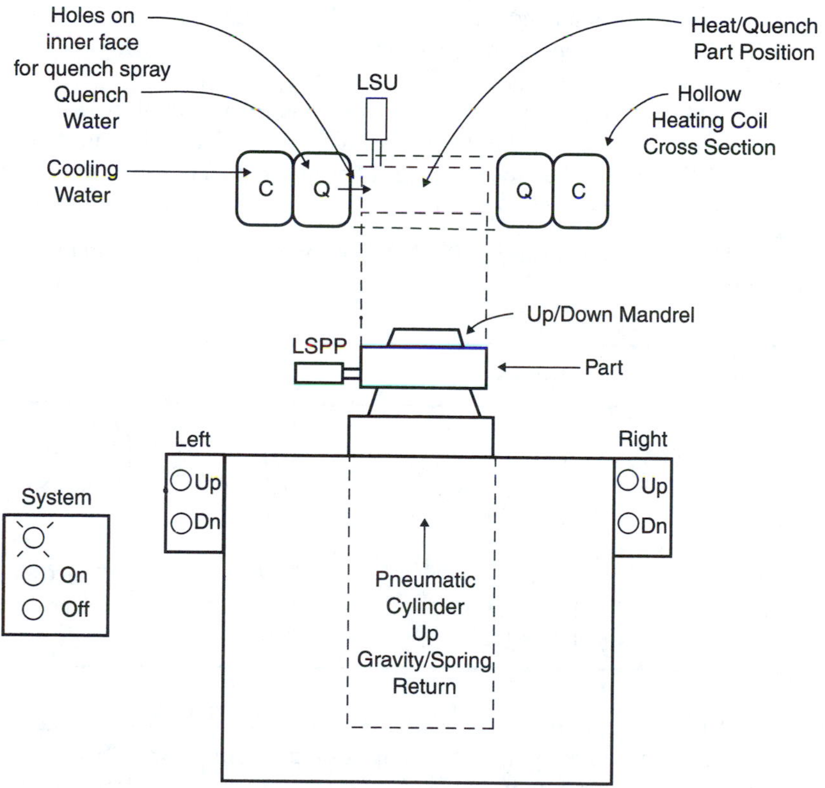

FIGURE 9–14
Heat/Quench Station Layout.

5. The part is raised from bottom to top by pneumatic air action. A solenoid valve supplies this air to a pneumatic elevating cylinder. The lower-limit switch must be actuated before the part will rise. Lower-limit switch actuation indicates that there is a part on the mandrel and that the mandrel is down. Note that the limit switch opens as the part leaves the bottom position.
6. The mandrel makes contact with a limit switch at the top of the travel.
7. Heat comes on for 10 seconds and goes off.
8. Quench comes on for 8 seconds and goes off.
9. The part returns down by gravity and spring action. The upper-limit switch becomes inactivated when the mandrel starts down.
10. The part and mandrel reach the bottom. The down-limit switch is again actuated.
11. The system should reset.
12. The part is removed.

Some optional features not included in this program are:

- If you assume the heat generator and both water pumps are on, interlocks could be added to insure they are running throughout the process.
- The same ring part could be processed two or more times. We could require the ring to be removed after step 12 before resetting takes place.
- Is proper temperature reached? A thermocouple sensor could be incorporated to monitor temperature.
- Manual controls for setup could be added. These are Up, Heat, and Quench.
- Safety features could be added, such as a safety shield that lowers during heat. Where does the process restart after interrupted power is restored?
- Other features as required.

The next step is to assign PLC register or address numbers to the various inputs and outputs.

Inputs		Outputs	
0001	Master Shop	0019	Solenoid Valve—Up
0002	Master Start	0021	Heat On Contactor Coil
0003	Left Stop-Up	0023	Quench Spray Water Solenoid
0004	Left Start-Down		
0005	Right Stop-Up		**Options**
0006	Right Start-Down	0017	System On Pilot Light
0007	Limit Switch Down	0018	Machine On/Up Pilot Light
0008	Limit Switch Up		

A ladder diagram to carry out the process is then developed, as shown in figure 9–15.

The next step is to draw the connection diagram for the PLC. There are eight input connections and five output connections, three to process actuators and two to pilot lights. The connection diagram is illustrated in figure 9–16.

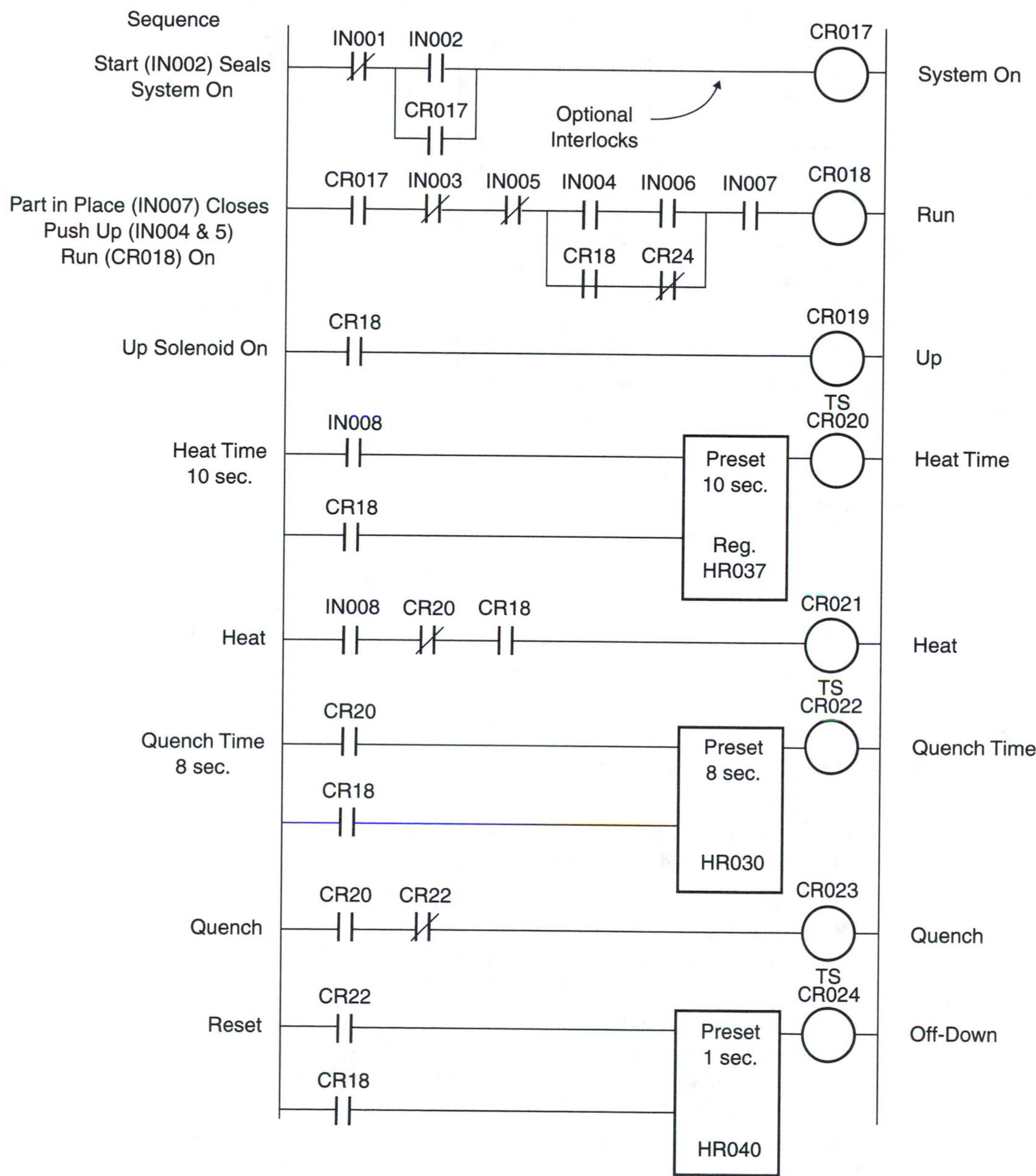

FIGURE 9–15
Heat/Quench Machine Program.

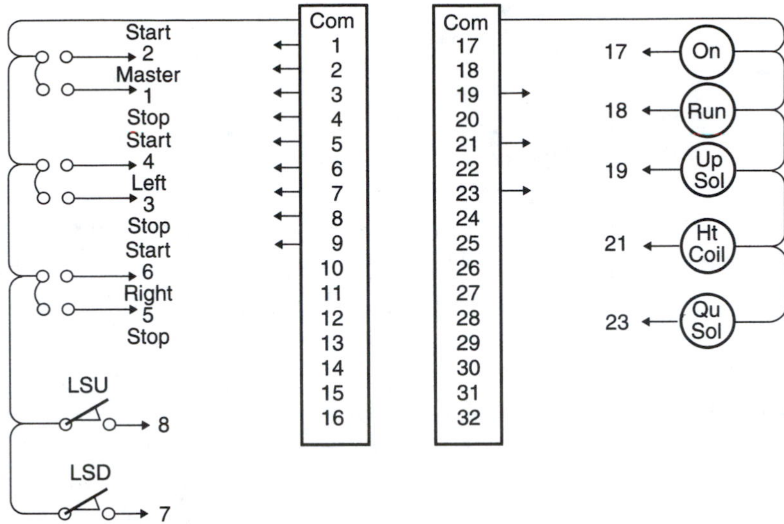

FIGURE 9–16
Connection Diagram for PLC Module, Inputs, and Outputs.

The final two developmental steps are to program the PLC for the process and make modifications as required.

TROUBLESHOOTING QUESTIONS

The PLC has been programmed to operate as shown in the figures referred to. However, the circuit has the malfunction(s) noted. What misprogramming or other factor or factors could cause the malfunction?

TS 9–1 Refer to the figure 9–4 (example 9–1) PLC circuit.

1. The timer times out in less than 1 second.
2. The timer will not reset when you deenergize IN001.

TS 9–2 Refer to figure 9–5 (example 9–2).

1. The lubricating pump does not start when the motor starts.
2. The lubricating pump goes off at the same time as the motor.
3. The lubricating pump never goes off.

TS 9–3 Refer to figure 9–7 (example 9–3–2).

1. Both outputs go on after 8 seconds.
2. The timers will not reset.
3. Both outputs go on and stay on.
4. Neither output will go on.

TS 9–4 Refer to figure 9–8 (example 9–4). The pulse circuit works once and stops.

TS 9–5 Refer to figure 9–9 (example 9–5). The output goes off after 17 seconds, then comes back on immediately for 17 more seconds, and repeats this pattern.

TS 9–6 Refer to figure 9–10 (example 9–6). Each light flashes once and the system stops.

TS 9–7 Refer to figure 9–11 (example 9–7).
1. The second output, CR019, does not go on.
2. Only the first timer resets.

EXERCISES

Write a PLC program for these exercises, insert them into a PLC, and test the programs for correct operation.

1. A timer is to turn on a fan switch 8.6 seconds after a wall switch is turned on. If the wall switch is turned off during the 8.6 second time interval, the timer is to reset to zero seconds, so that when the wall switch is again turned on, the delay is the full 8.6 seconds.
2. When a switch is turned on, C goes on immediately and D goes on 9 seconds later. Opening the switch turns both C and D off.
3. E and F are turned on by a switch. When the switch is turned off, E goes off immediately. F remains on for another 7 seconds and then goes off.
4. G and H both go on when an input is energized. G turns off after 4 seconds. H continues running until the input is deenergized. Turning input off at any time turns both outputs off.
5. Two pulsers start at the same time. Pulse output J is to pulse every 12 seconds. Pulse output K is to pulse every 4 seconds.
6. When L is turned on, M is to go on 11 seconds later. M goes on after 11 seconds, no matter how long L is turned on.
7. A. Two lights are to flash on and off at different intervals. One is on 5 seconds and off 5 seconds. The other is on 8 seconds and off 8 seconds. B. Two lights are to flash alternately, one for 5 seconds, one for 8 seconds.
8. There are four outputs: R, S, T, and U. R starts immediately when an input is energized. S starts 4 seconds later. T starts 5 seconds later than S. U goes on 1.9 seconds after S. One switch turns all outputs off.
9. Repeat exercise 7 for turning off delay on. S goes off 4 seconds after R. T goes off 6 seconds after R. U goes off 2.5 seconds after S.
10. An output pulse, V, is to go on 3.5 seconds after an input, W, is turned on. The V time-on interval is to last 7.5 seconds only. V is to go on again 3 seconds later for 5.3 seconds.
11. There are three mixing devices on a processing line: A, B, and C. After the process begins, mixer A is to start after 7 seconds elapse. Next, mixer B is to start 3.6 seconds after A. Mixer C is to start 5 seconds after B. All then remain on until a master Enable switch is turned off.
12. When a start button is depressed, M goes on. Five seconds later, N goes on. When Stop is pushed, both M and N go off. In addition, 6 seconds after M and N go off, fan F, which had previously been off, goes on. F remains on until the start button is again depressed, at which time it goes off.
13. A wood saw, W, a fan, F, and a lubrication pump, P, all go on when a start button is pushed. A stop button stops the saw only. The fan is to run an additional 5 seconds to blow the chips away.

The lube pump is to run for 8 seconds after shutdown of W. Additionally, if the saw has run more than one minute, the fan should stay on indefinitely. The fan may then be turned off by pushing a separate fan reset button. If the saw has run less than one minute, the pump should go off when the saw is turned off. The 8-second time delay off does not take place for a running time of less than 1 minute.

10

PLC Counter Functions

OUTLINE

OBJECTIVES

At the end of this chapter, you will be able to

- □ Describe the PLC counter functions.
- □ List some of the major counting functions used in circuits and processes.
- □ Apply the PLC counter function and associated circuitry to process control.
- □ Apply combinations of counters and timers to process control.

10–1 INTRODUCTION

PLC counters have programming formats which are similar to timer formats. One counter input furnishes count pulses which the PLC function analyses. Another input usually carries out Enable/Reset. Alternatively, there can be separate inputs for Enable and Reset, depending on the application requirements. Conventional counters replaced by the PLC counter function include mechanical, electrical, and electronic types. Typical examples of these counters can be found in various manufacturers' manuals and are described in various other texts.

Most PLCs contain both up and down counters which function similarly. Some PLCs also include a combination up/down counter in one function. Others contain special high-speed counters for high-frequency counting capability. The up counter counts from 0 up to a preset count, where some indicating action takes place. The down counter starts from the preset number and counts down to 0, where the indicating action takes place.

10–2 PLC COUNTERS

Figure 10–1 shows four configurations of typical PLC counter functions. Figure 10–1A is the basic PLC counter function, which is the same for up and down counters except for the designation of UC or DC. Figure 10–1a1 is one format, the block format, and is the one we use in the chapter. The count at which the counter output is to go on is entered into the preset space. The count can be a constant as shown, 21, or can be a register which has a varying value from the process. For the up counter, the count starts at 0 and increments by 1 each time IN001 is pulsed on. When the preset value is reached the output, CR17 in our example, goes on. As the count goes on beyond the preset value, the output stays on. Opening IN002 at any time resets the counter to 0. The down counter operates in a similar manner. The count starts at the preset value and increments down by one each time the input IN001 goes on. When the count reaches 0, the output goes on.

Figure 10–1a2 shows a coil format found in some PLC models. The counter output is internal logic. The counter, 32 in our example, counts to 21 and turns on. To actuate an output, a program line uses a contact from the count logic, 32, to actuate the output, 74. Some PLCs contain an additional up/down counter such as that shown in figure 10–1b. There are three inputs: one for up counts, one for down counts, and an enable/reset input. The up/down counter can be used as a single function when the process uses an up and a down counter in tandem. The preset can be set to a value as shown or can be set at the maximum count if you only wish to observe the net count in the designated register, HR102. One precaution when using this function is to know whether the function will accept negative numbers in case the net count goes negative.

Some PLCs contain a high-speed counter, as shown in figure 10–1c. Refer to the PLC operating manuals for the frequencies at which their high-speed counter operates. A number of auxiliary functions work with these counters. Again, see the PLC manuals for the system setup.

In some applications, you may wish to stop the count for a time without it resetting to 0 and start counting again later where the count left off. This is accomplished as shown in figure 10–1d. The Enable and Reset inputs are separate for this counter function.

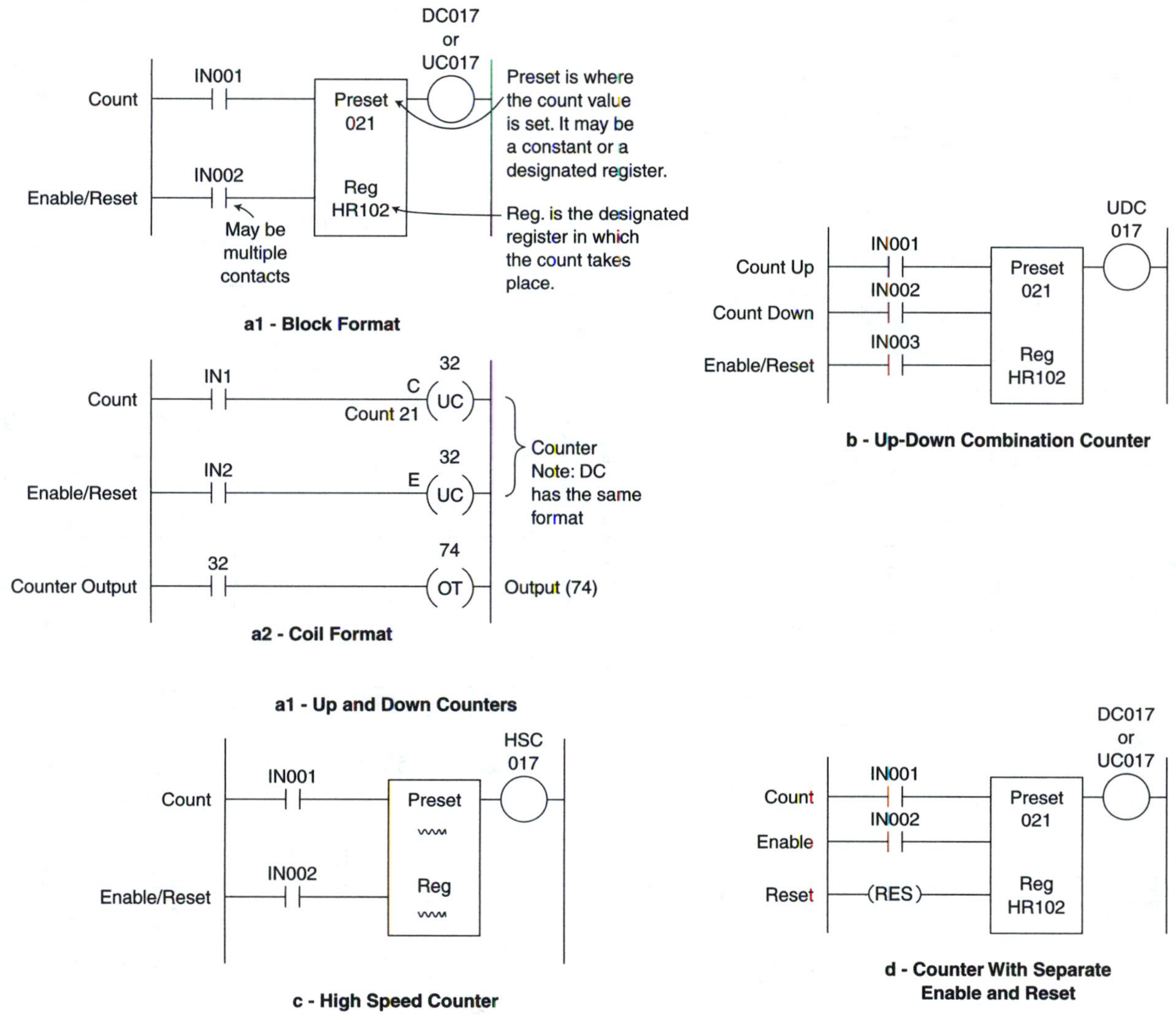

DC017
or
UC017
Count
IN001
Preset
021
Enable/Reset
IN002
Reg
HR102
May be multiple contacts
Preset is where the count value is set. It may be a constant or a designated register.
Reg. is the designated register in which the count takes place.
a1 - Block Format
Count
IN1
C
32
UC
Count 21
Enable/Reset
IN2
E
32
UC
Counter Note: DC has the same format
Counter Output
32
74
OT
Output (74)
a2 - Coil Format
a1 - Up and Down Counters
Count Up
IN001
Count Down
IN002
Enable/Reset
IN003
Preset
021
Reg
HR102
UDC
017
b - Up-Down Combination Counter
Count
IN001
Enable/Reset
IN002
Preset
Reg
HSC
017
c - High Speed Counter
Count
IN001
Enable
IN002
Reset
(RES)
Preset
021
Reg
HR102
DC017
or
UC017
d - Counter With Separate Enable and Reset

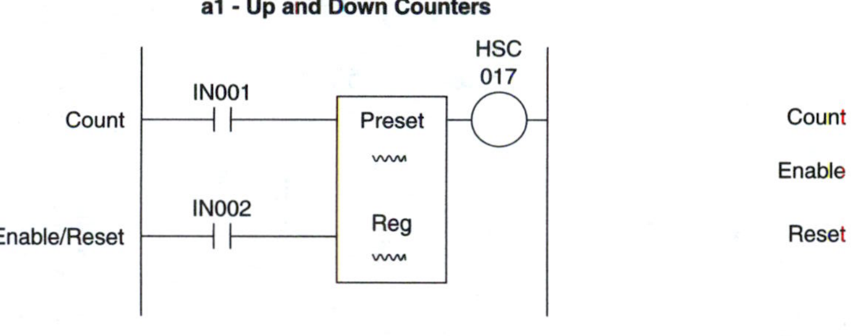

FIGURE 10–1
PLC Counter Functions

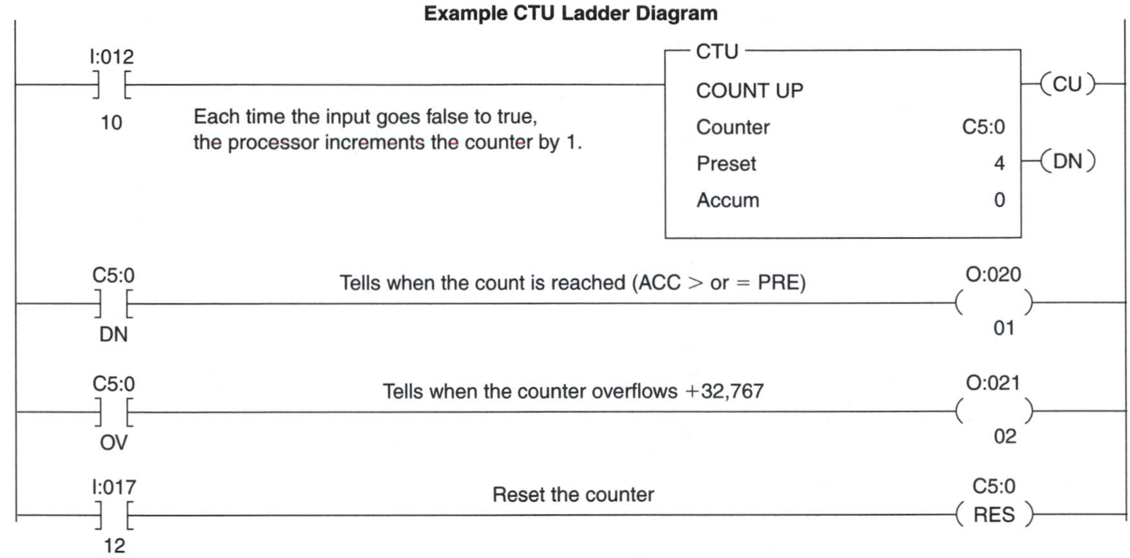

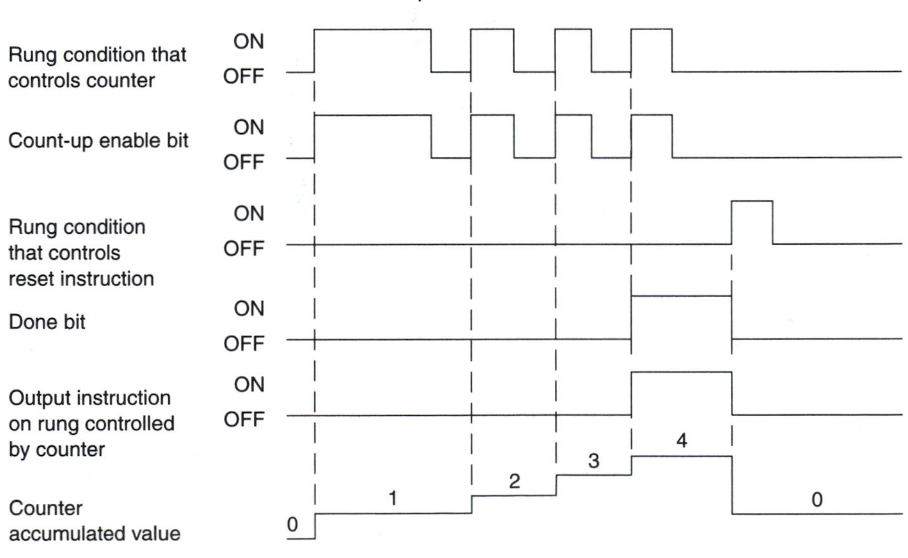

e Allen-Bradley Counter Format

FIGURE 10–1 (continued)

Figure 10–1e illustrates the Allen-Bradley up-counter function along with a functional timing diagram. The Allen-Bradley down-counter function is similar.

10–3 EXAMPLES OF COUNTER FUNCTION INDUSTRIAL APPLICATION

We show six examples of the use of the PLC counter. The first is for a basic application for counting events. The second and third examples use more than one counter for process control counts. The fourth through sixth examples use the counter function in conjunction with the timer function, which was described in chapter 9.

Example 10–1 Straight counting in a process. The counter output goes on after the set count is received by repetitive pulses to the counter input.

Example 10–2 Two counters used with a common register to give the sum of two counts.

Example 10–3 Two counters used with a common register to give the difference between two counts.

Example 10–4 A process where a timed interval is started when a count reaches a preset value.

Example 10–5 A process where a count of events is to start after a fixed time interval.

Example 10–6 A process where a rate is determined by dividing a count by a time interval.

EXAMPLE 10-1 Figure 10–2 illustrates the fundamental use of a PLC counter. After a certain number of counts occurs, the output goes on. The output can be used to energize an indicator. The output status could also be utilized in the ladder diagram logic in the form of a contact. The counter function is shown for either an up counter or a down counter. They both perform the same function in this illustration. Either counter will function if its enable line is energized. After the count input receives 18 pulses, the CR output will energize.

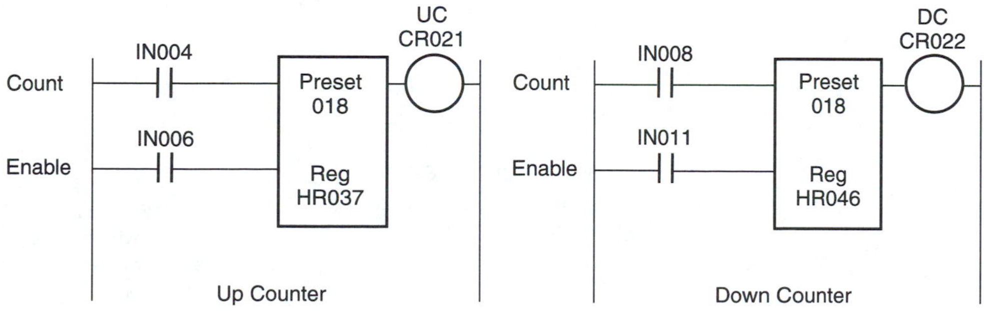

FIGURE 10–2
Example 10–1: Basic PLC Counter Operation

EXAMPLE 10-2 Figure 10–3 illustrates the use of a combination of two counters. Suppose that we wanted an output indicator to go on when six of part C and eight of part D are on a conveyor. This circuit would monitor the proper counts. IN002 and IN003 are proximity devices that pulse on when a part goes by them. Note that the circuit would not indicate more than six or eight parts; it would only indicate when there are enough parts.

To repeat the process, turn IN001 off to reset the system. Then reclose IN001.

EXAMPLE 10-3 The third example, shown in figure 10–4, concerns keeping track of the net number of parts on a conveyor. The number of parts going on the conveyor is counted by one proximity device's count. The number leaving the conveyor is counted by a second proximity device's

FIGURE 10–3
Example 10–2: Dual-Counter Application

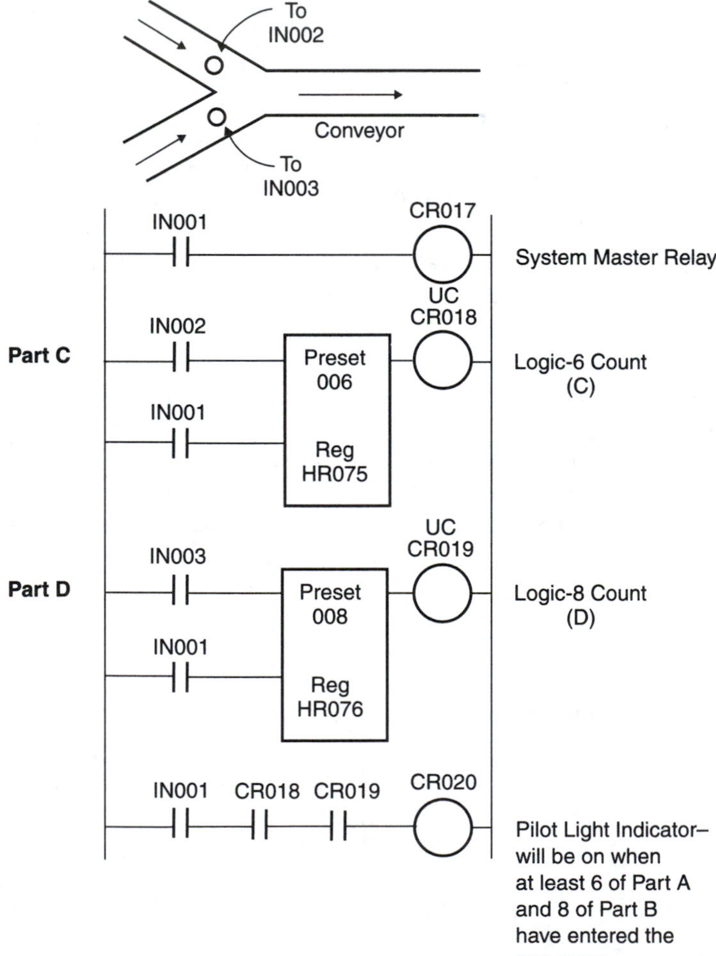

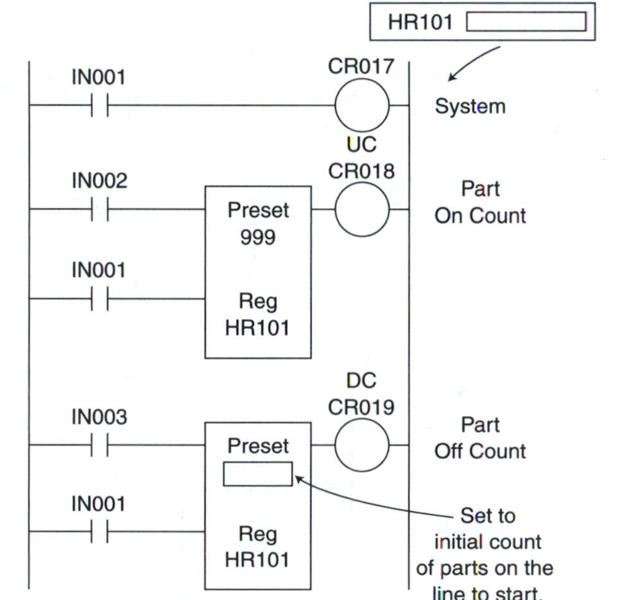

a - Using an Up and a Down Counter

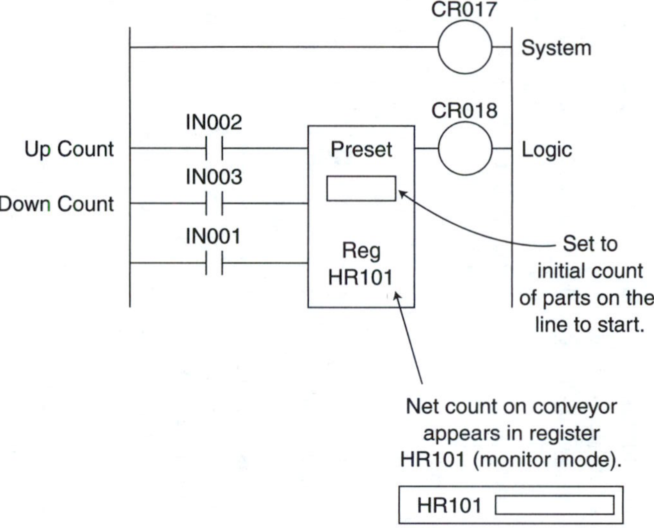

b - Using an Up/Down Counter

FIGURE 10–4
Example 10–3: Counters Used for a Net Count

count. Each proximity device feeds information into its own counter function. The total net count is kept in a holding register common to both counters.

A program using two counters, one up and one down, for this application is shown in figure 10–4a. An accurate initial count is needed. When starting the operation, the number of parts on the conveyor must be determined. This count number is programmed into the common register, HR101. It is normally necessary to put this count number in the down counter as the preset number. Then, any parts going onto the conveyor pulse the up counter. The counter's register (which is common to both counters) will have its value increased by one for each entering part. Similarly, the parts leaving decrease the common register's count through the down counter. The number value in register HR101 represents the number of parts on the conveyor. We are assuming that no parts fall off the conveyor and none are added along the way.

The up counter preset value is irrelevant in this application. It does not matter whether the counter outputs are on or off. The output on–off logic is not used. We have arbitrarily set the up counter's preset values to the maximum.

If the PLC you are using has an up/down counter, the program shown in figure 10–4b may be used. The operation is the same as for figure 10–4a.

EXAMPLE 10-4 This example is for a timed process that occurs after a certain process count is reached. After a count of 15 from a sensor, a paint spray is to run for 25 seconds. Figure 10–5 shows a program that accomplishes the count and time operation.

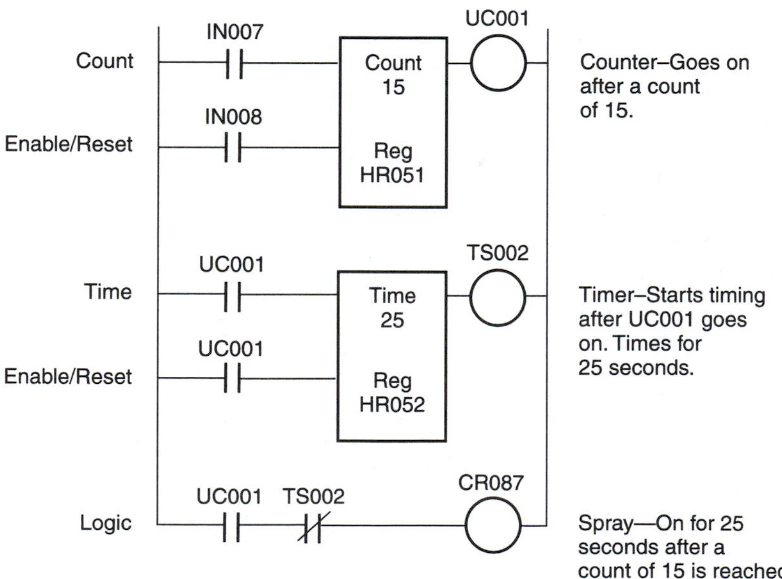

FIGURE 10–5
Example 10–4: Count and Time Program

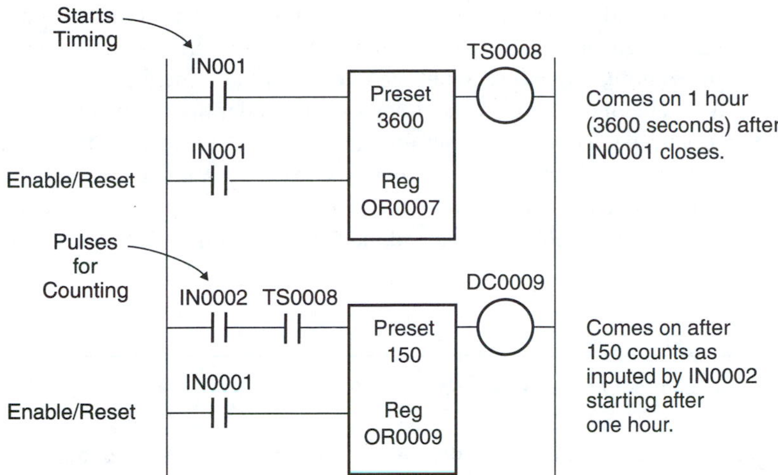

FIGURE 10–6
Example 10–5: Delay of the Start of the Counting Process

EXAMPLE 10-5 The fifth example, shown in figure 10–6, is for a delayed counting period. In this process we do not wish to start counting until one hour after the process starts. A timer output contact in the timer run line closes after the time period. The closure then enables the counter to start counting input pulses. After a count of 150, the output comes on.

EXAMPLE 10-6 How many parts per minute are going past a certain process point? This example addresses this problem. Figure 10–7 is a ladder diagram scheme for obtaining the product part flow

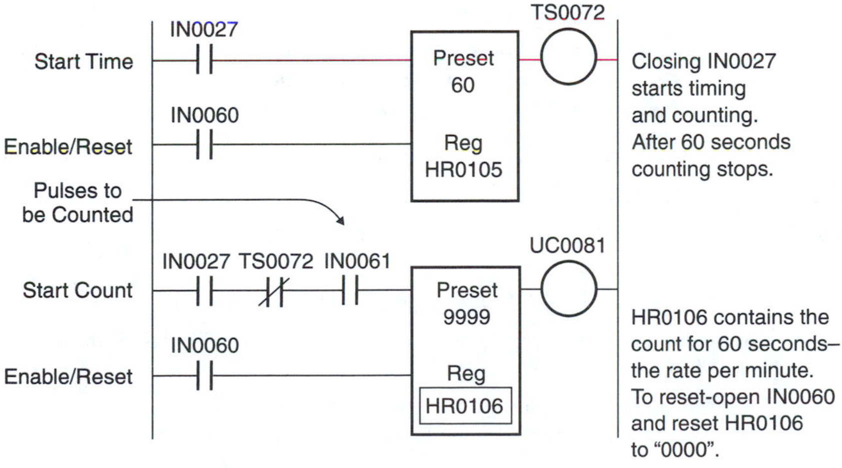

FIGURE 10–7
Example 10–6: Rate-per-Time-Period Program

rate. The timer and counter are enabled at the same time. The counter is pulsed for each part going past its sensor, which is connected to IN0027. The counting begins and the timer starts timing through its 60-second time interval at the same time.

At the end of 60 seconds, the timer's count ladder line is opened by a normally closed timer-output-related contact. Pulses continue but do not affect the PLC counter. The number of parts for the past minute are now recorded in counter register HR0106. The part count for the minute can be recorded manually or by a computer technique and will remain in HR0106 until IN0060 is opened and the counter and timer are reset. After IN0060 is reclosed, the process starts over.

TROUBLESHOOTING QUESTIONS

The PLC has been programmed to operate as shown in the figures referred to. However, the circuit has the malfunction noted. What misprogramming or other factor or factors could cause the malfunction?

TS 10–1 Refer to figure 10–2 (example 10–1). The counter counts to 1 and resets as the input goes on and then off.

TS 10–2 Refer to figure 10–3 (example 10–2).

1. The pilot light never goes on, even when the designated count is reached.
2. The pilot light never goes out.

TS 10–3 Refer to figure 10–4 (example 10–3).

1. As the process progresses, the count displayed in HR101 does not change.
2. The count displayed in HR101 goes negative.

TS 10–4 Refer to figure 10–5 (example 10–4).

1. The spray goes on when the control circuit is turned on.
2. The spray does not go off after the 25-second interval.
3. The spray never goes on.

TS 10–5 Refer to figure 10–6 (example 10–5).

1. The count starts after 6 minutes, not 1 hour.
2. The counter does not count.

EXERCISES

Design, construct, and test PLC circuits for the following processes:

1. An indicating light is to go on when a count reaches 23. The light is then to go off when a count of 31 is reached.
2. A machine, M, is to be turned on either when count A goes up to 11 or when count B goes up to 16. One stop button or switch resets the entire process.
3. A fan, F, is to be turned on when count L goes from 7 down to 0 and when either count M goes up to 14 or count N has not gone all the way from 14 down to 0. One switch or stop button resets the entire process.

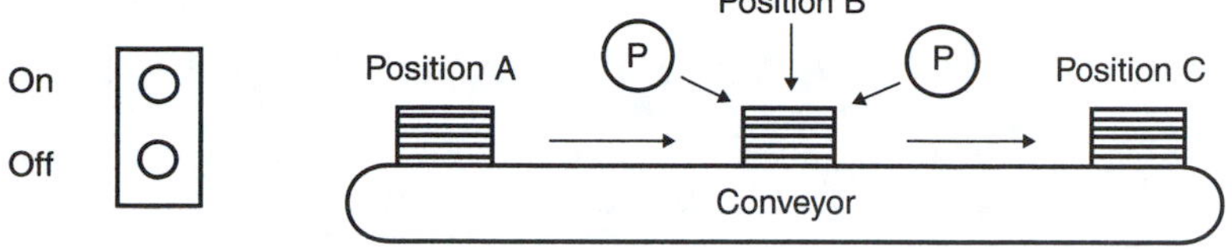

FIGURE 10–8
Diagram for Exercise 10

4. A solenoid, S, is to go on when count C goes up to 22, and when count D goes down from 37 to 0, and when count E goes up to 8. Furthermore, if count F goes down from 17 to 0 at any time, the solenoid is to be kept from operating. One stop button resets the entire process, including the solenoid's being off.

5. Repeat exercise 3 except that when F goes from 17 down to 0, the entire process is to be reset by a start–stop–seal system.

6. A bottling process for 12 bottles operates as follows:
   Bottles are counted until all 12 are in position for filling.
   When in position in the carton, the 12 bottles are filled simultaneously for 6.3 seconds.
   After filling, there is a pause of 3.8 seconds for foam to subside.
   The 12 caps are then put on and counted as they are installed.
   A solenoid then pushes the completed carton of 12 on to a conveyer. The system is reset for a new group (to be restarted manually) of 12 bottles by a limit switch that indicates that the carton is out of the "fill" position and on the conveyor.

7. A stacking and banding system (S) requires a spacer to be inserted (I) in a stack of panels after 14 sheets are stacked. After 14 more (28 total), the stack is to be banded (B). Add sensors and assumed output devices as required.

8. Refer to exercise 2 and add the following additional steps to the process. After banding is completed, there is a 2-second delay for the bander to pull back. Then, an identification spray color dot (P) is to be applied to the stack. Spray time is 4 seconds.

9. Two feeder conveyors (F1 and F2) feed a part onto one main conveyor (M). A proximity device is at the end of each feeder conveyor. The proximity device outputs are fed as pulses to counters. Each counter then shows the count of parts being put onto the main conveyor. In addition, another proximity device at the end of the conveyor pulses in response to parts leaving and then sends the pulses to another counter.

   Develop a program to have a single register showing the number counter of parts on the conveyor. Assume that the register is initially set to the same count as the count of parts on the conveyor.

10. Program an automatic control for the system shown in figure 10–8.

When the on button is pushed, a stacker (S) starts stacking plywood sheets at position A. Stack height is controlled by a PLC counter function, not a height sensor. When 12 parts are stacked, the conveyor (CV) goes on and moves the stack to position B. A sensor is used to stop the conveyor at B. At B, paint (P) is applied for 12.5 seconds. After painting is complete, the conveyor is restarted manually. The conveyor then moves parts to position C. At position C the stack stops automatically and the stack is removed manually. The stop button stops the process any time it is depressed. Assume that only one stack is on the conveyor at a time. Add limit switches and other devices as required.

Intermediate Functions

11

PLC Arithmetic Functions

OUTLINE

OBJECTIVES

At the end of this chapter, you will be able to

- Add and subtract numbers using the PLC ADD and SUBTRACT functions.
- Explain and demonstrate how the PLC handles overflow and negative numbers for the ADD and SUBTRACT functions.
- Explain the programming, operation, and utilization of the repetitive clock.
- Multiply and divide numbers using the PLC MULTIPLY and PLC DIVIDE functions.
- Find square roots by using the PLC SQUARE ROOT function.
- Describe the various PLC trigonometric and log functions.
- Describe other major PLC arithmetic functions.
- Apply arithmetic functions to process control operations.

11–1 INTRODUCTION

Most PLCs have basic arithmetic functions. Arithmetic functions are sometimes designated as math functions. These basic functions—ADD, SUBTRACT, MULTIPLY, DIVIDE, and SQR (square root)—are covered in sections 11–2 and 11–4. A program for performing these functions (and others), continuously, the Repetitive Clock, is covered in section 11–3. *Continuously* is defined as operating quickly in one or two scan times between updates.

Larger-capability PLCs have an extended number of advanced arithmetic functions. Section 12–5 illustrates the trigonometric and logarithmic functions found in most advanced PLCs. Various other major advanced functions are covered in section 12–6.

11–2 PLC ADDITION AND SUBTRACTION

Figure 11–1 shows the PLC ADD function. For figure 11–1a, it adds only when the enable line changes from off to on. The addition will not take place continuously just because Enable is on. When enabled, the numerical value in the operand-2 register is added to the numerical value in the operand-1 register. The resulting value then appears in the specified destination register. In many PLCs, operand 2 can be a number or the value in a designated register.

Figure 11–1b shows another format that operates similarly. The two numbers to be added are entered in locations 30 and 31. The result of the addition then appears in location 32 when the input 121 is energized. Figure 11–1c shows another format found on some PLCs. You can add more than two numbers using one function. Otherwise you would have to add two numbers and then add the third number to the result, requiring two lines of pro-

FIGURE 11–1
PLC ADD Function

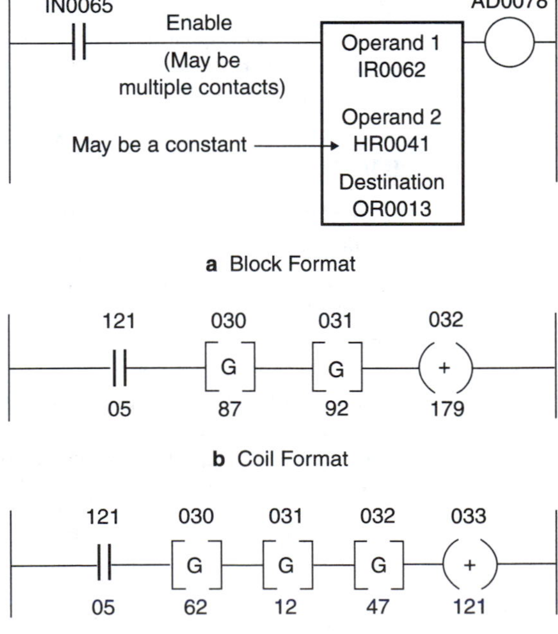

gram. Throughout the ensuing examples in this chapter, we use the block format only for the ADD function and for the other functions shown.

When does the output coil come on? In some PLCs its status is irrelevant. In other PLC systems, the coil's energization indicates register overflow or negative value. For overflow, the coil only comes on when the resulting number exceeds the register counting capability; otherwise, it remains off. For example, assume the decimal register limit is 9999. If you add 643 plus 568, the sum, which equals 1211, is within the register limit, and the coil will not come on. The sum, 1211, will appear in the destination register. On the other hand, the sum of 8973 plus 8632 is 17,605, which exceeds 9999, so the coil will come on. In this case, the destination register will read 7605, the excess over 10,000.

Note that we are using decimal numbers throughout this chapter. The arithmetic functions also work for binary numbers and other numbering systems. You may choose any numbering system, depending on the requirements of your particular PLC model. The only precaution is to use the same numbering system throughout the arithmetic function.

To find out which register types, HR, OR, and so on, can be used as operands, consult the manufacturer's manual for your PLC.

Figure 11–2 illustrates the ADD function for two examples. One example is for a result less than 9999, and the other is for a result greater than 9999.

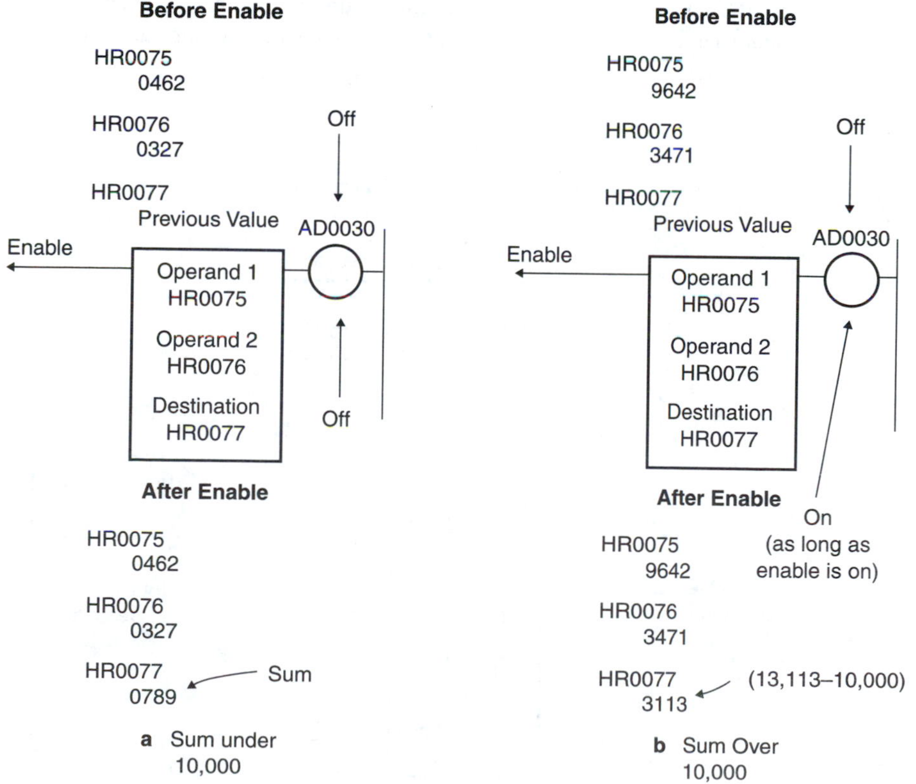

FIGURE 11–2
Two Addition Examples

A sample industrial problem for the ADD function is shown in figure 11–3. Two conveyors feed a main conveyor. For some reason we cannot get to the main conveyor to make a count. The main conveyor count is determined from the count of parts entering from the other two conveyors. The count on each feeder-conveyor is determined by a counter (not shown). The counters on each feeder-conveyor are input-pulsed by a proximity detector once for each part leaving the conveyors. The count of total parts entering the main conveyor is then determined by adding the two feeder conveyor counts using the ADD function. For illustration, we monitor the total count every 30 seconds. The input of the ADD function is pulsed on and immediately off by pulsing the ADD function enable. The count could be printed out as shown in figure 11–3 every 30 seconds.

Note that this addition method is an alternate solution to a similar problem in chapter 10 using counters and a common register.

The PLC format for SUBTRACT is the same as for ADD, and the function operation is similar. For subtraction, operand 2 is subtracted from operand 1. The result is found in the destination register. Figure 11–4 shows the SUBTRACT function.

For subtraction, we discuss the block formats only. As with addition, the SUBTRACT function operates only when the enable line goes on. In some PLCs the direct answer will appear. In other PLCs, the coil status is significant for complete answer description. The coil on-off operation for subtraction differ from that of addition. When the result is positive, the coil is off and the result is found in the destination register. When the result is negative, the coil is on and the resulting negative number value is found in the destination register.

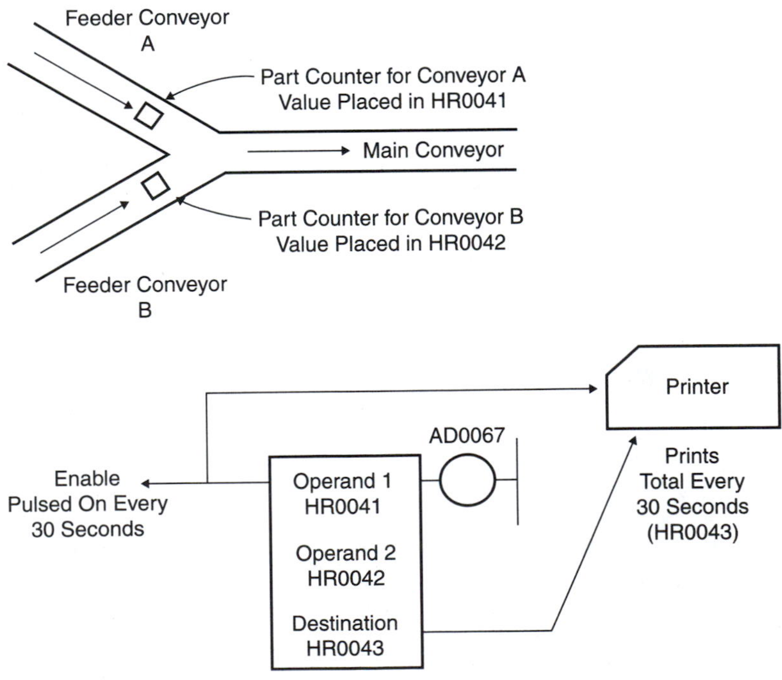

FIGURE 11–3
Using the ADD Function for a Conveyor Part Count

FIGURE 11–4
PLC SUBTRACT Function

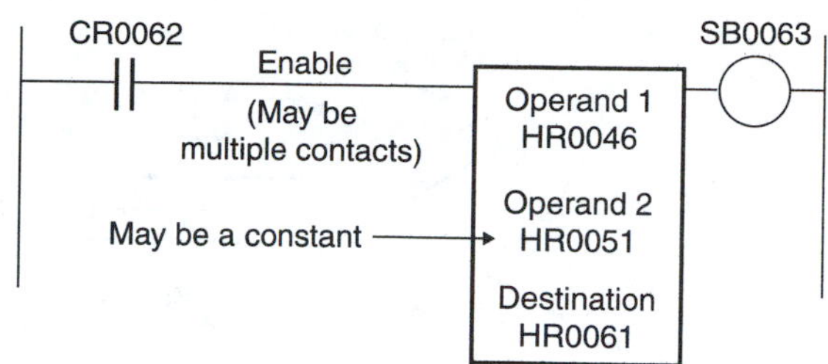

a Block Format

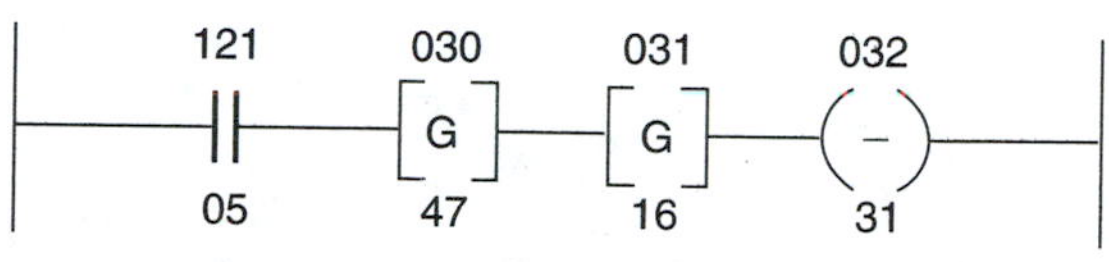

b Coil Format

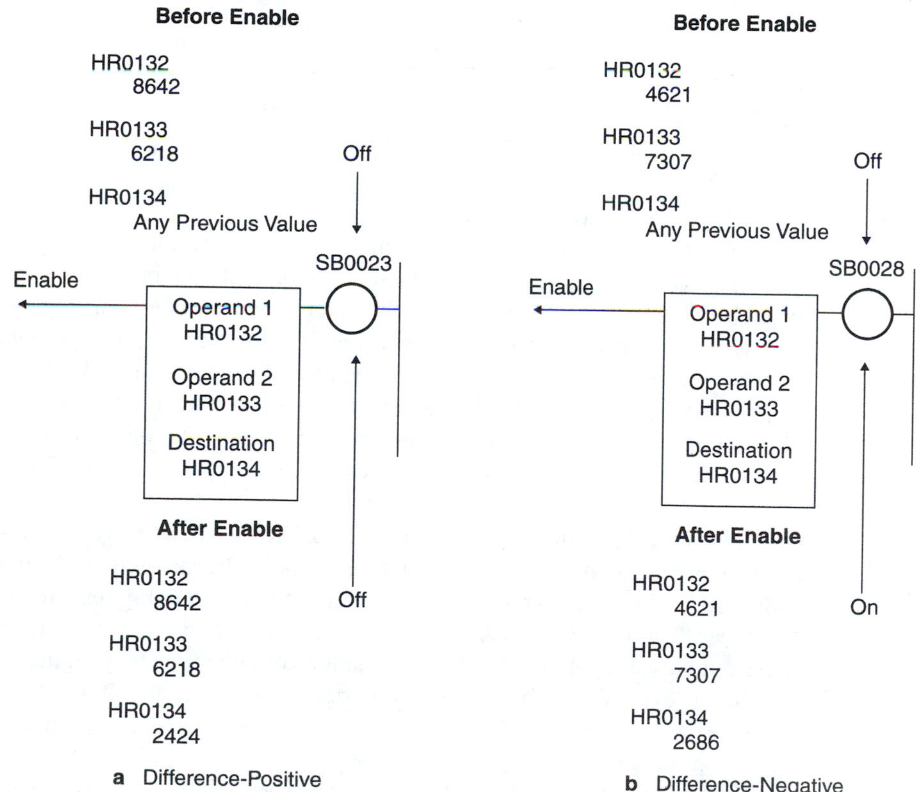

a Difference-Positive

b Difference-Negative

FIGURE 11–5
Two Subtraction Examples

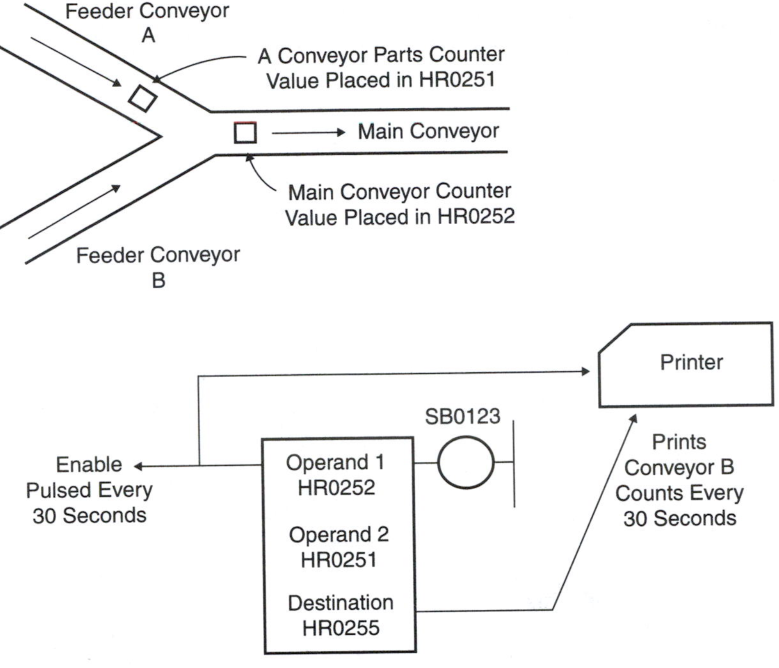

FIGURE 11–6
Using the SUBTRACT Function for a Conveyor Count

Figure 11–5 shows the function operation for both a positive and a negative answer. The coil is off for a positive result and on for a negative result.

A possible industrial problem for subtraction is shown in figure 11–6. It is similar to the addition example. In this example, the output count and only one input count are available. One of the conveyor inputs is not accessible for some reason. To obtain the input A count value, subtract the input B count value from the output count. The result is the A conveyor count. The count is again determined by a 30-second counting interval. The count is taken for 30 seconds, used, and then reset to zero. Again, initializing is needed periodically for accurate operation. Initialization figures the initial number of parts on conveyor B into the counting results.

One useful function of the ADD and SUBTRACT functions is to set a range. For example, there is an inspection system with a periodically changing base dimension and periodically changing tolerances. The PLC can easily and quickly reset the dimensions and tolerances. The preset dimensional values are transmitted to two specified PLC outputs. These output values are used to set the positions of an automatic gauging system.

In the example, the base dimension, or set point, is set at 6.250 inches. The allowable tolerances chosen for this illustration are +0.025 and −0.035 inch. Figure 11–7 shows the range in graphic form.

Figure 11–8 illustrates how the settings can be accomplished by programming a PLC. For three eternally fixed values, a PLC is not needed. Our system is valuable when the di-

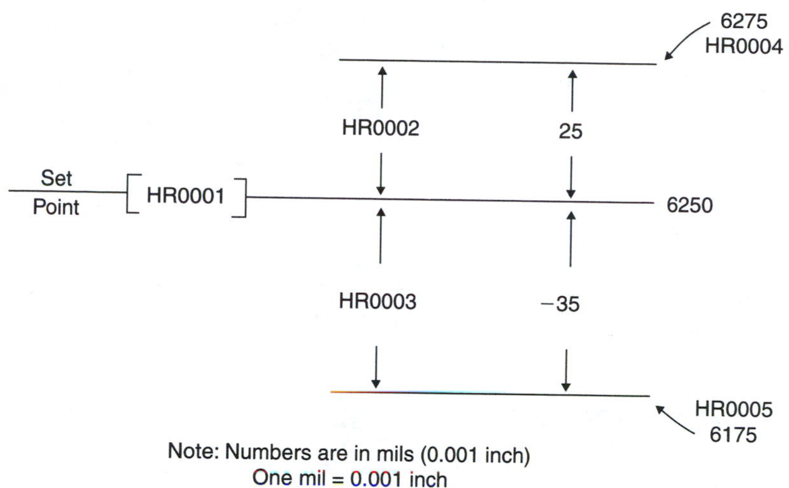

FIGURE 11–7
Graphic Representation of Dimensions

mensions are varying quickly as production varies. The set point is entered into HR0001. HR0002 receives the upper tolerance value, and HR0003 receives the lower tolerance value. When the circuit in figure 11–9 is enabled, the upper and lower limits are calculated. They appear in HR0004 and HR0005, respectively.

To revise dimensions or tolerances, the mathematical values in HR0001, HR0002, and HR0003 are changed. When PLC functions are re-enabled, the new values appear im-

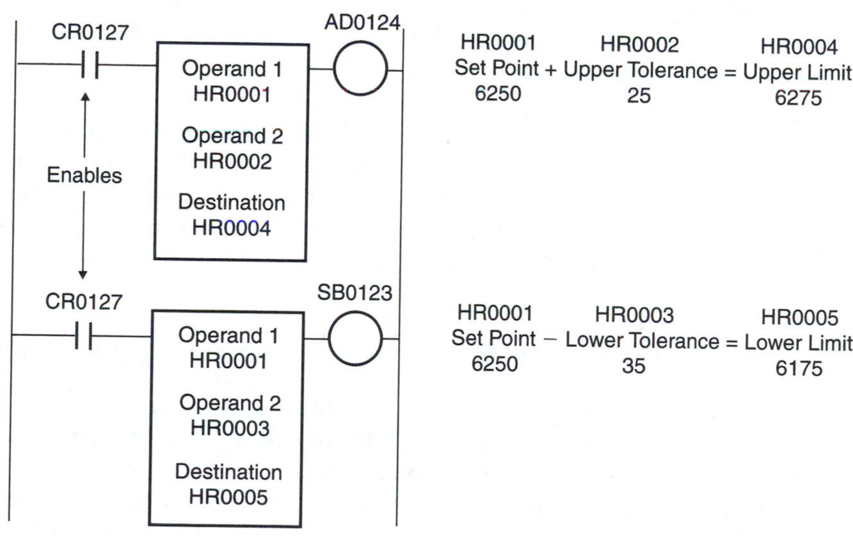

FIGURE 11–8
PLC Operation with Set Points and Tolerances

FIGURE 11–9
Repetitive Clock Circuit

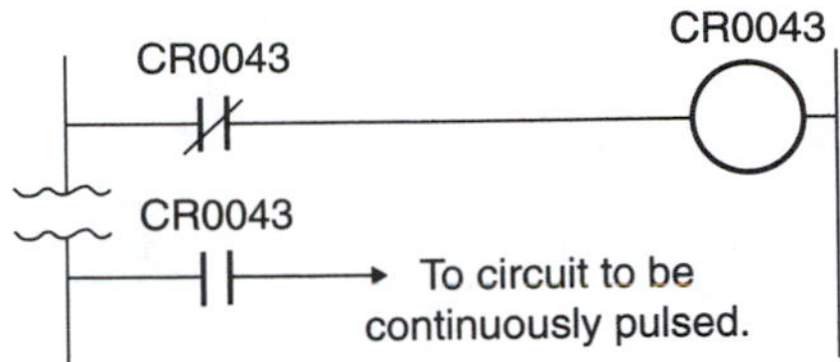

mediately (on the next scan) at the PLC output. Comparing the actual dimensions with the upper and lower limits we have set is discussed in chapter 12, which covers comparison functions. Moving different dimensional numbers into the set point and tolerance registers is discussed in chapter 16, which covers the MOVE functions.

11–3 THE PLC REPETITIVE CLOCK

The ADD and SUBTRACT functions discussed so far in this chapter do not usually operate continuously. They only perform the addition or subtraction operation once when the function is enabled. If Enable remains on, nothing else happens, even though the operands change. When Enable is off, nothing happens, either. Therefore, a repetitive on-off enable is needed for continuous operation.

Figure 11–9 shows a repetitive clock arrangement. A coil turns itself off and on at a very fast rate, about two times the scan-time rate. If this is used as an enable, the operation of an arithmetic function is essentially continuous—if you consider every millisecond or so to be continuous.

The sequence for the repetitive clock is this: On the first scan, the relay coil is turned on through its own contact. When the coil goes on, it opens its own enabling contact, which is normally closed. At the end of the first scan, the CR0043 NC contact is updated and opened. On the next scan, the coil is turned off. Update at the end of the second scan recloses the self-enabling contact. The process then repeats itself continuously.

When fixed, longer intervals are needed, timers are used instead of the repetitive clock. These interval situations were covered in chapter 9. The repetitive clock can be used with other arithmetic functions besides ADD and SUBTRACT. This includes all functions in the remainder of this chapter. The repetitive clock can also be used with many other functions which need periodic updating or actuation. Number comparison functions of the next chapter are examples of further use of the repetitive clock.

11–4 PLC MULTIPLICATION, DIVISION, AND SQUARE ROOT

The multiplication format is similar to the addition and subtraction formats previously discussed. Figure 11–10 shows the MULTIPLY function. We discuss the block function only. Operand 1 is assigned a register number. Operand 2 can be another register, or it may be a constant. The result of the multiplication appears in the destination when the function is enabled. The destination is two registers wide by necessity. Multiplying 0034 by 0086 would require only one register, four numbers wide, for the answer (2064). However, multiplying

FIGURE 11–10
MULTIPLY Function

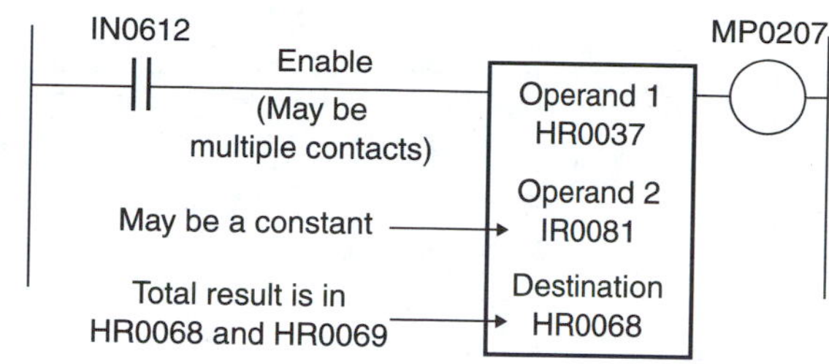

A Block Format

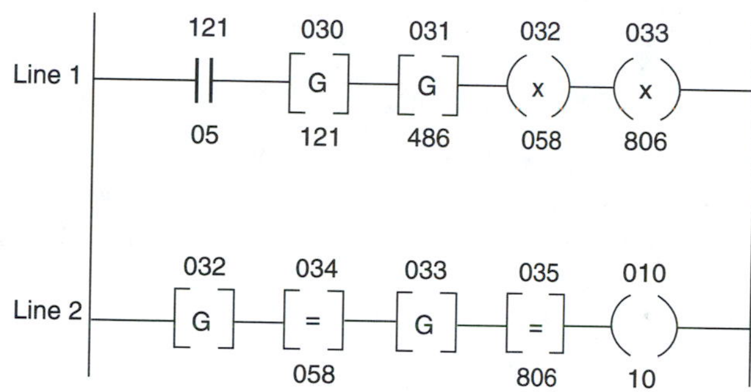

B Coil Format

Note: For result of less than 6 digits, use Line 1 only

6453 by 8933 (57,644,649) would require an eight-bit-wide slot or two registers to accommodate the answer.

The multiplication takes place only when enable comes on. Normally, the coil comes on when the multiplication is completed. The coil on-off state has no numerical significance as it did in ADD and SUBTRACT.

A simple process application for counting cartons is shown in figure 11–11. The count from a carton counter enters the PLC and is put into IR0001 and then into operand 1. Each carton contains 12 bottles; therefore, 12 is entered as a constant in operand 2. To keep

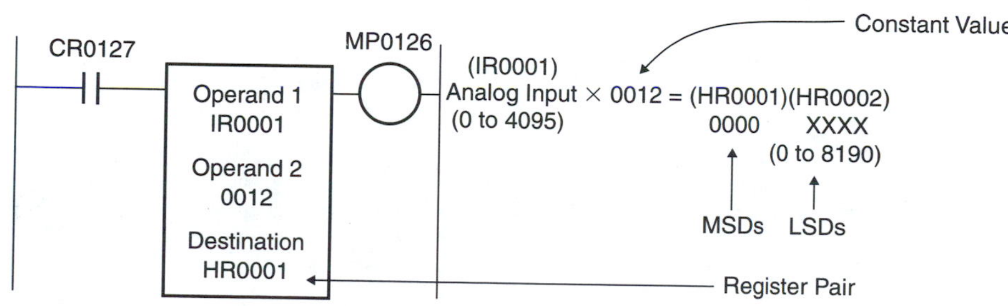

FIGURE 11–11
Multiplication Example

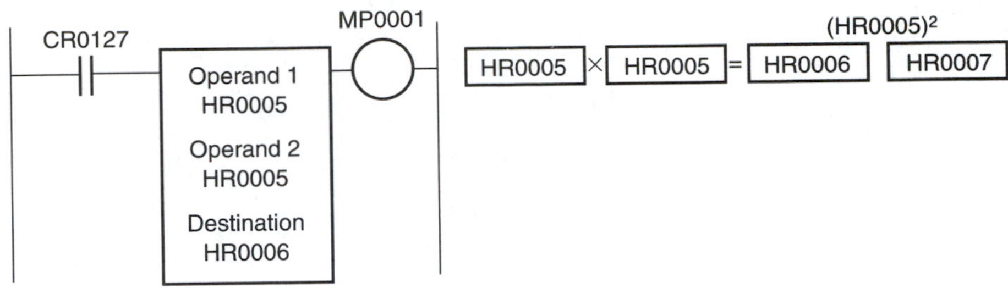

FIGURE 11–12
Squaring by the MULTIPLY Function

a constant count of bottle output, the PLC would constantly multiply the carton count by 12. The destination register, HR0001, will show the number of bottles output each time the function is enabled. A number of these MULTIPLY functions could be combined to give a total plant unit output. The constants in each function would be bottles per carton for that particular count. A PLC addition program of all individual counts would then give the total plant bottle count.

FIGURE 11–13
DIVIDE Function

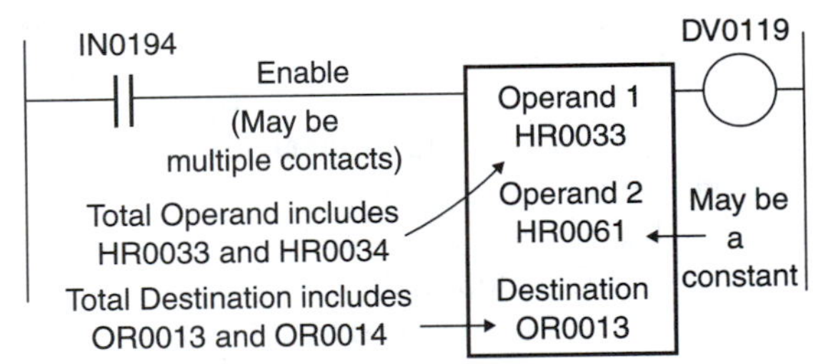

A Block Format

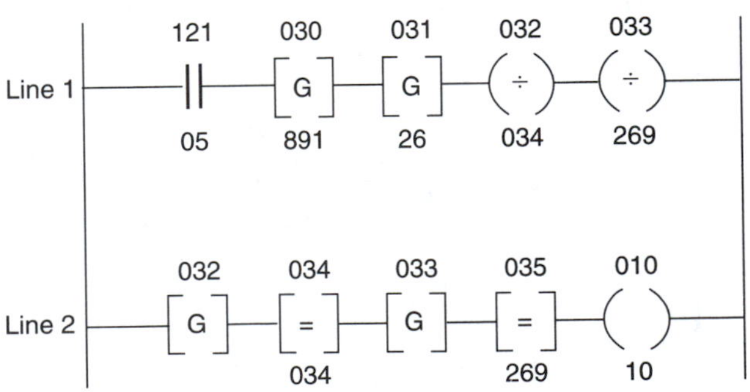

B Coil Format

Note: Value is in 032 and Decimal is in 033.

FIGURE 11–14
DIVIDE Destination Register Content Determination

			Remainder
Operand 1 Value in Two Registers	648,127	Operand 1 minus	648,127
divided by	÷	267	−
Operand 2 Constant Value	2,421	× 2421	646,407
	=		=
equals	267.71045	equals	1720

Destination Value in the Two Registers: 0267 | 1720

Value is *not* 7104

There is normally no squaring function in a PLC format. Squaring is simply a matter of putting the number to be squared into both operand 1 and operand 2 of a MULTIPLY function. The square of the original number then appears in the destination register. The squaring function is shown in figure 11–12. We show the block format only.

The division function, which is shown in figure 11–13, is similar to the multiplication function. We discuss the block format only. Operand 1, the dividend, is divided by operand 2, the divisor. The numerical result of the division appears in the destination register when the function is enabled. Again, the division takes place only at the time Enable is energized. To facilitate division, operand 1 is two registers wide and operand 2 is only one register wide. Operand 2 may normally be a register or a constant numerical value.

In PLCs, the destination is almost always two registers wide. The first destination register is the numerical result of the division. The second register value is the remainder in numerical form. The remainder is not a decimal value. A numerical example of a division is illustrated in figure 11–14. The determination of number value in the second destination register is explained in the figure.

Figure 11–15 is an example of scaling by division. An analog measurement numerical value, in inches, is fed into a PLC input register. The measurement value is transferred (not shown) within the PLC to input register IN0078. To get the value in feet to output from the PLC into an indicator, divide by 12 with a DIVIDE function. The result, now in the required dimension of feet, appears in register OR0124.

FIGURE 11–15
Example of the Process Use of the DIVIDE Function

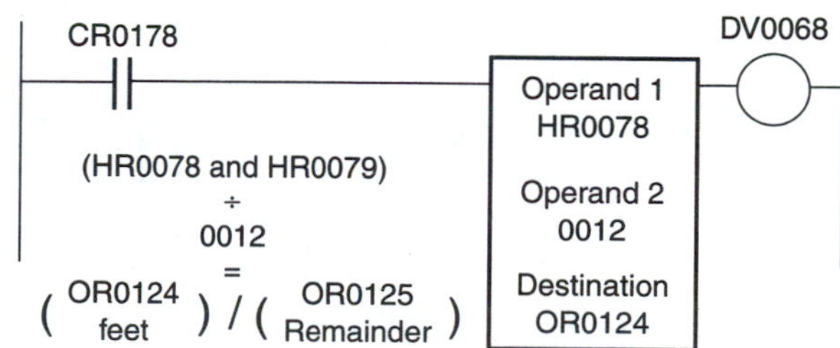

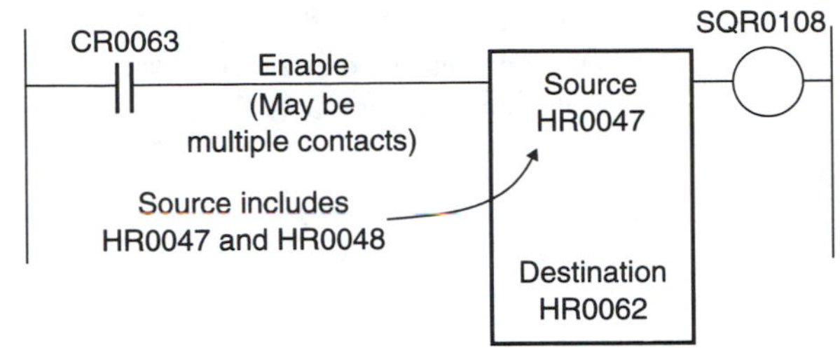

FIGURE 11–16
SQUARE ROOT Function

If the required output accuracy is less than whole feet, convert the remainder into decimal form. This would be accomplished by an added program step (not shown): Divide the numerical remainder in OR0125 by 12 and place the result in another output register, for example, OR0126. You would then have four-decimal-place accuracy when the OR0126 output register value is recognized as ten-thousandths of an inch.

Figure 11–16 shows the SQUARE ROOT function found on some PLCs (block format only). The number whose square root we want to determine is placed in the source. The source input number is contained in two registers so that it may be up to 99,999,999 in value. When enabled, the function calculates the square root and places it in the destination. The destination is one register wide, up to 9999 in value. There is usually no remainder register.

11–5 PLC TRIGONOMETRIC AND LOG FUNCTIONS

A number of functions perform trigonometric and logarithmic operations in general in mid-size PLCs. A general block function for these functions is shown in figure 11–17. Input go-

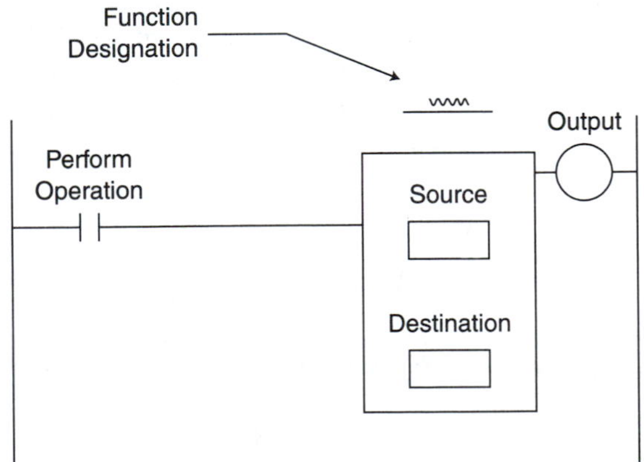

FIGURE 11–17
Arithmetical General Block Diagram

ing from off to on causes the mathematical operation to take place. The source register value is analyzed appropriately and the calculated value appears in the destination register. The angles involved in trigonometric function are normally stated in radians. There are upper limits on how big a number the log functions will handle, as listed in operating manuals. Some PLCs have an antilog function, where a conversion is made from a log to its number. These various functions are:

Function	Symbol	Source	Destination
Arc Cosine	ACS	Cosine	Degrees
Arc Sine	ASN	Sine	Degrees
Arc Tangent	ATN	Tangent	Degrees
Cosine	COS	Degrees	Cosine
Sine	SIN	Degrees	Sine
Tangent	TAN	Degrees	Tangent
Convert Radians to Degrees	DEG	Radians	Degrees
Convert Degrees to Radians	RAD	Degrees	Radians
Natural Log	LN	Plus Number	Natural Log
Log to Base 10	LOG	Plus Number	Log

11-6 OTHER PLC ARITHMETIC FUNCTIONS

Some additional, nontrigonometric and nonlogarithmic arithmetic functions are available in midsized and larger PLCs. The function blocks are identical to the one shown in figure 11–17. These functions are:

Function	Symbol	Source	Destination
Compute—Using a Formula	CPT	A formula	The answer
Average a File	AVE	A series of numbers	The average
Negative	NEG	Plus or minus	Minus or plus
Standard Deviation	STD	A series of numbers	Value of std. dev.
X to the power of Y	XPY	X and Y values	The answer

The first function, CPT, can take complex expressions within the limits of the PLC, which are spelled out in the operating manual. For example, the expression (A + B)* (Sine C − Square root of D) could be entered. Instead of programming a number of arithmetic steps, you program only one function. A table of the mathematical functions usable in the Allen-Bradley system is shown in figure 11–18.

Table 4.C
Valid Operations for Use in a CPT Expression

Type	Operator	Description	Example Operation
Copy	none	copy from A to B	enter source address in the expression; enter destination address in destination
Clear	none	set a value to zero	0 (enter 0 for the expression)
Arithmetic	+	add	2 + 3 2 + 3 + 7 (Enhanced PLC-5 processors)
	−	subtract	12 − 5 (12 − 5) − 7 (Enhanced PLC-5 processors)
	*	multiply	5 * 2 6 * (5 * 2) (Enhanced PLC-5 processors)
	| (vertical bar)	divide	24 | 6 (24 | 6) * 2 (Enhanced PLC-5 processors)
	−	negate	− N7:0
	SQR	square root	SQR N7:0
	**	exponential * (x to the power of y)	10**3
	LN	natural log *	LN F8:20
	LOG	log to the base 10 *	LOG F8:3
Trigonometric	ACS	arc cosine *	ACS F8:18
	ASN	arc sine *	ASN F8:20
	ATN	arc tangent *	ATN F8:22
	COS	cosine *	COS F8:14
	SIN	sine *	SIN F8:12
	TAN	tangent *	TAN F8:16
Bitwise	AND	bitwise AND	D9:3 AND D10:4
	OR	bitwise OR	D10:4 OR D10:5
	XOR	bitwise exclusive OR	D9:5 XOR D10:4
	NOT	bitwise complement	NOT D9:3
Conversion	FRD	convert from BCD to binary	FRD N7:0
	TOD	convert from binary to BCD	TOD N7:0
	DEG	convert radians to degrees *	DEG F8:8
	RAD	convert degrees to radians *	RAD F8:10

* Available in Enhanced PLC-5 processors only.

FIGURE 11–18
Valid Operations for use in a CPT Function (courtesy of Allen Bradley).

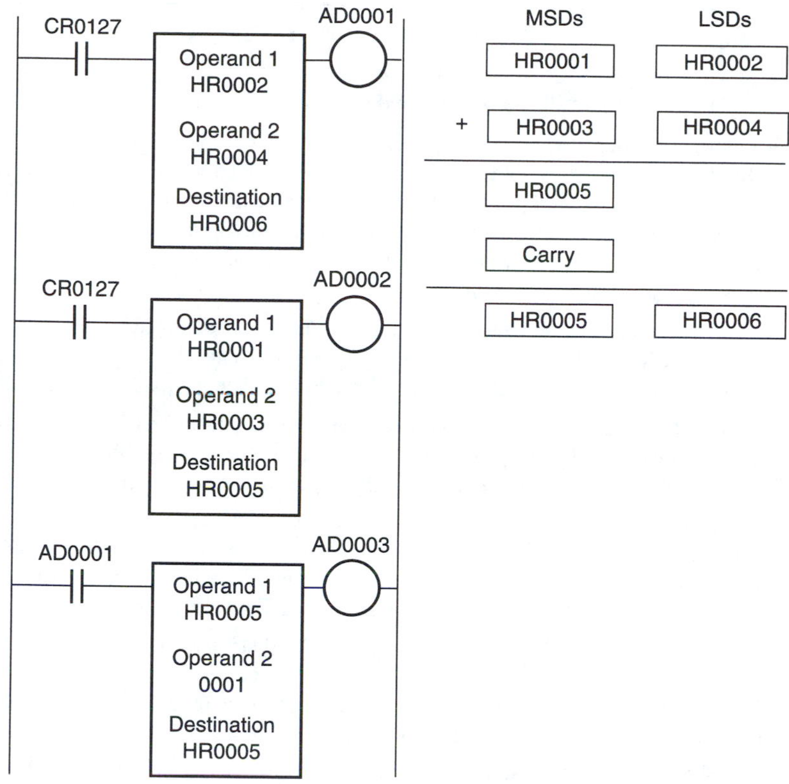

FIGURE 11–19
Typical DOUBLE PRECISION Function Format

Another arithmetic process often needed is the DOUBLE PRECISION procedure. Suppose a process needs seven- or eight-place figure accuracy, not just the normally calculated four figures. Some advanced PLCs have the capability to double the number of output decimal digits, for example, from four digits to eight. For more precise processes, this increased accuracy may be required. The PLC's system to increase accuracy is called DOUBLE PRECISION. Figure 11–19 illustrates one manufacturer's system for carrying out DOUBLE PRECISION for the ADD function. Consult your user's manual for how your particular PLC does this, or, indeed, whether it can do it at all.

TROUBLESHOOTING PROBLEMS

The PLC has been programmed to operate as shown in the figures referred to. However, the circuit has the malfunction noted. What misprogramming or other factor or factors could cause the malfunction?

TS 11–1 Refer to figure 11–3. The printed count is for one conveyer only.

TS 11–2 Refer to figure 11–6. The printer starts printing negative numbers.

EXERCISES

All exercises assume that all numbers are in decimal form.

1. Construct a basic PLC ADD function. Use IN0001 as the enable circuit. Use HR0001 and HR0002 as the registers holding the numbers to be added and HR0003 as the sum. Insert relatively small numbers in HR0001 and HR0002, enable the function, and verify that the resulting sum in HR0003 is correct. Use the appropriate numbering system to accomplish the addition. Next, insert large numbers whose sum exceeds the capability of HR0003 and observe the result. Does the result correspond to figure 11–2?
2. Three conveyors feed a main conveyor. The count from each feeder conveyor is fed into an input register in the PLC. Construct a PLC program to obtain the total count of parts on the main conveyor. As an additional exercise, use a timer to update the total every 15 seconds.
3. Construct a SUBTRACT function in a manner similar to exercise 1. Insert numbers in the operands that result in a positive destination number. Next, use numbers that give a negative answer. Does the negative answer produce the results as shown in figure 11–5?
4. Two conveyors, A and B, feed a main conveyor, C. A third conveyor, R, removes rejects a short distance down the main conveyor. The counts for conveyors A, B, and R are each input into holding registers in the PLC. Construct a PLC program to obtain the total output, C, part count. As an additional exercise, use a timer to update the total at a time interval of your choice.
5. Set up a process range PLC program following the system shown in figures 11–7 and 11–8. The nominal value or set point is 15.35 inches. The tolerances are +0.27 and −0.27 inch. Show that the resulting PLC-calculated limits are correct. To test the program's validity further, change the set point and tolerances to different values and check the results.
6. Repeat exercise 5 with different tolerances. Both tolerances are negative, −0.05 to −0.20.
7. A main conveyor has two conveyors feeding it. One feeder puts 6-packs on the main conveyor; the other feeds 8-packs. Both feeder conveyors have counters that count the number of packs leaving them. Design a program to give a total can count on the main conveyor.
8. A conveyor has 6-, 8-, and 12-packs of canned soda entering it. Each size of an entering pack has an individual pack quantity counter feeding a PLC register. To know how many total cans are entering the conveyor, set up a program for multiplying and then adding to give a total can count.
9. We have an output that gives us a dimension in inches. We wish to have the dimension displayed in feet and yards. Develop a PLC program to output all three dimensions. (*Hint:* Use two divisions for two outputs and one direct-access output.)
10. Set up a PLC program to obtain an output, P, in register OR0055. The output is to give a value based on two inputs, M and N. P equals the square of M plus the square root of N.
11. Develop programs for other math equations of your choice. Example: $N = (J + K - L)/M$.

12

PLC Number Comparison Functions

OUTLINE

OBJECTIVES

At the end of this chapter, you will be able to

- □ List and define the six basic COMPARE functions.
- □ Apply each of the basic COMPARE functions to application processes.
- □ Describe advanced COMPARE functions and their use.
- □ Apply combinations of COMPARE functions to do multiple comparison analysis.

12–1 INTRODUCTION

Medium and large PLCs have number comparison capabilities. The number comparisons are performed internally in a manner similar to microcomputers and microprocessors. With the PLC, there is no internal programming necessary for the operator. The PLC programming is set up for direct keyboard/screen arithmetical logic. This chapter illustrates how to perform number comparisons of all types of PLC programming.

What kind of number comparisons can be made by a PLC? We may wish to compare two numbers. We might compare a varying count to a fixed value. We might wish to compare two varying input values every five seconds.

In an even more complicated situation, we might wish to determine whether a periodically varying number is between two limits. These limits of comparison might be fixed, or one or both limits could be variable.

Some PLC models have two basic comparison functions, with four other comparison functions derived from the two basic functions. Other PLCs have all six functions as individual functions. These are covered in section 12–2. Other PLCs have more advanced compare functions, covered in section 12–3. Applications of comparison functions are illustrated in section 12–4.

12–2 PLC BASIC COMPARISON FUNCTIONS

Many PLCs have only two COMPARE functions: equal, and greater than or equal to. To perform any one of the other four functions (not equal, less than greater than, and less than or equal to), combinations of the basic two are used. Some PLCs have all six individual functions, which makes programming easier. Of course, some less-expensive PLCs do not have COMPARE functions at all.

This chapter uses the PLC comparison system with two basic functions for illustrations. The other four PLC COMPARE functions use the inverse or combinations of the two basic functions.

Figure 12–1 shows a table of comparison functions. Functions 1 and 3 are the two basic functions that we have discussed. The other four are derived functions. The six direct functions for PLCs having them in their programming capability are listed on the right side of the table.

Let's take an example of each COMPARE function. Assume that A, the standard for comparison, is placed in operand 2. A is set at 182. Then B, the number to be compared to A, will be placed in operand 1. We are therefore comparing the value of B to the value of A, 182.

1. Equal (EQ) is true only if B is exactly 182 also.
2. Not equal (NE) is true if B is 181 or less, or if B is 183 or more.
3. Greater than or equal to (GE) is true only when B is 182 or less.
4. Less than (LT) is true only when B is 183 or more.
5. Greater than (GT) is true only when B is 181 or less.
6. Less than or equal to (LE) is true only when B is 182 or more.

Comparison	Function	Equation	Circuit (conducts when equation is true)	Direct Function
*1	Equal (EQ)	A = B	EQ —| |—	EQU
2	Not equal	A ≠ B	EQ —|/|—	NEQ
*3	Greater than or equal to (GE)	A ⩾ B	GE —| |—	GEQ
4	Less than	A < B	GE —|/|—	LES
5	Greater than	A > B	GE —| |— EQ —|/|—	GRT
6	Less than or equal to	A ⩽ B	GE —|/|— EQ —| |—	LEQ

* Basic Functions

FIGURE 12–1
Six COMPARE Functions

In actual operation, A might be a varying number, not a fixed value of 182. Later chapter examples illustrate how it may be changed periodically.

In figure 12–2 we show three typical programming formats. A, the block format, is the type that we will use for illustration in this chapter. B is one type of coil format. For B, the type of comparison—for example, equal to (=)—is entered in the 37 coil center (not shown). Logic on (numbers OP1 and OP2 equal) and IN 0071 on energizes output 106 through a 37 input contact. C is another coil-type format similar to B but on one line only. For C, we show a function for equal to for illustration. If the number in location 030 is equal to 225, output 021 is energized.

Figure 12–2 shows three basic comparison function layouts. The two numbers being compared are operand 1 and operand 2. One operand can be a constant, the other a register. Both operands may also be registers that contain numerical values. The identification number of the register would be specified in the functional block.

When the function is enabled by the input contact, the comparison is made. If the comparison is true, the output goes on. If the comparison is not true, the output goes off, or stays off. The comparison in some PLCs is made continuously as long as the enable is on. It makes the comparison on each scan. In some other PLCs the comparison is made only at the time the enable goes on. To make another numerical comparison, the input must go off and then back on.

The patterns of the two basic COMPARE functions are normally similar. Figure 12–2 shows three formats of layouts that may be used for any COMPARE function. The

FIGURE 12–2
Typical PLC COMPARE Functions

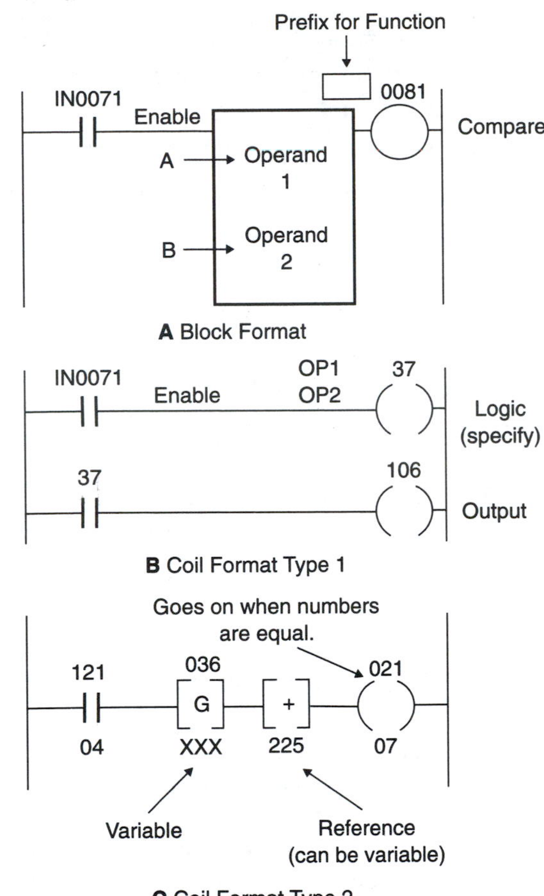

only difference between the two basic functions would be the coil designation (and the mathematical manipulation by the PLC CPU).

12–3 PLC BASIC COMPARISON FUNCTION APPLICATIONS

This section illustrates the use of each of the six COMPARE functions in PLC process control situations. Both direct and derived functions are shown for each example. Note that only the equal-to and greater-than-or-equal-to COMPARE functions are the same for derived and direct function systems. Additionally, it shows one process application with two COMPARE functions used in combination and an industrial COMPARE application.

EXAMPLE 12-1 This example consists of two illustrations of the use of the equal-to COMPARE function. We are banding dowels into bundles of 40. A counter function (not shown) keeps track of the count of the number of dowels in the bundle as they are added. The dowel counter's count

FIGURE 12–3
Example 12–1: Equal-To COMPARE Function

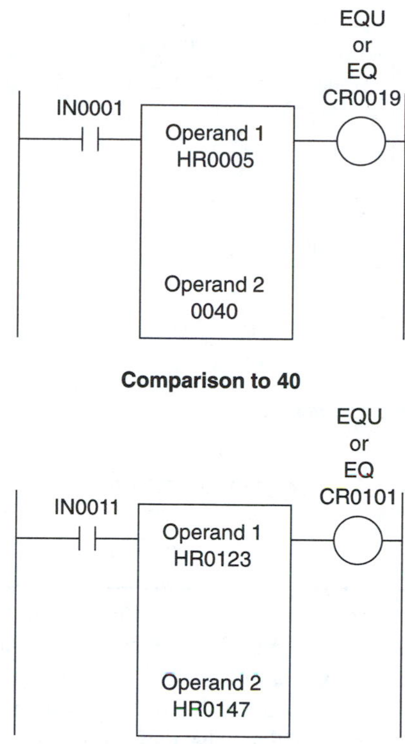

number is kept in HR0005. The running count is compared to 40, as shown in figure 12–3. When the count reaches 40, the comparison is true, and the output, CR0019, goes on. Output CR0019 is connected to a bander that operates when 40 is reached. The bundle removal and system reset systems are not shown. The other system would involve other functions.

The count in IR0071 must be in the correct numbering system. We are comparing to a decimal 40, so IR0071 must be in decimal for proper operation. Some PLCs convert to the proper numbering system automatically and some do not. Appropriate number conversions may be needed, as shown in chapter 13.

Another equal-to application is also shown in figure 12–3. In this case, an output must go on when two numbers are equal. The number's values do not matter, except that they are equal. The two numbers to be compared are fed from the outside process into HR0123 and HR0147 for the illustration. When enabled, output CR0101 will come on any time the numbers in the two registers are exactly equal.

EXAMPLE 12-2 This example is for the not-equal-to function. Figure 12–4 shows its programming. In the example, the output is to be on except when an input count is exactly 87. The input count is tracked in IR0062. Operand 2 can be programmed as the number 0087. It could also be programmed as a register, HR0183. The number 87 would then be inserted into HR0183.

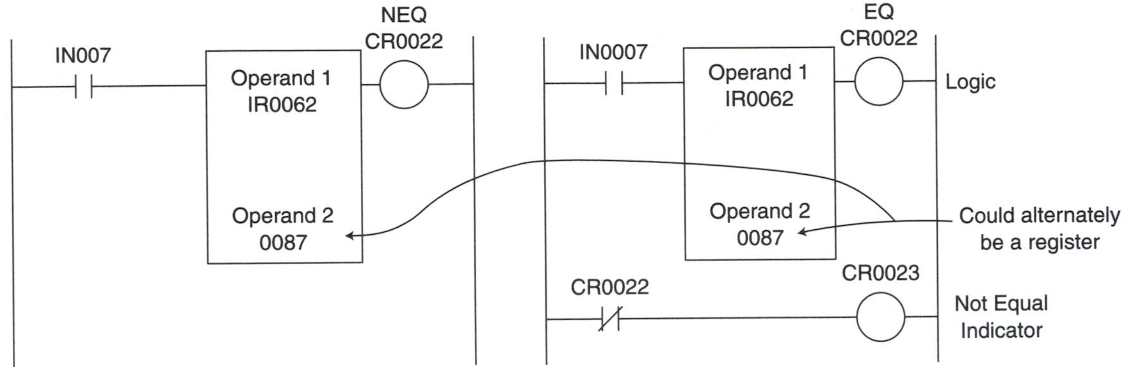

FIGURE 12–4
Example 12–2: Not-Equal-To COMPARE Function

EXAMPLE 12-3 An automatic pill-bottle-filling operation has two possible bottle sizes. One bottle is to be filled to a count of 225 or more. The other is to have 475 or more. This example uses a greater-than-or-equal-to COMPARE function. Figure 12–5 shows the PLC function to control the pill counts. The pill count (counter not shown) is fed from an input to IR 0142 as the bottle is filled. The appropriate minimum number of pills for proper filling, 225 or 475, is inserted into HR 0128. A bottle is put under the pill dispenser (not shown). For a small bottle, the 225 limit is put into HR 0128; for a large bottle, 475 would be entered into HR 0128.

FIGURE 12–5
Example 12–3: Greater-Than-or-Equal-To COMPARE Function

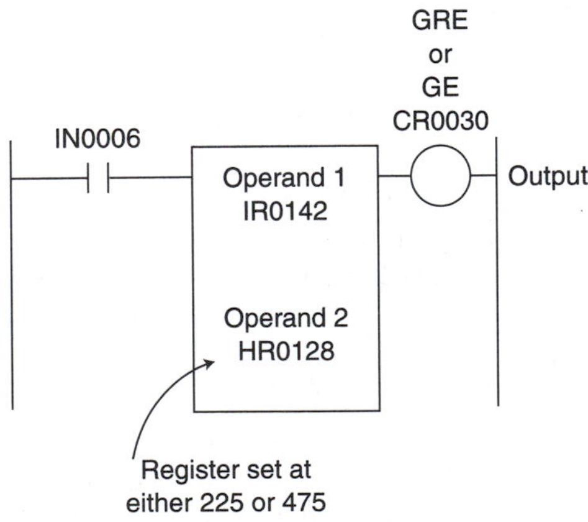

As the bottle starts filling, Enable is continuously pulsed. The comparison is untrue and output CR0030 is off. Once the pill count reaches 225 for the small bottle, CR0030 goes on. Output CR0030 is connected to a cap-and-remove operation (not shown). The bottle is capped and removed, and the process is reset and can be repeated. The same sequence would be carried out for the large bottle with HR0128 set at 475.

Why not use an equal-to function for example C? EQ would probably work, but what if the process overshoots? Suppose the count somehow got to 226 for the small bottle. The fill would go on unabated. If the count got to 226 (or 476 for the large size), however, the fill process would not erroneously continue if you use the GE function.

EXAMPLE 12-4 Completed assemblies flow off of a production line as illustrated in figure 12–6. If an assembly is removed for rework somewhere along the line, the part to be reworked is automatically counted. If there are more than 18 assemblies removed in an hour, a light in the foreman's office is to turn off. If there are fewer than 18 rejects per hour, the light will remain on. The required hourly reset-to-0 system is not shown. (The only part of this program that is shown is the comparison portion.) The defect count is kept in HR0063, operand 1. The allowable hourly number of defects, 18, is inserted as operand 2.

EXAMPLE 12-5 The greater-than COMPARE function is illustrated in figure 12-7. Two comparison functions are required for this example. This production operation requires a count greater than 348 for the output to turn on. The number 348 is inserted as operand 2 in both functions. For a count of 347, the EQ function keeps the output off. Below a count of 348, the GE function keeps the output off. For 349 or more, the output is on, as both contacts will be closed for the indicator ladder line.

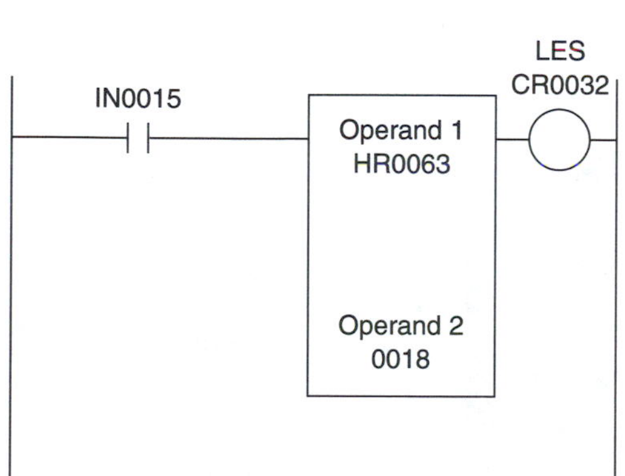

a - Direct Function

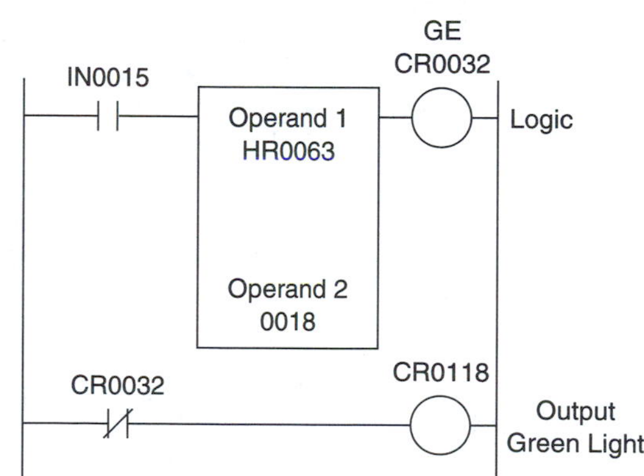

b - Derived Function

FIGURE 12–6
Example 12–4: Less-Than COMPARE Function

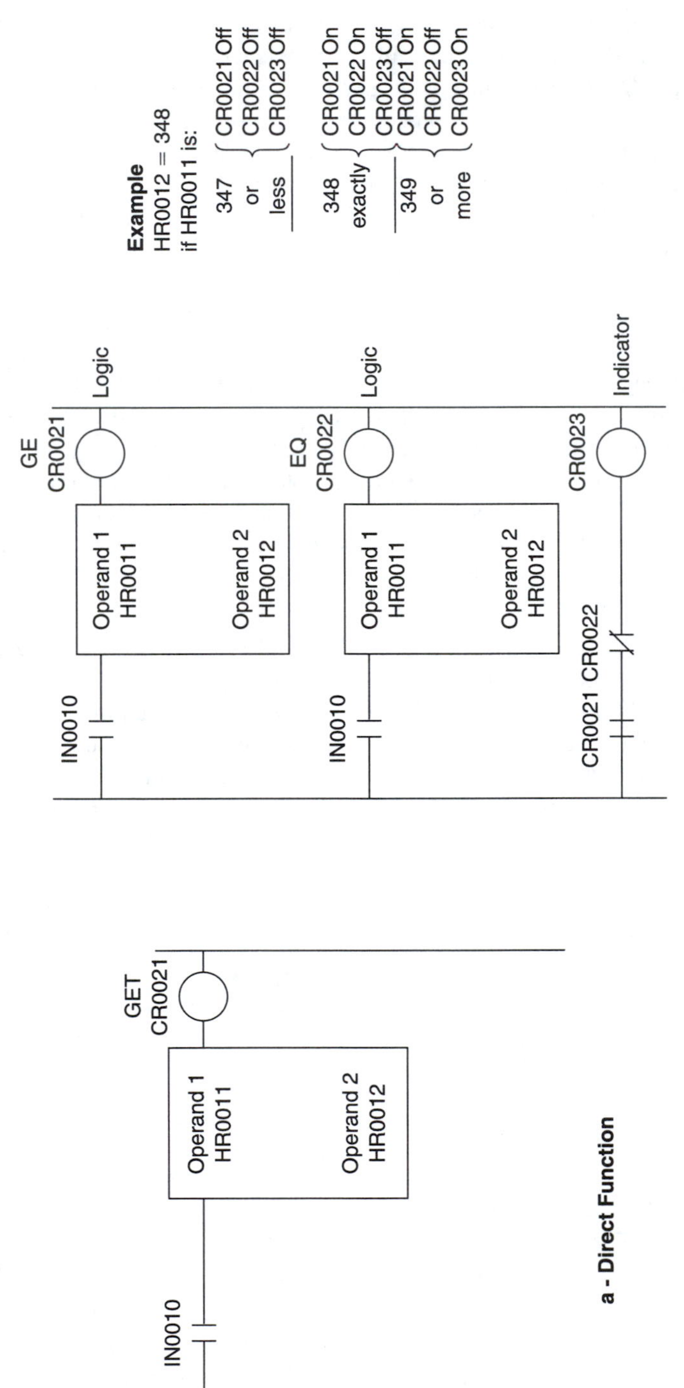

FIGURE 12–7
Example 12–5: Greater-Than COMPARE Function

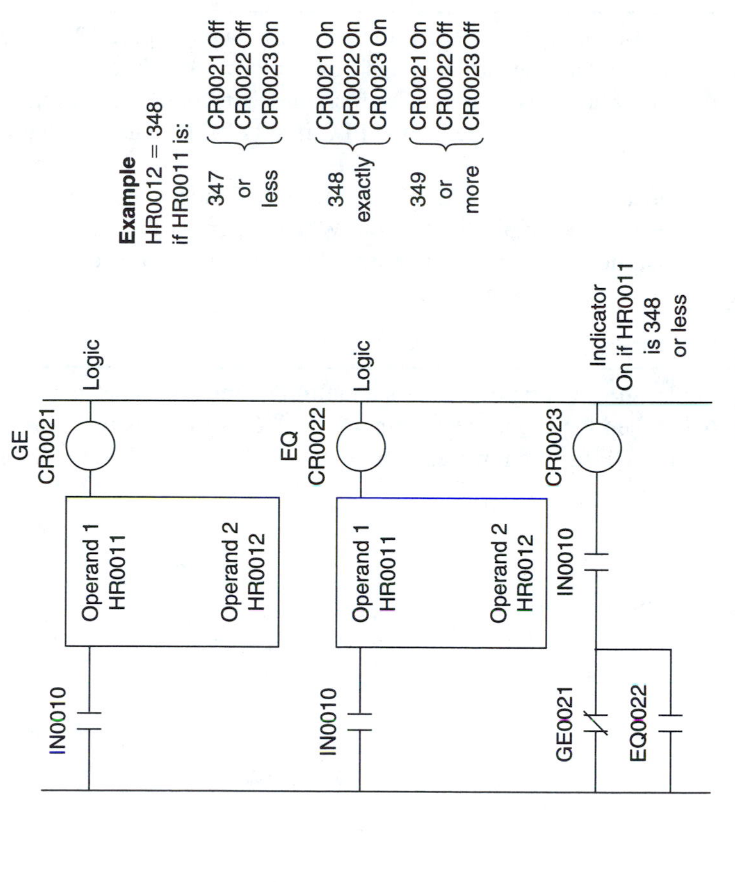

FIGURE 12–8
Example 12–6: Less-Than-or-Equal-To Function

EXAMPLE 12-6 The less-than-or-equal-to COMPARE function is shown in figure 12–8. A production system produces a product that can be one of three colors: red, white, or blue. The production is limited to 348 blue units per day. The blue units are counted by using a color-sensitive detector. The detector count is fed to the PLC into HR0111. The maximum desired count, 348, is inserted into HR0012.

The indicator is on for counts below 348. When the count reaches 348, GE stays on, and EQ comes on. The output remains on. When the count goes up one more to 349, EQ goes off, indicating that the production limit for blue has been reached. The output is now off and will remain off for higher counts.

EXAMPLE 12-7 This example, shown in figure 12–9, is a multiple-comparison program for lighting an indicator only when the count is between 15 and 22. A GE function is used for the lower count. Another GE function is used as an indicator for 22 and is set at 23. IN0016 enables

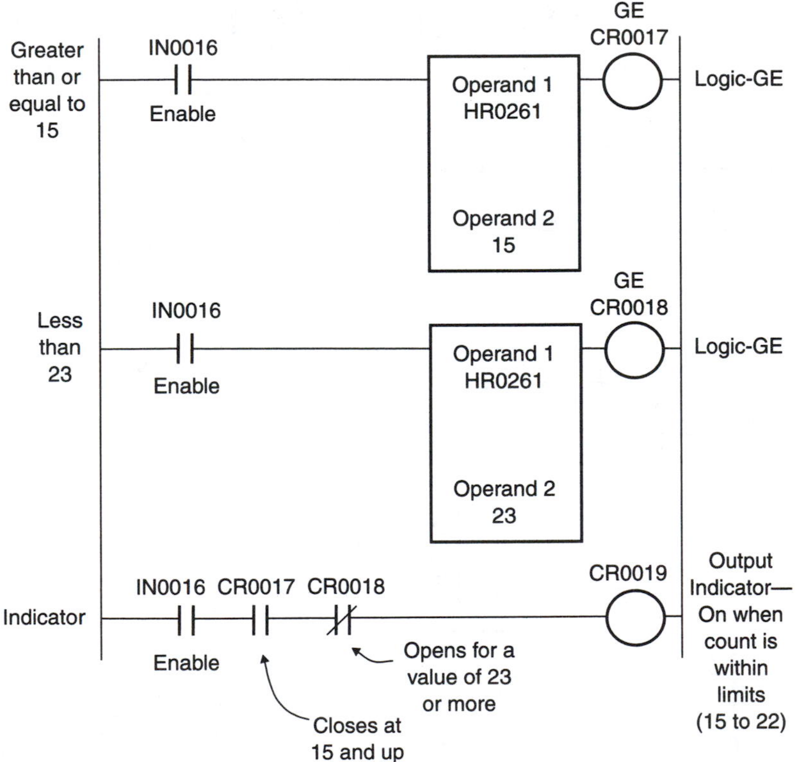

FIGURE 12–9
Example 12–7: Multiple-Comparison Program

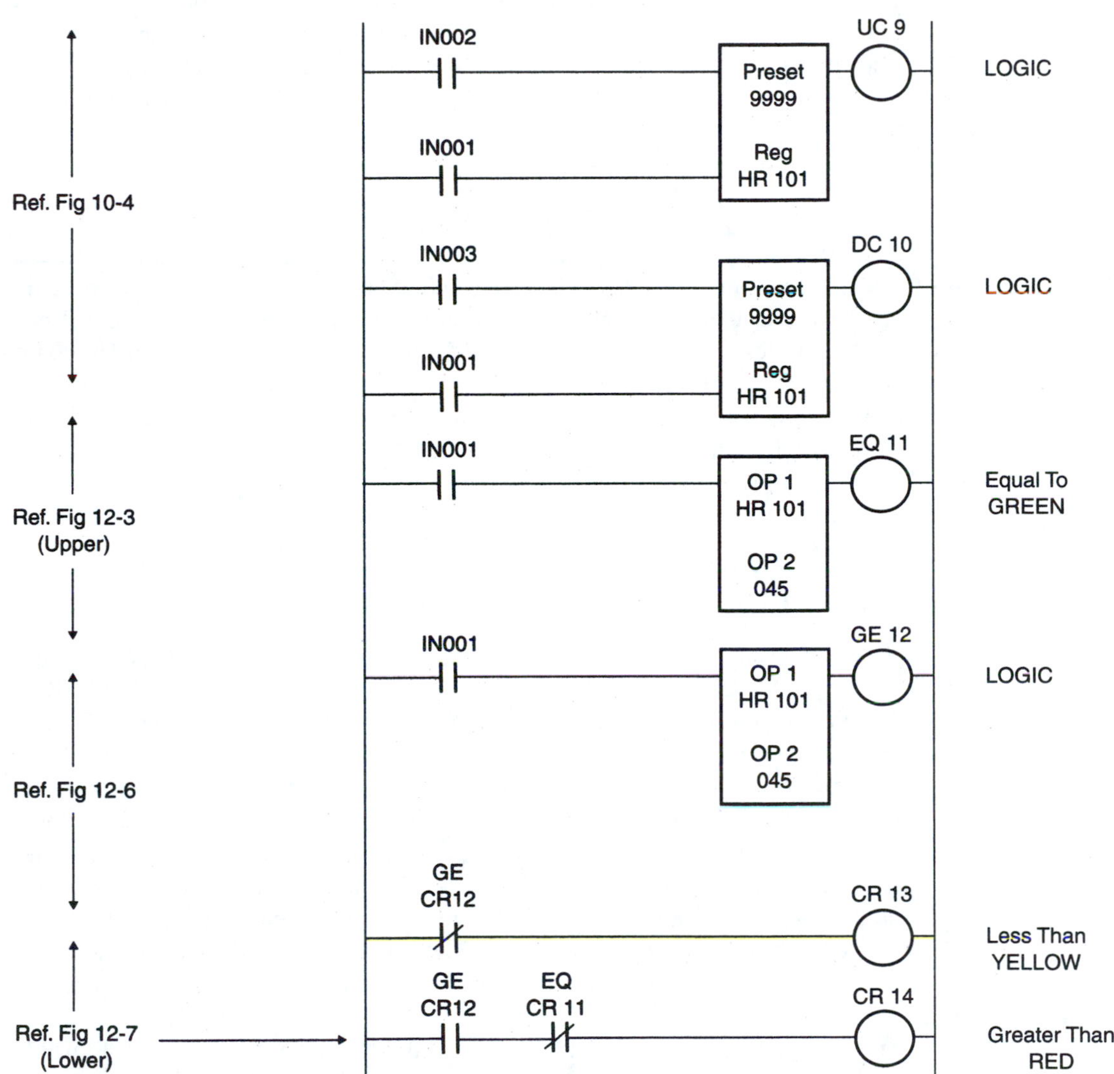

FIGURE 12–10
Example 12–8: PLC Program for Pilot Light Indication of Conveyor Part Count

the functions. Below 15, the top GE function is off, keeping the output in the lower ladder line off.

When the count, starting from 0, reaches 15, the top GE function goes on. The output, CR0019, turns on when the CR0017 contact closes. The other contact in the bottom line, CR0018, remains closed because the lower GE function has not yet come on.

When the up count reaches 22, the output is still on. When the count reaches 23, the lower GE function goes on. Its contact in the lower ladder line opens. The output is therefore turned off at 23 and beyond.

EXAMPLE 12-8

This problem shows how COMPARE functions might be used in industry. A conveyor is supposed to have exactly 45 parts on it. You have three indicating lights to indicate the conveyor count status: less than 45, yellow; exactly 45, green; and more than 45, red. The count of parts on the conveyor is set at 45 each morning by an actual count of parts. There are two sensors on the conveyor. One is actuated by parts entering the conveyor, and the other is actuated by parts leaving. A PLC program to carry out this function is shown in figure 12–10.

12–4 PLC ADVANCED COMPARISON FUNCTIONS

Three often useful comparison functions included in some advanced PLCs are LIMIT TEST (LIM), MASK COMPARE EQUAL TO (MEQ), and COMPARE EXPRESSION (CMP) function. In this section we describe the use of each of these advanced COMPARE functions.

Suppose that you want to see if a varying value is between two limits, which are either fixed or varying. You could set up a program like that shown in figure 9–2, perform two comparisons, and do a logic step. With the LIM function you need to program only one function. The function block is shown in figure 12–11a. There are three inputs: the varying value and the two limits. Figure 12–11b shows the two applications of the LIM function. The first determines if the value is between the high and low limits. The second, with the low limit greater than the high limit, shows if the value being evaluated is outside the limits.

The second advanced function sometimes used is the MASK COMPARE EQUAL TO (MEQ) function. The function compares bits in two registers selectively. The mask is used to denote which bits to compare and which not to compare. The function is shown in figure 12–12a. The use of the mask is illustrated in figure 12–12b. Where a mask bit is 1, the comparison is made. Where the mask bit is 0, no comparison is made. When enabled, the function goes on (true) if all unmasked bits pairs match. If one or more pair do not match, the function remains off (false).

Another advanced function sometimes used is the COMPARE EXPRESSION

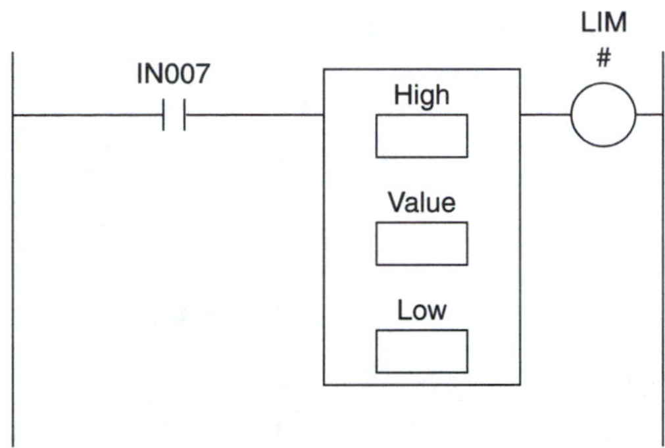

a - Function

LIM Example Using Integer:

- **If value Low Limit ⩽ value High Limit:** When the processor detects that the value of B (Test) is equal to or between limits, the instruction is true; if value Test is outside the limits, the instruction is false.

	false <	------- true ------	> false	
from −32,768	. . . .	A C	. . . .	to +32,767
		< value B >		

- **If value Low Limit ⩾ value High Limit:** When the processor detects that the value of Test is equal to or outside the limits, the instruction is true; if value Test is between, but not equal to either limit, the instruction is false.

	true <	------ false -----	> true	
from −32,768	 C		A	to +32,767
	value B <		< value B	

b - Function Operation

FIGURE 12–11
Operation of the Limit Test (LIM) Function (Courtesy of Allen-Bradley)

FIGURE 12–12
Operation of the Masked Comparison (MEQ) Function

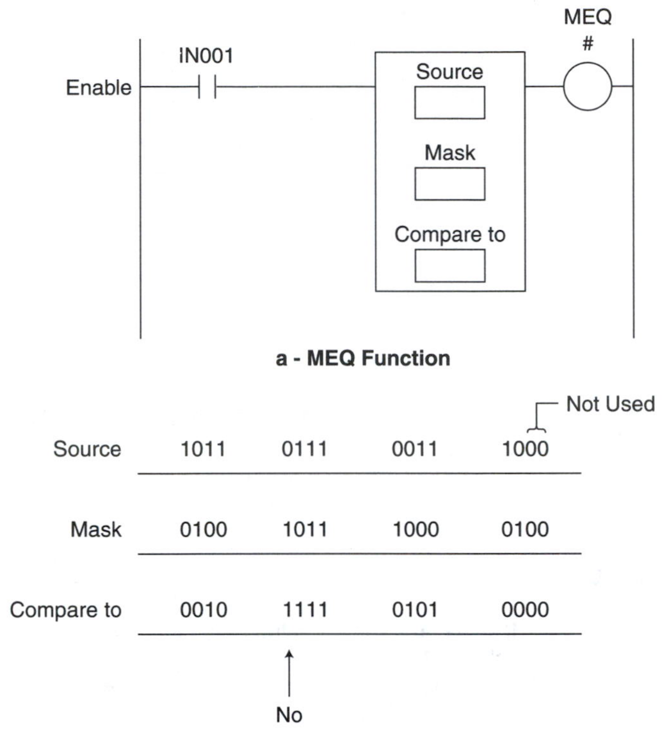

Masked Comparison is False

Comparison of 14 bits for a 16 bit register
Comparison is masked so that only bits 2, 5, 7, 8, 9 & 14 are compared

b - MEQ Operation

(CMP) function. This function compares values of complex expressions. For example, if you wished to determine whether one side of an equation was equal to the other side, you would program

$$((\text{HR107}*72) - 81) = (\text{SQR HR006} + 23)*(\text{HR } 056|0.68$$

Any of the other five comparison functions can also be applied. The valid operations for advanced Allen-Bradley PLCs is shown in figure 12–13. When enabled, the CMP function checks the equation values. If the equation is true, the output goes on. If false, the output remains off.

Valid Operations for Use in a CMP Expression

Type	Operator	Description	Example Operation
Comparison	=	equal to	if A = B, then ···
	<>	not equal to	if A <> B, then ···
	<	less than	if A < B, then ···
	<=	less than or equal to	if A <= B, then ···
	>	greater than	if A > B, then ···
	>=	greater than or equal to	if A >= B, then ···
Arithmetic	+	add	2 + 3 Enhanced PLC-5 processor: 2 + 3 + 7
	−	subtract	12 − 5
	*	multiply	5 * 2 PLC-5/30, -5/40, -5/60, -5/80: 6 * (5 * 2)
	| (vertical bar)	divide	24 | 6
	−	negate	− N7:0
	SQR	square root	SQR N7:0
	**	exponential * (x to the power of y)	10**3 Enhanced PLC-5 processors only)
Conversion	FRD	convert from BCD to binary	FRD N7:0
	TOD	convert from binary to BCD	TOD N7:0

FIGURE 12–13
Valid operations for Use in a CMP Expression (Courtesy of Allen-Bradley)

TROUBLESHOOTING PROBLEMS

The PLC has been programmed to operate as shown in the figures referred to. However, the circuit has the malfunction noted. What misprogramming or other factor or factors could cause the malfunction?

TS 12–1 Refer to figure 12–9 (example 12–7).

1. The output indicator never goes on.
2. The output indicator does not go off when a count of 23 is reached.

TS 12–2 Refer to figure 12–10 (example 12–8).

1. Yellow and green lights both go on.
2. Green and red lights both go on.

EXERCISES

Construct and test PLC COMPARE function programs for the following exercises.

1. A light is to come on only if a PLC counter has a value of 45 or 78. (*Hint:* Two EQ functions with outputs in parallel.)
2. A light is to be on if a PLC counter does not have values of either 23 or 31.
3. A light is to come on if three input numbers have the same value. (*Hint:* Use two functions with the same registers and two contacts controlling an output coil.)
4. An output is on if the input count is less than 34 or more than 41.
5. Same as exercise 4, but if the count is 37, the output is also on.
6. An output is to be on if the count is between 34 and 41. The count includes 34 and 41.
7. Same as exercise 6, but if the count is 37, the output is to be off.
8. There are two conveyors, each with sensors to count input and output of parts entering and leaving the conveyors, as in example 12–8 for one conveyor. There are to be three indicating lights for the process as follows:

Number of parts on the conveyors equal	White light
Number of parts on conveyor 1 is greater	Green light
Number of parts on conveyor 2 is greater	Blue light

9. Repeat exercise 8 except that the desired situation is that conveyor 1 have exactly 12 more units on it than conveyor 2. When this situation is true, a white light goes on. If conveyor 1 has less than 12 more parts on it than conveyor 2, a green light goes on. If conveyor 1 has more than 12 more parts on it than conveyor 2, a blue light goes on.
10. Use two or more LIM functions to check a varying value for the following values of true or false.

Less than 52	True
52 to 59	False
More than 59 and less than 71	True
71 to and including 86	False
More than 86 and less than 99	True
99 and above	False

11. For the MEQ function, what would the mask look like to check all the even-number (only) bits in a source for a match to the COMPARE bits?

13

Numbering Systems and PLC Number Conversion Functions

OUTLINE

OBJECTIVES

At the end of this chapter, you will be able to

- □ Explain the basis of the binary counting system.
- □ Convert numbers from binary to decimal, and vice versa.
- □ Explain the basis of the BCD system.
- □ Convert numbers from BCD to binary, and vice versa.
- □ Program a PLC for binary-to-BCD and BCD-to-binary conversions.
- □ Explain the basis of the octal and hex systems.
- □ Convert among the three numbering systems—decimal, octal, and hex.
- □ Describe the three coding systems—Gray, ASCII, and EBCDIC.

13–1 INTRODUCTION

Most PLC functional and programming operations can be handled in the decimal numbering system. Some PLC models and individual PLC functions use other numbering systems. This chapter deals with some of these numbering systems, including binary, BCD, octal, hexadecimal, Gray, ASCII, and EBCDIC. The basics of each system are explained, and conversions from one system to another are illustrated.

If your PLC works entirely in decimal, this chapter is optional. The purpose of the chapter is to give you enough background to handle any numbering system that you may encounter on larger PLCs or in the future. Various texts are available that detail the use of all numbering systems.

Note that this chapter does all conversions "long hand." Actual conversions can be made easily with a moderately priced, scientific, hand-held calculator. Calculators normally handle only up to 511 decimal. Various personal computer programs for larger numbers are available.

13–2 NUMBERING SYSTEMS: DECIMAL, BINARY, AND BCD

The first part of this section describes the binary numbering system, how it differs from decimal, and how to convert from one to the other.

The decimal numbering system, which uses a base of 10, is the one we use every day. The decimal system is said to have developed by tens because the originators had ten fingers and so could easily count to ten. Above ten, another person or means had to be used to keep track of the tens. If the count was over 99 (and less than 1000) a third means was needed to keep track of the 100s. Thus, the decimal system was developed.

When digital computers were developed, they were designed to count by 2's; their counters were either off or on. Therefore, the base of the computer counting system is 2, instead of the 10 used in the decimal system. This base-two binary system, which had been around for a long time, became the basis for all computer systems. With the advent of integrated circuit (IC) chips, hundreds of binary, on-off switches can now be found on a single IC chip in computers. All PLCs work internally in the binary system because they are IC-chip based. Since outside information is decimal or BCD (BCD is defined later in this section), a conversion to and from the binary used in the PLC CPU is needed.

Figure 13–1 shows a comparison of the interpretation of a decimal and a binary number. The numbers are arbitrarily picked for this illustration. There are 10 possible count values (0 through 9) in each decimal position. For the binary counter, there are only 2 possible counts, 0 or 1. As a result, for binary we must move to the next bit or *slot* after we reach 1, not 9, as in decimal. The binary number shown has a decimal equivalent value of 4 + 1, or 5.

FIGURE 13–1
Decimal and Binary System Comparison

	100s	10s	1s
Typical Decimal Number 976 ⟶	↓	↓	↓
	9	7	6
	4s	2s	1s
Typical Binary Number 101 ⟶	↓	↓	↓
	1	None	1

FIGURE 13–2
Decimal and Binary Value Comparison

Decimal-Base 10	Binary-Base 2
1	1
2	10
3	11
4	100
5	101
6	110
7	111
8	1000
9	1001
10	1010
11	1011
12	1100
13	1101
14	1110
15	1111
16	10000
17	10001
18	10010
19	10011
20	10100
21	10101
22	10110
23	10111
24	11000
25	11001

Figure 13–2 compares the value of a given quantity for decimal and binary systems.

Figure 13–3 illustrates the counting slots or bit status for some typical binary numbers. The more bits you use, the higher you can count. The maximum decimal equivalent count is given for each example (1 in all bits). The equivalent decimal values for the given binary bit patterns are given also. Typical PLCs work in 8- or 16-bit form. Other PLCs may have 4, 6, 12, or other numbers of bits as a base for each address or register.

Bits	Maximum Value - IF All Bits Are 1	Typical Binary Number	Equivalent Decimal Value
4	15	8s 4s 2s 1s 1 1 0 1	= 13
8	255	128s 64s 32s 16s 8s 4s 2s 1s 1 0 0 1 1 1 0 1	= 157
16	4095	2048s 1024s 512s 256s 128s 64s 32s 16s 8s 4s 2s 1s 1 0 1 1 0 1 1 0 0 1 0 1	= 2917
32	65,553	32786 16384 8192 4096 2048 1024 512 256 128 64 32 16 8 4 2 1 1 0 1 0 1 1 0 0 1 1 1 0 1 1 1 0	= 44,288

FIGURE 13–3
Typical Binary Numbers and Decimal Equivalents

Digital Number			Conversion	Decimal Result
		1011	(8 + 0 + 2 + 1)	11
	0101	0101	(0 + 64 + 0 + 16) + (0 + 4 + 0 + 1)	85
	0011	1110	(0 + 0 + 32 + 16) + (8 + 4 + 2 + 0)	62
	1101	0011	(128 + 64 + 0 + 16) + (0 + 0 + 2 + 1)	211
1010	1010	1111	(2048 + 0 + 512 + 0) + (128 + 0 + 32 + 0) + (8 + 4 + 2 + 1)	2735

FIGURE 13–4
Binary-to-Decimal Conversions

Figure 13–4 gives some examples of conversions from binary to decimal. Four-, 8-, and 12-bit numbers are shown in the example. Different numbers of bits are converted similarly, starting from right to left.

Figure 13–5 illustrates the opposite conversion, from decimal to binary. The conversion is not as easily accomplished as binary to decimal, especially as the numbers get larger. Basically, the procedure is to find the largest number in the digital columns that is less than the decimal number being converted. A 1 is placed in that binary column. Next, subtract the column's binary/decimal value number from the decimal number being converted. Then move one column to the right and apply the same procedure to the remainder. Keep repeating the procedure until the remainder is zero.

A related number system used often in PLCs is the binary-coded decimal, or BCD, system. The key to the BCD system is to utilize four digital bits for each single decimal number of output or input. A commonly used BCD system contains four decimal numbers and can display decimal values from 0000 through 9999. The four-digit decimal number is then represented by 4 times 4, or 16 bits. Figure 13–6 shows some typical decimal numbers and their BCD equivalents. For comparison, the binary values are also shown in the table.

PLC CPUs function in binary, not BCD or decimal. If the PLC received a BCD number from an input thumbwheel (a small rotary device set to 0 through 9 by rotating it with

Decimal Number →	Binary Number							
	128	64	32	16	8	4	2	1
1	0	0	0	0	0	0	0	0
3	0	0	0	0	0	0	1	1
7	0	0	0	0	0	1	1	1
22	0	0	0	1 22 – 16 = 6	0	1 6 – 4 = 2	1	0
87	0	1 87 – 64 = 23	0	1 23 – 16 = 7	0	1	1	1
125	0	1 125 – 64 = 61	1 61 – 32 = 29	1 29 – 16 = 13	1 13 – 8 = 5	1 5 – 4 = 1	0	1
164	1 164–128 = 36	0	1 36 – 32 = 4	0	0	1	0	0

FIGURE 13–5
Decimal-to-Binary Conversion

FIGURE 13–6
Decimal and BCD Equivalents

Decimal	BCD	Binary (Ref.)
1	0001	1
2	0010	10
3	0011	11
4	0100	100
5	0101	101
6	0110	110
7	0111	111
8	1000	1000
9	1001	1001
10	0001 0000	1010
11	0001 0001	1011
12	0001 0010	1100
13	0001 0011	1101
14	0001 0100	1110
15	0001 0101	1111
16	0001 0110	10000
17	0001 0111	10001
18	0001 1000	10010
19	0001 1001	10011
20	0010 0000	10100
21	0010 0001	10101
22	0010 0010	10110
23	0010 0011	10111
24	0010 0100	11000
25	0010 0101	11001

Table of Values

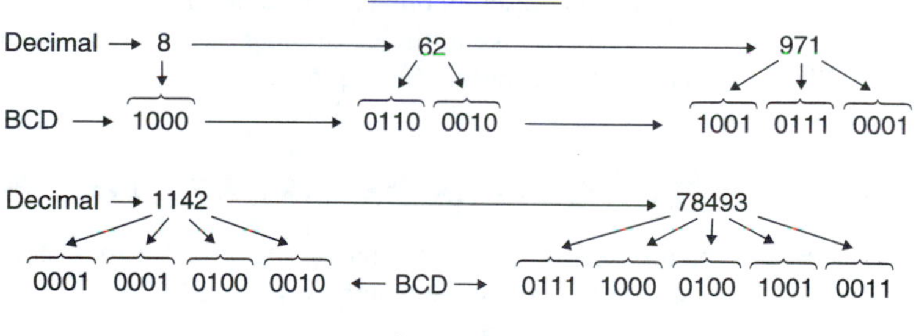

Conversions

the thumb or finger), it would interpret the number as a binary number. The BCD input number must be converted to its binary equivalent for correct PLC CPU operation. We could convert the number manually, as shown in Figure 13–6; however, if it must be done 25 times per second, for example, manual conversion would be impossible. For fast operation, the conversion must be performed by the PLC. A PLC function for the conversion is available. Your PLC may do the conversion automatically; however, in most cases, you must program in the BCD-to-binary conversion for inputs. For outputs, you program binary-to-BCD conversions.

Figure 13–7 shows a simple example of a required conversion. The PLC is required to take an input value in BCD and multiply it by 0.5. The resulting number is then displayed

FIGURE 13–7
BCD-to-Binary and Binary-to-BCD PLC Processing

8	6	4	2	← Thumbwheel Setting in Decimal

↓

1000 0110 0100 0010 BCD (8642)

↓

0010 0001 1100 0010 Binary (for 8642 decimal)

↓ (x .5) ← Internally scaled 1 to 2 ($^1/_2$ value)

0001 0000 1110 0001 Binary (for 4321 decimal)

↓

0010 0011 0010 0001 BCD (4321)

↓

(4) (3) (2) (1) ← Output reading on LED read out in decimal

in BCD on an output LED readout. Since the PLC functions in binary, the input BCD value must be converted to binary before internal processing can take place. Then, the converted binary number is multiplied in binary by 0.5. This resulting binary value is then converted to BCD for output to the display.

13–3 PLC CONVERSIONS BETWEEN DECIMAL AND BCD

In this section of the chapter we show the formats of the PLC number conversion functions and use the conversion functions in a PLC program. Many PLCs include these functions

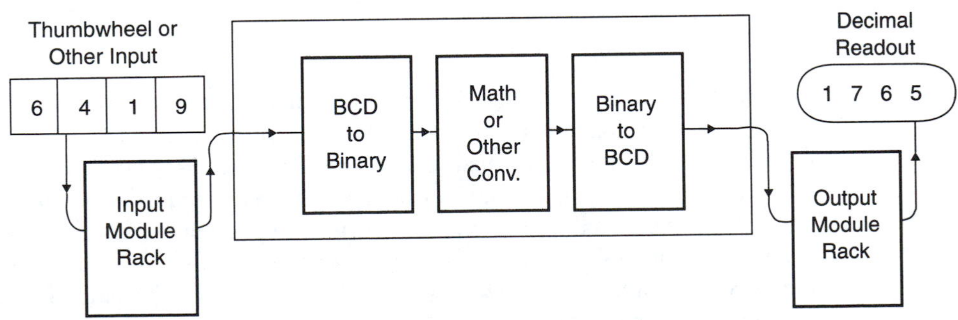

FIGURE 13–8
Block Diagram of a PLC's BCD-to-Binary Conversion Program

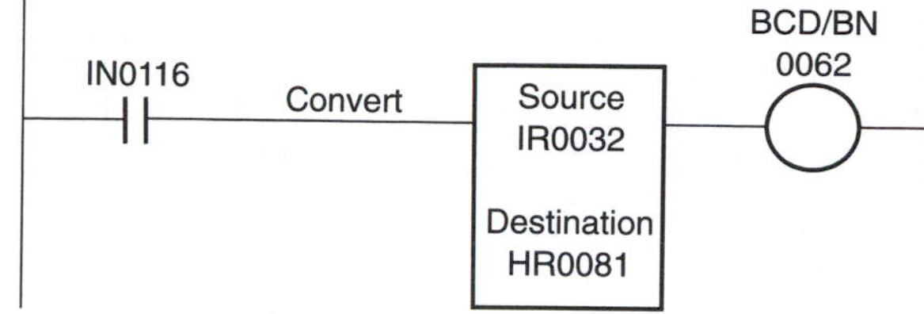

FIGURE 13–9
BCD-to-Binary Conversion Function

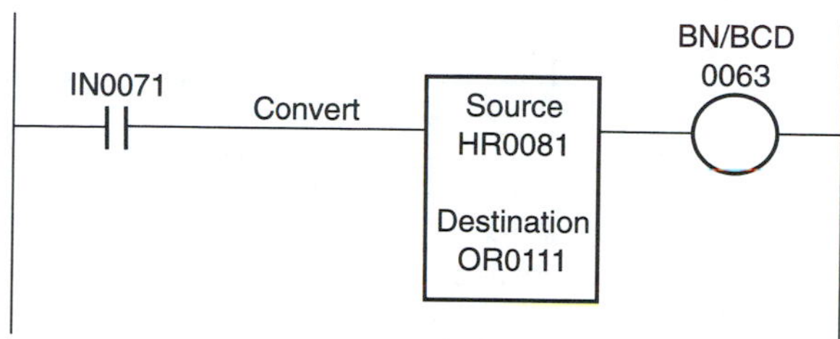

FIGURE 13–10
Binary-to-BCD Conversion Function

for converting BCD numbers to binary for use by the CPU. They also can contain the inverse conversion, binary to BCD. Figure 13–8 is a typical PLC application problem requiring number conversions. It illustrates in block diagram form how the conversions are accomplished. The mathematical manipulation in the middle block can take any form.

Figure 13–9 shows the layout of a typical BCD-to-binary conversion function, which is usually used for input data conversions. It converts the BCD value found in the source register to a binary number in the destination register for PLC CPU use. When the function's input line is energized, values in the source register are converted from BCD to binary. The resulting number is put into the destination register. Typically, the coil comes on only if one of the BCD digits exceeds 9 during mathematical conversion.

Figure 13–10 shows how the reverse of the conversion in figure 13–9 is accomplished. A binary value in the source is converted to a BCD value in the destination. This function's primary use is for feeding an output display. Typically for this function, the coil comes on only if the binary value exceeds 9999 in decimal during operation.

To learn how the individual functions of figures 13–9 and 13–10 function, you may use the MONITOR mode (covered in chapter 26). Program the function to be observed in the EDIT mode, place the PLC in the MONITOR mode, and then call up the two registers involved in the conversion function. Insert appropriate values into the first register. Enable the conversion function being used and observe the resulting value in the second register. Verify the conversion function's correct mathematical operation manually or by calculator conversion computation.

13–4 OCTAL AND HEXADECIMAL NUMBERING SYSTEMS

Two other number codes often encountered in dealing with PLCs and computers are the octal code and the hexadecimal. Hexadecimal is normally shortened to hex. The octal code uses a base of 8, and the hex uses a base of 16. These bases are in contrast to decimal and

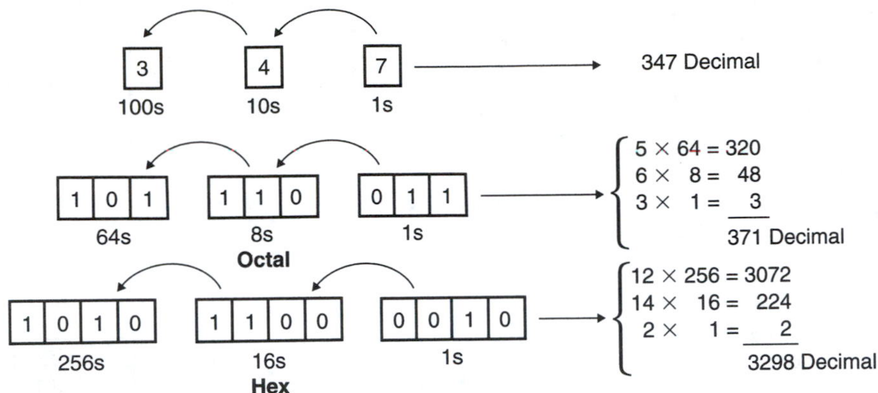

FIGURE 13–11
Decimal–Octal–Hex System Comparison

FIGURE 13–12
Octal Code Table

Decimal Base (10)	Octal Base (8)	(Ref.) Binary Base (2)
1	1	1
2	2	10
3	3	11
4	4	100
5	5	101
6	6	110
7	7	111
8	10	1000
9	11	1001
10	12	1010
11	13	1011
12	14	1100
13	15	1101
14	16	1110
15	17	1111
16	20	10000
17	21	10001
18	22	10010
19	23	10011
20	24	10100
21	25	10101
22	26	10110
23	27	10111
24	30	11000
25	31	11001

Octal Number	Conversion	Decimal Result
0203	$(0 \times 512) + (2 \times 64) + (0 \times 8) + (3 \times 1)$	131
1111	$(1 \times 512) + (1 \times 64) + (1 \times 8) + (1 \times 1)$	585
2673	$(2 \times 512) + (6 \times 64) + (7 \times 8) + (3 \times 1)$	1467
1071	$(1 \times 512) + (0 \times 64) + (7 \times 8) + (1 \times 1)$	569
4918 ⟶	Not a valid octal number –	(9s & 8s impossible)

$$\text{base} \begin{cases} 8 - 8 \\ 8 \times 8 = 64 \\ 8 \times 8 \times 8 = 512 \end{cases}$$

FIGURE 13–13
Octal-to-Decimal Conversions

BCD, which have a base of 10. If the computer's numbers are 3 bits long, it is in the octal numbering system. Three bits can count up to 7 as a maximum and start over at 8. If 4 bits are used for each number, the numbering system is in hex. Four bits can count up to 15 and then start over at 16. Figure 13–11 shows the comparison of octal- and hex-to-decimal equivalents.

Figure 13–12 gives a table comparing values for decimal numbers with their octal equivalents (and binary for reference). When we reach 8, octal starts over and a 1 goes in the slot to the left.

How do we convert between octal and decimal? Figure 13–13 shows some typical octal-to-decimal conversions, and figure 13–14 illustrates some decimal-to-octal conversions. For decimal-to-octal conversions, repetitive division is used.

As stated previously, the hex system uses 4 bits per number, compared to octal's 3. Therefore, hex is base 16, as was shown in figure 13–11. Since we must count beyond 9 in a single column, we use sequential letters of the alphabet for numbers 10 through 15. The hex system is shown in figure 13–15.

Figure 13–16 shows how hex numbers are converted to decimal. The conversion uses straightforward multiplication, as does the octal-to-decimal conversion.

The conversion of decimal to hex is a little more involved than hex to decimal. We use repetitive division similar to that used for decimal-to-octal conversions, as shown in figure 13–17.

Decimal Number	Conversion	Octal Number
5	None Needed	5
11	11/8 = 1 with 3 Remainder	13
28	28/8 = 3 with 4 Remainder	34
85	85/8 = 10 with 5 Remainder	
	↳ 10/8 = 1 with 2 Remainder	125
116	116/8 = 14 with 4 Remainder	
	↳ 14/8 = 1 with 6 Remainder	164
982	982/8 = 122 with 6 Remainder	
	↳ 122/8 = 15 with 2 Remainder	1726
	↳ 1 with 7 Remainder	

FIGURE 13–14
Decimal-to-Octal Conversions

FIGURE 13–15
Decimal–Hex–Binary Comparison

Decimal Base 10	Hex Base 16	Binary Base 2
0	0	0000
1	1	0001
2	2	0010
3	3	0011
4	4	0100
5	5	0101
6	6	0110
7	7	0111
8	8	1000
9	9	1001
10	A	1010
11	B	1011
12	C	1100
13	D	1101
14	E	1110
15	F	1111
16	10	10000
17	11	10001

Hex Number	Conversion	Decimal Result
13	$(1 \times 16) + (3 \times 1)$	19
BC	$(11 \times 16) + (12 \times 1)$	188
F4D	$(15 \times 256) + (4 \times 16) + (13 \times 1)$	3917
C1B7	$(12 \times 4096) + (1 \times 256) + (11 \times 16) + (7 \times 1)$	49591

$$\text{Base}\begin{cases} 16 \\ 16 \times 16 = 256 \\ 16 \times 16 \times 16 = 4096 \end{cases}$$

FIGURE 13–16
Hex-to-Decimal Conversions

Decimal Numbers	Conversion	Hex Number
7	—	7
12	—	C
21	21/16 = 1 with 5 Remainder	15
111	111/16 = 6 with 15 Remainder	6F
247	247/16 = 15 with 7 Remainder	F7
398	398/16 = 24 with 14 Remainder ↳ 24/16 = 1 with 8 Remainder	18E

FIGURE 13–17
Decimal-to-Hex Conversions

13–5 OTHER NUMBERING AND CODE SYSTEMS

Three other numbering/coding systems are often encountered in PLC work: Gray, ASCII, and EBCDIC. The Gray code involves only numbers; ASCII and EBCDIC codes involve both numbers and letters or symbols.

The Gray code's structure, which is shown in figure 13–18, is constructed so that only one digit changes as you go down one step. You will note this arrangement by following down the left Gray code column. Note the step from the fourth step to the fifth step. Only one number changes for the Gray code. The third digit changes from a 0 to a 1. For the same binary step, 3 bits' digits change.

The single-digit change is important when certain mechanical and photoelectric encoders are used. Details of these encoding systems may be found in various texts on control systems. If you follow the binary code in the right-hand column, you see that one, two, three, or more digits change from one number to the next. The Gray code has only one change per step. A change of more than one digit is difficult to handle when using encoders.

The Gray code and standard binary are different and are not interchangeable. Caution must be exercised not to mix the two. If your PLC works in binary, putting Gray code values into it will give false input information. The inverse is also true for output; a Gray code output will not function properly in binary.

ASCII stands for *American Standard Code for Information Interchange*. As you can see from its listing in figure 13–19, it covers numbers, letters, symbols, and abbreviations. The ASCII code requires 6 or 7 memory bits, depending on the system used. It is used to interface the PLC CPU with alphanumeric keyboards and printers.

The EBCDIC code is similar to the ASCII code in function. It is shown in figure 13–20. EBCDIC uses 8 bits, compared to the 6 or 7 bits used for the ASCII code. Whether ASCII or EBCDIC is used depends on the particular PLC's operational requirements and capabilities.

FIGURE 13–18
Gray Code Compared to Binary Code

Gray Code	Binary
0000	0000
0001	0001
0011	0010
0010	0011
0110	0100
0111	0101
0101	0110
0100	0111
1100	1000
1101	1001
1111	1010
1110	1011
1010	1100
1011	1101
1001	1110
1000	1111

Hexadecimal	Decimal	Octal	Binary	Character	Description
00	0	000	0000000	NUL	Null
01	1	001	0000001	SOH	Start of heading
02	2	002	0000010	STX	Start of text
03	3	003	0000011	ETX	End of text
04	4	004	0000100	EOT	End of transmission
05	5	005	0000101	ENQ	Enquiry
06	6	006	0000110	ACK	Acknowledge
07	7	007	0000111	BEL	Bell
08	8	010	0001000	BS	Back space
09	9	011	0001001	HT	Horizontal tab
0A	10	012	0001010	LF	Line feed
0B	11	013	0001011	VT	Vertical tab
0C	12	014	0001100	FF	Form feed
0D	13	015	0001101	CR	Carriage return
0E	14	016	0001110	SO	Shift out
0F	15	017	0001111	SI	Shift in
10	16	020	0010000	DLE	Data link escape
11	17	021	0010001	DC1	Device Control 1
12	18	022	0010010	DC2	Device Control 2
13	19	023	0010011	DC3	Device Control 3
14	20	024	0010100	DC4	Device Control 4
15	21	025	0010101	NAK	Negative acknowledge
16	22	026	0010110	SYN	Synchronize
17	23	027	0010111	ETB	End of transmission block
18	24	030	0011000	CAN	Cancel
19	25	031	0011001	EM	End of media
1A	26	032	0011010	SUB	Substitute
1B	27	033	0011011	ESC	Escape
1C	28	034	0011100	FS	File separator
1D	29	035	0011101	GS	Group separator
1E	30	036	0011110	RS	Record separator
1F	31	037	0011111	US	Unit separator
20	32	040	0100000	SP	Space
21	33	041	0100001	!	Exclamation

FIGURE 13–19
ASCII Code

Hexadecimal	Decimal	Octal	Binary	Character	Description
22	34	042	0100010	”	Double quote
23	35	043	0100011	#	Number or pound
24	36	044	0100100	$	Dollar sign
25	37	045	0100101	%	Percentage
26	38	046	0100110	&	Ampersand
27	39	047	0100111	’	Apostrophe or single quote
28	40	050	0101000	(	Left parenthesis
29	41	051	0101001	)	Right parenthesis
2A	42	052	0101010	*	Asterisk
2B	43	053	0101011	+	Plus
2C	44	054	0101100	,	Comma
2D	45	055	0101101	–	Minus
2E	46	056	0101110	.	Period
2F	47	057	0101111	/	Slash
30	48	060	0110000	0	Zero
31	49	061	0110001	1	One
32	50	062	0110010	2	Two
33	51	063	0110011	3	Three
34	52	064	0110100	4	Four
35	53	065	0110101	5	Five
36	54	066	0110110	6	Six
37	55	067	0110111	7	Seven
38	56	070	0111000	8	Eight
39	57	071	0111001	9	Nine
3A	58	072	0111010	:	Colon
3B	59	073	0111011	;	Semi-colon
3C	60	074	0111100	<	Less than
3D	61	075	0111101	=	Equal
3E	62	076	0111110	>	Greater than
3F	63	077	0111111	?	Question
40	64	100	1000000	@	At sign
41	65	101	1000001	A	Letter A
42	66	102	1000010	B	Letter B
43	67	103	1000011	C	Letter C
44	68	104	1000100	D	Letter D
45	69	105	1000101	E	Letter E
46	70	106	1000110	F	Letter F
47	71	107	1000111	G	Letter G
48	72	110	1001000	H	Letter H
49	73	111	1001001	I	Letter I
4A	74	112	1001010	J	Letter J
4B	75	113	1001011	K	Letter K
4C	76	114	1001100	L	Letter L
4D	77	115	1001101	M	Letter M
4E	78	116	1001110	N	Letter N
4F	79	117	1001111	O	Letter O
50	80	120	1010000	P	Letter P
51	81	121	1010001	Q	Letter Q
52	82	122	1010010	R	Letter R
53	83	123	1010011	S	Letter S
54	84	124	1010100	T	Letter T
55	85	125	1010101	U	Letter U

FIGURE 13–19
(continued)

Hexadecimal	Decimal	Octal	Binary	Character	Description
56	86	126	1010110	V	Letter V
57	87	127	1010111	W	Letter W
58	88	130	1011000	X	Letter X
59	89	131	1011001	Y	Letter Y
5A	90	132	1011010	Z	Letter Z
5B	91	133	1011011	[	Left bracket
5C	92	134	1011100	\	Back slash
5D	93	135	1011101	]	Right bracket
5E	94	136	1011110	↑	Up arrow
5F	95	137	1011111	←	Back arrow
60	96	140	1100000	'	Back quote or accent mark
61	97	141	1100001	a	Small letter a
62	98	142	1100010	b	Small letter b
63	99	143	1100011	c	Small letter c
64	100	144	1100100	d	Small letter d
65	101	145	1100101	e	Small letter e
66	102	146	1100110	f	Small letter f
67	103	147	1100111	g	Small letter g
68	104	150	1101000	h	Small letter h
69	105	151	1101001	i	Small letter i
6A	106	152	1101010	j	Small letter j
6B	107	153	1101011	k	Small letter k
6C	108	154	1101100	l	Small letter l
6D	109	155	1101101	m	Small letter m
6E	110	156	1101110	n	Small letter n
6F	111	157	1101111	o	Small letter o
70	112	160	1110000	p	Small letter p
71	113	161	1110001	q	Small letter q
72	114	162	1110010	r	Small letter r
73	115	163	1110011	s	Small letter s
74	116	164	1110100	t	Small letter t
75	117	165	1110101	u	Small letter u
76	118	166	1110110	v	Small letter v
77	119	167	1110111	w	Small letter w
78	120	170	1111000	x	Small letter x
79	121	171	1111001	y	Small letter y
7A	122	172	1111010	z	Small letter z
7B	123	173	1111011	{	Left brace
7C	124	174	1111100	|	Vertical bar
7D	125	175	1111101	}	Right brace
7E	126	176	1111110	~	Approximate or tilde
7F	127	177	1111111	DEL	Delete (rub out)

FIGURE 13–19
(continued)

FIGURE 13–20
EBCDIC Code

Character	ASCII	EBCDIC
0	011 0000	1111 0000
1	011 0001	1111 0001
2	011 0010	1111 0010
3	011 0011	1111 0011
4	011 0100	1111 0100
5	011 0101	1111 0101
6	011 0110	1111 0110
7	011 0111	1111 0111
8	011 1000	1111 1000
9	011 1001	1111 1001
A	100 0001	1100 0001
B	100 0010	1100 0010
C	100 0011	1100 0011
D	100 0100	1100 0100
E	100 0101	1100 0101
F	100 0110	1100 0110
G	100 0111	1100 0111
H	100 1000	1100 1000
I	100 1001	1100 1001
J	100 1010	1101 0001
K	100 1011	1101 0010
L	100 1100	1101 0011
M	100 1101	1101 0100
N	100 1110	1101 0101
O	100 1111	1101 0110
P	101 0000	1101 0111
Q	101 0001	1101 1000
R	101 0010	1101 1001
S	101 0011	1110 0010
T	101 0100	1110 0011
U	101 0101	1110 0100
V	101 0110	1110 0101
W	101 0111	1110 0110
X	101 1000	1110 0111
Y	101 1001	1110 1000
Z	101 1010	1110 1001

EXERCISES

In exercises 1 through 8, convert the given numbers from one system to the other system specified.

1. Binary to decimal:
111, 1011, 10 1010, 110 1101, 1011 0011, 1001 1011 1110

2. Decimal to binary:
14, 42, 85, 162, 207, 459

3. Decimal to BCD:
56, 381, 1121, 4583, 6666

4. BCD to decimal:
0110 0011 1011 1100 (*Hint:* Is this possible?)
0110 0011 0111 0110
1000 1001 0100 0001
0101 0000 1000 0101
0011 0011 0011 0111

5. Decimal to octal:
7, 15, 88, 327, 691, 1121

6. Octal to decimal:
6, 35, 77, 201, 847, 4464

7. Decimal to hex:
8, 14, 79, 410, 558, 1243

8. Hex to decimal:
11, 3C, 2A2, BCF, 1B4C, DDDDD

9. Octal uses 3 bits and hex uses 4 bits. (What would a quad number system, which utilizes only 2 bits, look like?) Construct a table of comparison values to decimal.

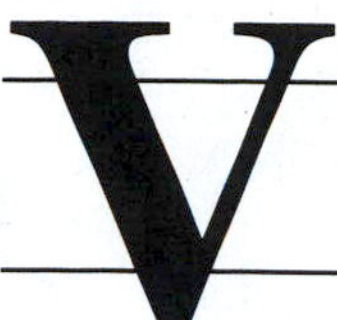

Data Handling Functions

14

The PLC SKIP and MASTER CONTROL RELAY Functions

OUTLINE

14–1 Introduction □ **14–2** The SKIP Function and Applications □ **14–3** The MASTER CONTROL RELAY Function and Applications

OBJECTIVES

At the end of this chapter, you will be able to

- □ Describe the operation of the SKIP function.
- □ Describe the operation of the MASTER CONTROL RELAY function.
- □ Apply the SK and MCR functions to operational applications.

14–1 INTRODUCTION

Both the SKIP (SK) and MASTER CONTROL RELAY (MCR) functions are powerful programming tools. The SK function allows us to skip, or bypass, a chosen portion of a ladder sequence. The coils and functions skipped remain in the state they were in during the last scan before SK was enabled. SKIP enables us to effectively branch to a different portion of the program. In some programming systems, SKIP is called ZCL (zone control last state). In many PLCs the SKIP function is carried out by a JUMP programming system, which is covered in chapter 15.

MCR operates similarly. When MCR is enabled on, the ladder diagram functions normally. When MCR is not enabled, a specified number of coils and functions are frozen in the off position. Coils in the frozen section will then stay off even if their individual enable lines are turned on. In some systems, MCR means Master Control Reset (not Relay).

The difference between the two functions is that SKIP leaves the next specified number of ladder lines in their previous on or off state. MCR turns the next specified number of ladder lines to the off state. Another difference is that SKIP is active when enabled and MCR is active when not enabled, which makes it fail-safe.

14–2 THE SKIP FUNCTION AND APPLICATIONS

The SKIP (SK) function, illustrated in figure 14–1, allows a portion of a PLC program to be bypassed when its coil is enabled. The enable line of the function is energized when the skip of one or more subsequent lines is desired. In addition to programming a coil number in the usual manner, the number of lines to be skipped is also specified and programmed as shown.

Figure 14–2 shows a basic application of the SKIP function in a program. The eight-line program used for illustration has seven lines with output functions. A SKIP function is included on the third line of the eight. When the SK function is off, the other seven functions operate in the normal manner. When the seven lines corresponding to inputs are on, their outputs are on, and when inputs are off, outputs are off. For this illustration, the value for number of lines to be skipped will be set at 3.

When the SKIP function (set at 3) is turned on, the first two lines will function as usual. However, the next three lines, 4 through 6, will stay on or off in their previous state. With SKIP on, changing the input on–off status feeding the coils on lines 4 through 6 will have no effect on output coils 4 through 6. Coils on lines 4 through 6 will retain their previous states. Lines 7 and 8 will continue to operate normally, unaffected by the SKIP func-

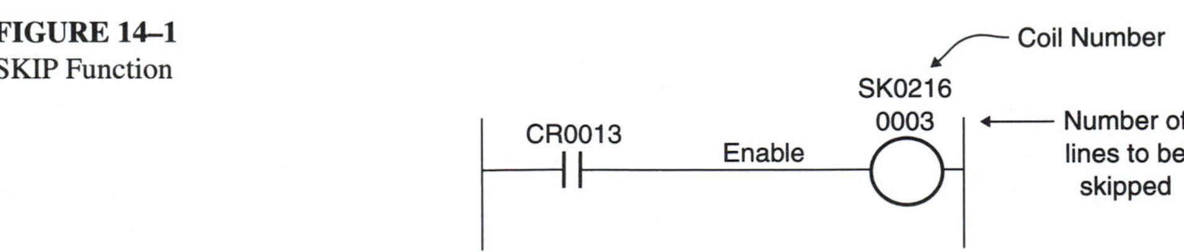

FIGURE 14–1
SKIP Function

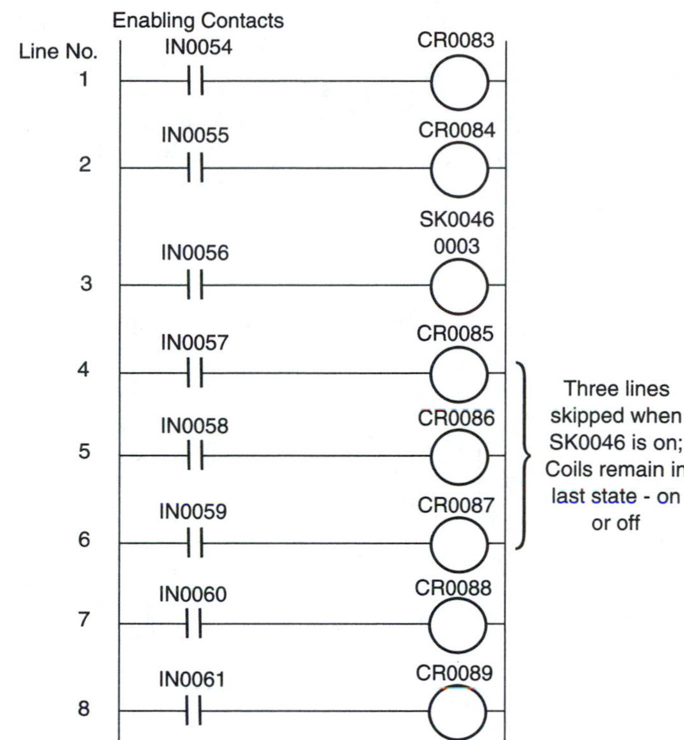

FIGURE 14–2
SKIP Function PLC Operation

tion's operation. Lines 7 and 8 could also be skipped if we had inserted a 5 in place of 3 as the number of lines to be skipped by the function. When SKIP is turned off, the ladder will operate normally.

For an illustration of the SKIP function, we apply it to a production line. The production line has eight stations, each of which can perform an assembly operation as the product comes down the line. Depending on the individual part number, all of the eight operations may or may not be set up and carried out. The pattern of whether or not the operations are to be set up and performed is stored in registers, as explained in detail in chapter 18. Each of the eight stations is set up to operate or not, according to register bit statuses. A bit of 1 says turn the setup on, and a 0 says turn the setup off.

At the third station, an inspection takes place. If the part is good, it continues down the operating line; if it is bad, it is shunted to a side conveyor and repaired. After repair, the part reenters at the beginning of the conveyor. The product flow and conveyor layout are shown in figure 14–3.

When a part arrives at the beginning of the line, a sensor (not shown) detects the presence of the parts at the beginning of the conveyor and causes the eight stations to be set up for operation. The sensor causes register contents to turn each of the eight setup switch contacts (the BP/IR contacts) on or off. Figure 14–4 shows this setup system on the left. The setup functions are CR0041 through CR0048. As the part proceeds down the conveyor,

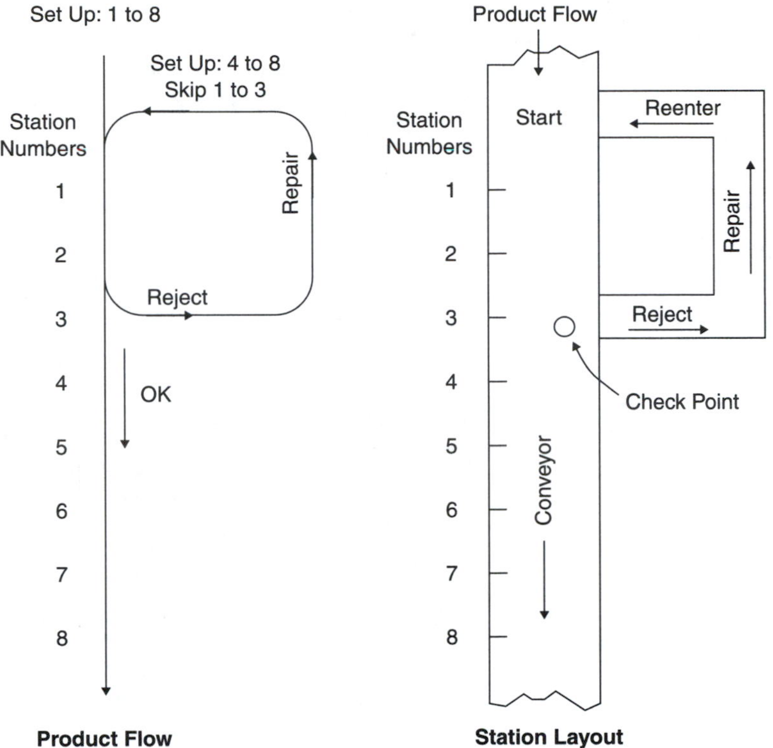

FIGURE 14–3
SKIP Function Application Layout

each operation is performed (if set up) when the part is detected by sensors at each station. These sensors are IN0021 through IN0028, as shown on the right of figure 14–4. The operations are CR0061 through CR0068.

If a part is rejected at station 3, it is shunted to repair. Later, when the repaired part reenters the conveyor, the setups of stations 1 through 3 do not have to be reset. Unnecessary reset is prevented by the two SK functions, 0011 and 0060. The two SK functions are turned on by a sensor at the repair reentry point.

14–3 THE MASTER CONTROL RELAY FUNCTION AND APPLICATIONS

The MASTER CONTROL RELAY (MCR) function operation is similar to the SK function. Figure 14–5 shows a typical MCR function. When its enable line is energized, it turns on. When MCR is off, the number of following ladder diagram lines specified are turned

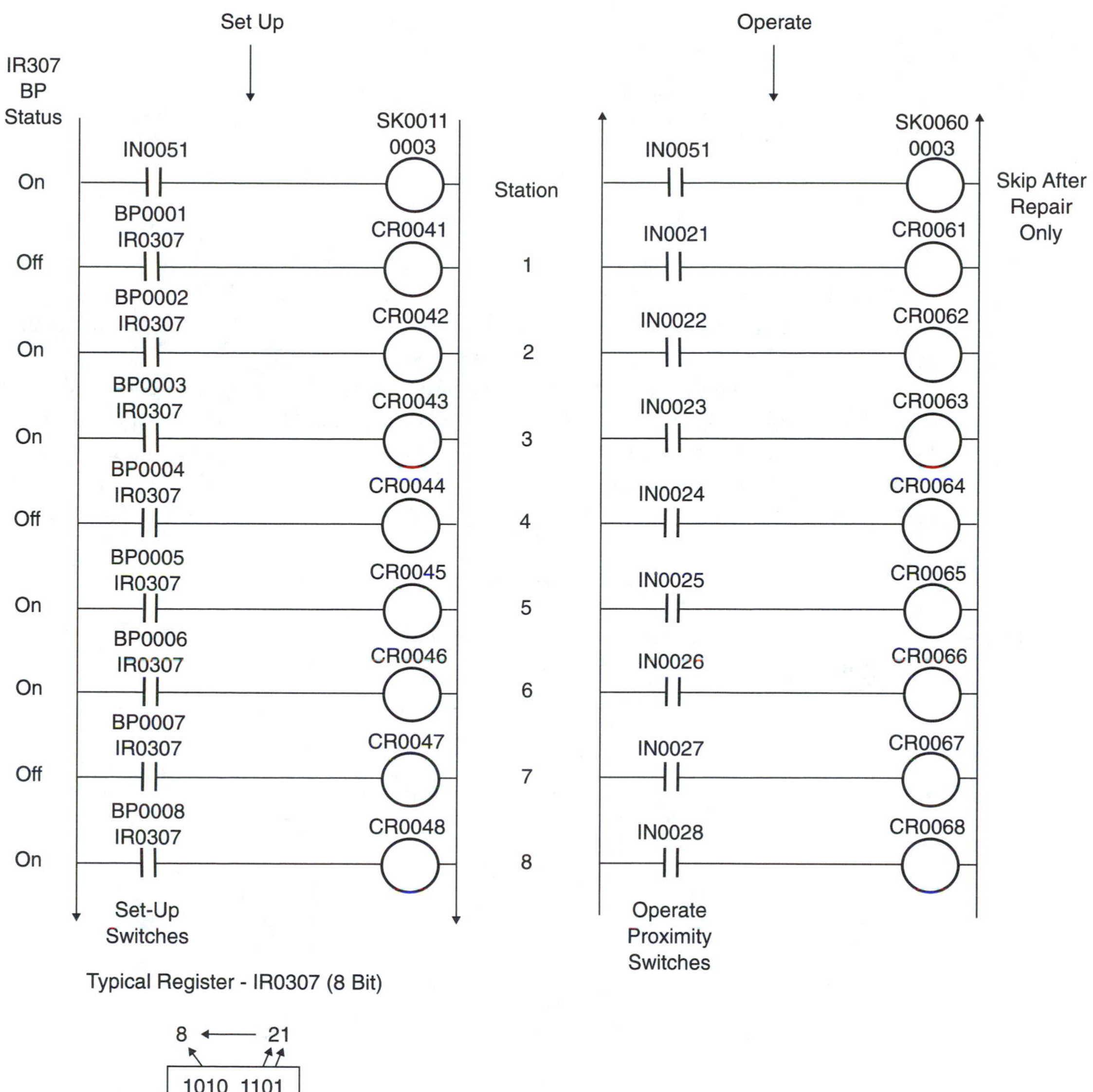

FIGURE 14–4
SKIP Application Program

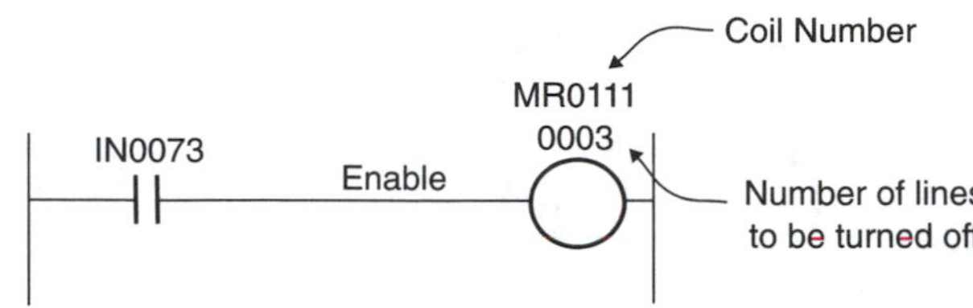

FIGURE 14–5
MCR Function

off. In contrast to the SK operation, where lines were skipped, the MCR turns the following specified number of lines to the off state. In many PLCs, the MCR function turns off only the nonretentive outputs in the zone. All retentive functions are unaffected under this system.

Figure 14–6 shows how the MCR function operates in a program. There are eight lines. The third line is the MCR function. The other seven lines are contact-coil functions. For fail-safe reasons, the MCR must be turned on to be inactive. If the function goes off for some reason, it is active and turns the specified lines off, also. When MCR is on, the other seven lines operate normally. When MCR is off, the next three lines, 4 through 6, are turned

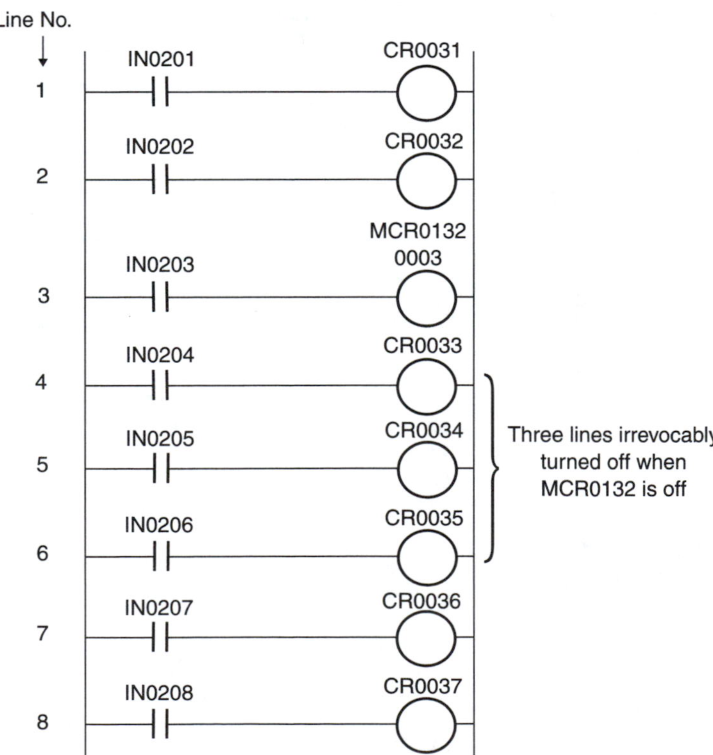

FIGURE 14–6
MCR Function PLC Operation

off. Lines 1, 2, 7, and 8 are unaffected. With MCR off, there is no way to turn on coils 4 through 6 by energizing their enable lines. When MCR is turned on, the ladder operates in the normal manner.

A production line example similar to the SK example will be used for the MCR application illustration. There are again eight production stations. Whether each station operates for a given part number as the part goes past depends on the setup (not shown). Each station's operation is initiated by proximity switches at each station. The proximity switches are IN0081 through IN0088. Figure 14–7 shows the production line layout and product flow.

Station 5 is an inspection station. Rejected parts are shunted to a repair conveyor. After repair, the part reenters the conveyor. When it reenters, it turns on IN0011, which turns on and seals an MCR relay, CR0021. The first five steps are therefore not repeated for the part, because the first five operations are prevented by the MCR. When the part gets to station 5, the MCR is unsealed, enabling stations 6 through 8 to function. These last three steps were not performed the first time through, but are now performed to complete the process. The MCR program for these operations is shown in figure 14–8.

FIGURE 14–7
MCR Application Layout

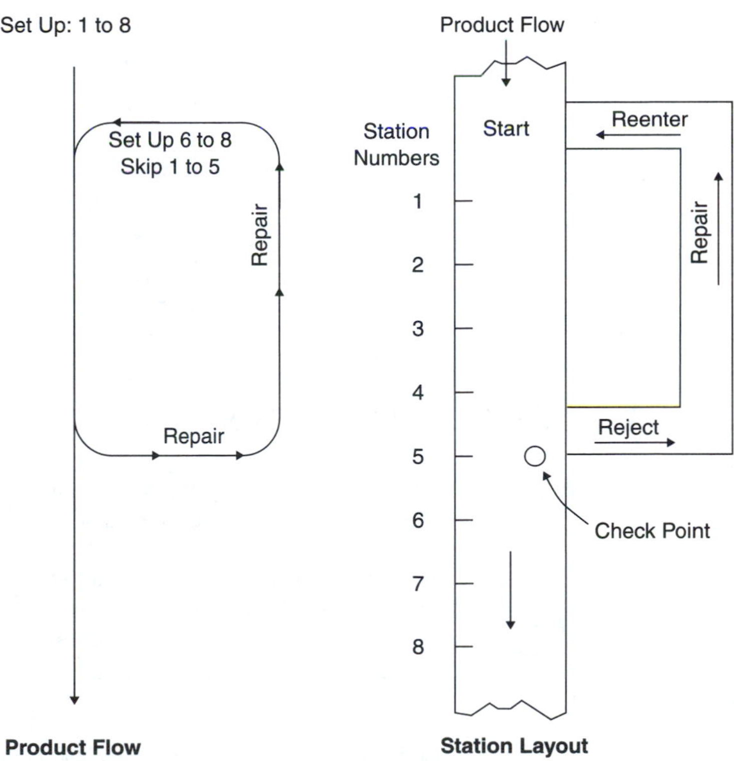

FIGURE 14–8
MCR Application Program

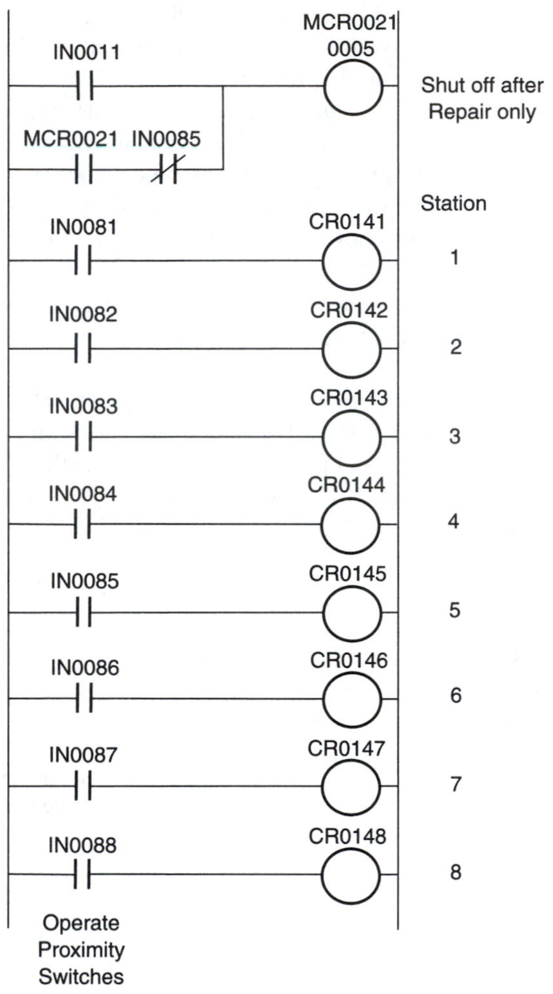

TROUBLESHOOTING PROBLEMS

TS 14–1 You have set up a SKIP program similar to that shown in figure 14–2. The three lines are skipped when the program is first turned on. What could be wrong?

TS 14–2 You have set up an MCR program similar to that shown in figure 14–6. The three lines are still on when the program is turned on. What could be wrong?

EXERCISES

1. For the 12-ladder line program in figure 14–9, insert three SK functions one at a time. The first problem is to skip lines 3 through 5. The second problem is to skip lines 8 and 9. The third is to skip lines 3 through 11. Program them and check out their operation. When the SK function is on,

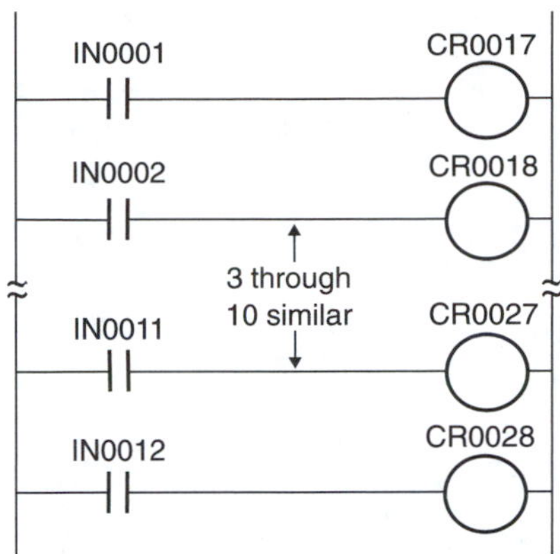

FIGURE 14–9
Diagram for Exercise 1

changing input status should not affect the previous status of outputs for the lines to be skipped. As an added problem, insert the line 8 and 9 skip and also the line 3 through 11 skip at the same time. What happens when either one or both are turned on?

2. Repeat exercise 1 using the MCR function instead of the SK function. When MCR is on, all MCR-designated lines should be off. Additionally, changing any control line's input status for MCR-controlled lines should have no effect on its output. Repeat the added problem in exercise 1, using MCR instead of SK.

3. Devise an MCR system to control the assembly line shown in figure 14–10. All 15 stations are to function as set up by one of two registers. Short and tall parts are sent down the line. Short parts get all 15 operations, if specified. Lines 6 through 9 are omitted for the tall parts only. Therefore, have operations 6 through 9 turned off by an MCR function when a tall parts goes by. Tall parts are detected by a limit switch just after station 5. After station 9, another limit switch unseals the MCR function, so the next part, whether large or small, is again set up for all operations.

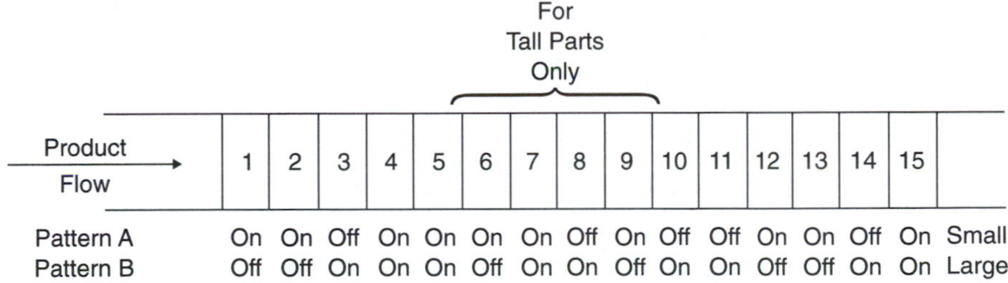

FIGURE 14–10
Diagram for Exercise 3

15

Jump Functions

OUTLINE

15–1 Introduction □ **15–2** Jump with Nonreturn □ **15–3** Jump with Return

OBJECTIVES

At the end of this chapter, you will be able to

- □ Explain the advantages of a jump instruction.
- □ Describe how a jump with nonreturn instruction works.
- □ Describe how a jump with return instruction works.
- □ Explain how subroutine nesting works.

15–1 INTRODUCTION

As with data processing computers, so with process control computers and PLCs, the jump programming function is quite useful. Simply put, the jump instruction is a command in a computer program that causes the sequence to go to, or branch to, a specified point other than the next line in the program sequence.

Two types of jump instructions are examined in this chapter. The first, the *jump with nonreturn*, is similar to the skip function in that it "leaps over" a certain portion of the main program when called upon to do so. The *jump with return* instruction, on the other hand, leaps to a subroutine when activated. It then returns from the subroutine to the main program.

Jump instructions reduce scanning time, programming effort, and memory space. Let's see how.

15–2 JUMP WITH NONRETURN

A jump with nonreturn instruction allows us to skip, or branch, to a different portion of the program, usually further down. Thus the scan leaps over a portion of the main program. All outputs in the skipped portion are frozen in their last state. This is because their inputs are not being scanned, they're being skipped.

The basic jump operation is illustrated in figure 15–1. Program scanning takes place from top to bottom, as usual. When a jump instruction is encountered, the scan leaps to the jump destination. Main program scanning continues from there.

For a jump to take place there must be a place to jump from and a place to jump to. With Allen-Bradley programmable controllers, a jump (JMP) and label (LBL) instructions go together. The former is placed in the rung where the leap is to occur; the latter is the target of the leap. Thus a jump jumps to a label. Both the jump and label must have the same address.

Figure 15–2 shows the nonreturn JUMP function. If input 121 (line 17) is on (true), the program will jump to the next line where an LBL is found (line 22). Since lines 18 through 21 are not scanned by the processor, input conditions are not examined and outputs controlled by these rungs remain in their last state. If input 121 is off (false), the program continues directly to line 18.

It is possible to jump to the same label from multiple jump locations. The concept is illustrated in figure 15–3. Here we see two jump instructions numbered 04. The single label is numbered 04, of course. The program scan will jump from either jump instruction to label 04, depending on which input, 02 or 03, is true. If input 02 is true, a jump to label 04 occurs. If input 02 is false, the program scan proceeds to input 03 and continues. If when the program encounters input 24 (line 81) and that input is true, a jump to label 04 takes place. If input 24 is false, however, program scanning continues to line 82.

15–3 JUMP WITH RETURN

Often, with PLC control, a machine or process goes through many repetitive subprocesses. For example, various specific time delays may be needed as parts are manufactured and assembled. Suppose that a process requires five 4-second delays in the assembly of an item.

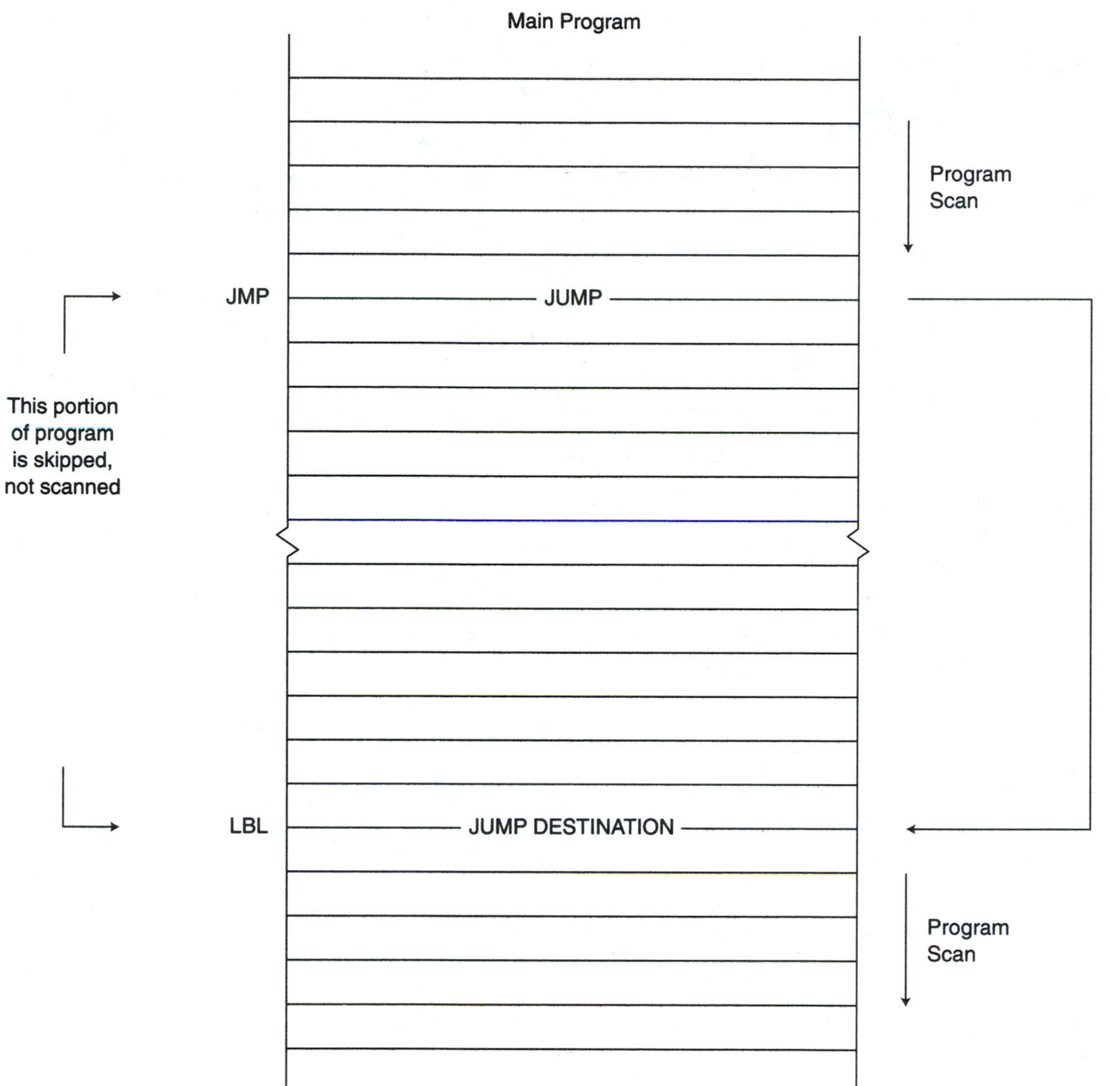

FIGURE 15–1
Basic Jump Operation

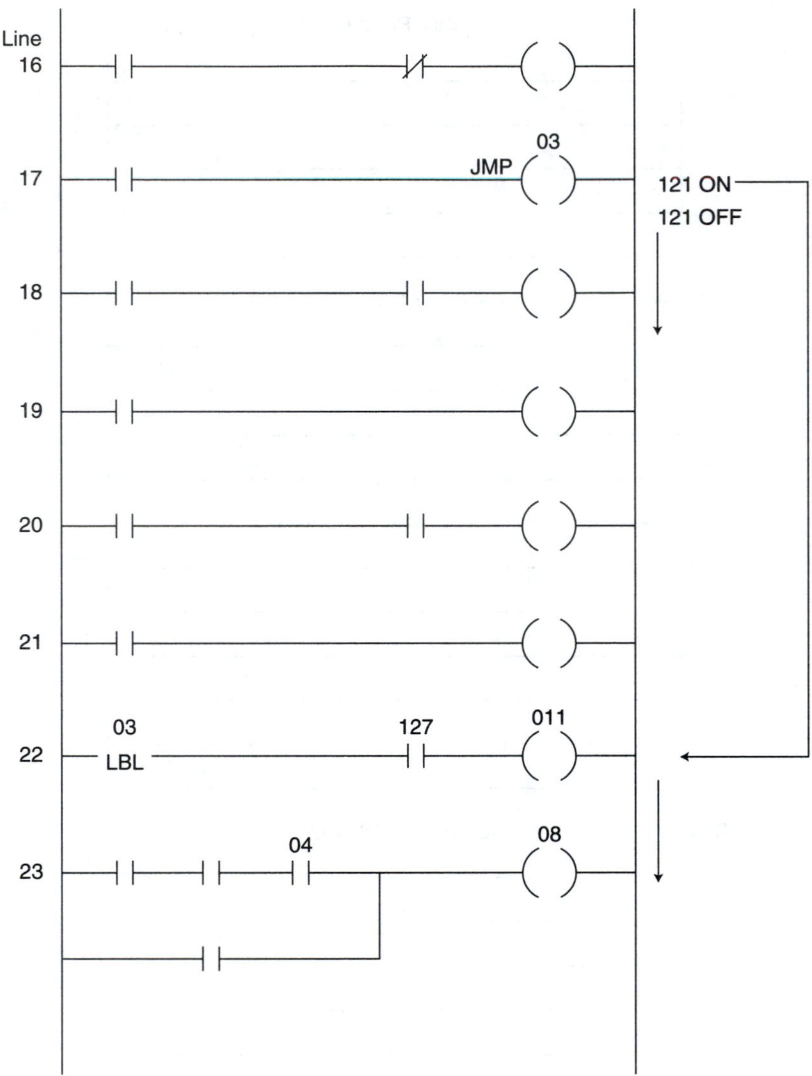

FIGURE 15–2
Jump Function

FIGURE 15–3
Jump-to-Label from Multiple Locations

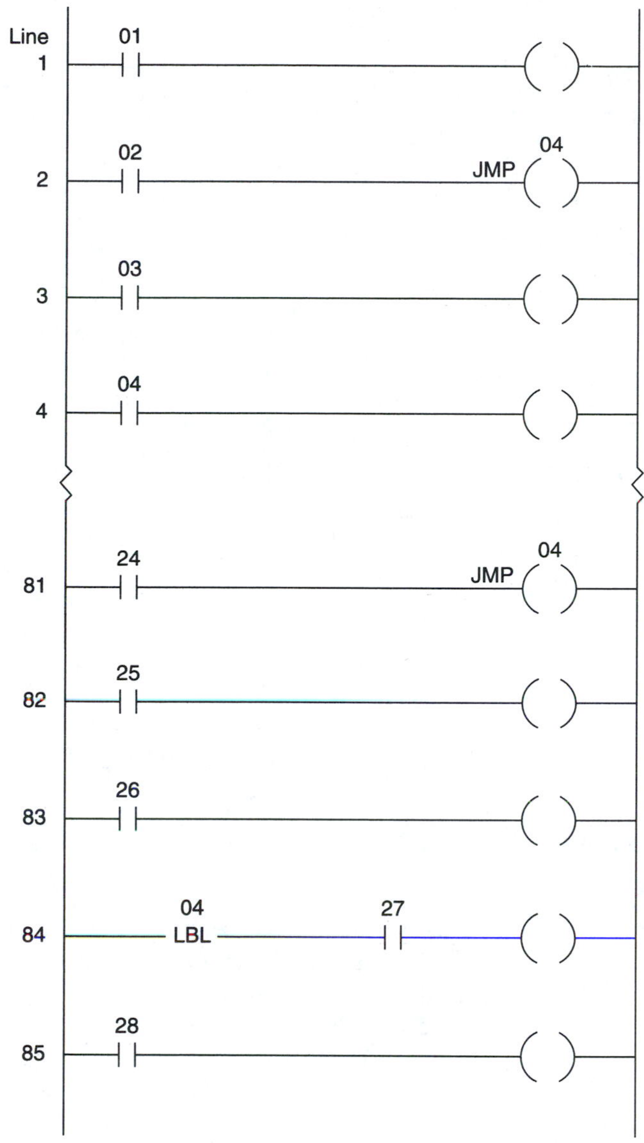

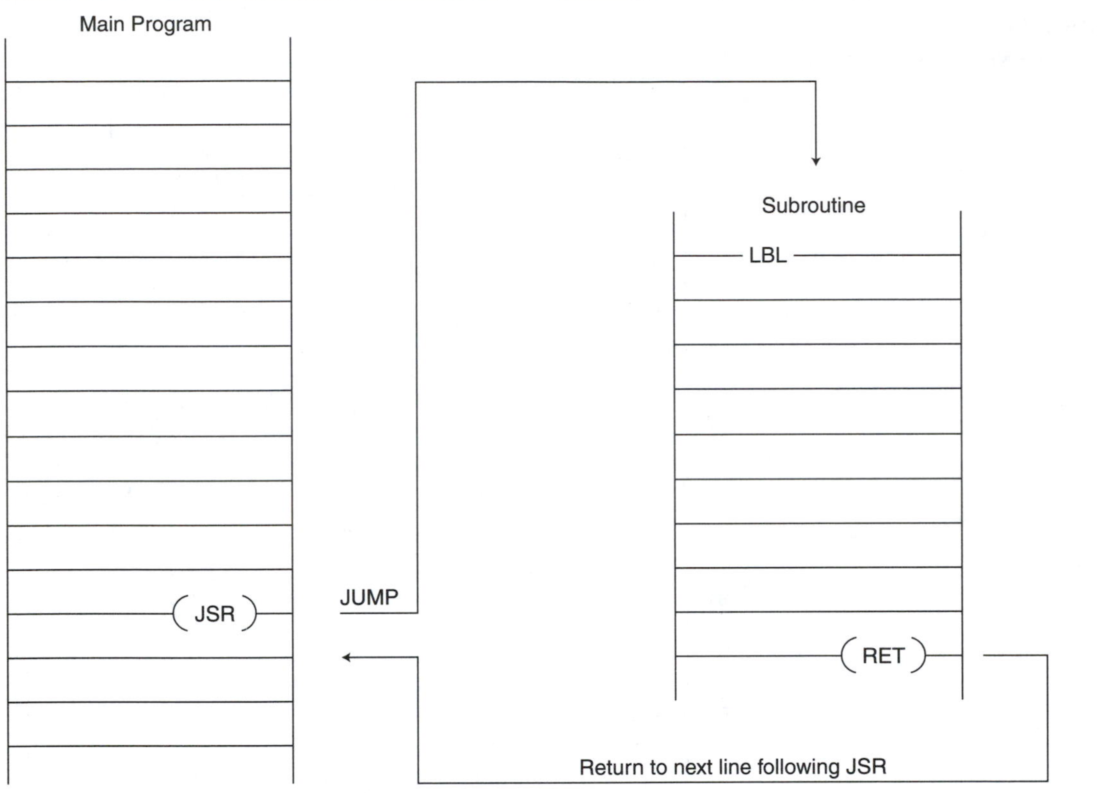

FIGURE 15–4
Jump-to-Subroutine

To write, at various points in the main program, five identical time delays would be a waste of effort and PLC memory. Better to write code for the time delay once and then call the code up every time it is needed. In our example, that would be four times. Such programming is easily accomplished with the use of subroutines.

A subroutine is a group of instructions written separately from the main program to perform a function that occurs repeatedly in the main program. Thus, the subroutine code is outside the main program, residing elsewhere in memory. In figure 15–4 we see the basic jump-to-subroutine operation. In PLC terms, this is often known as a jump with return.

Note in figure 15–4 that a jump-to-subroutine consists of a *call* operation and a *return* operation. When the rung containing the jump-to-subroutine instruction (JSR) is true, the subroutine is called. Where is the subroutine? At the label (LBL) in the subroutine area (see figure 15–5). Note that the JSR and LBL addresses are the same. When the subroutine is completed, an unconditional return to the main program must take place. The return is always to the rung following the JSR in the main program (figures 15–4 and 15–5).

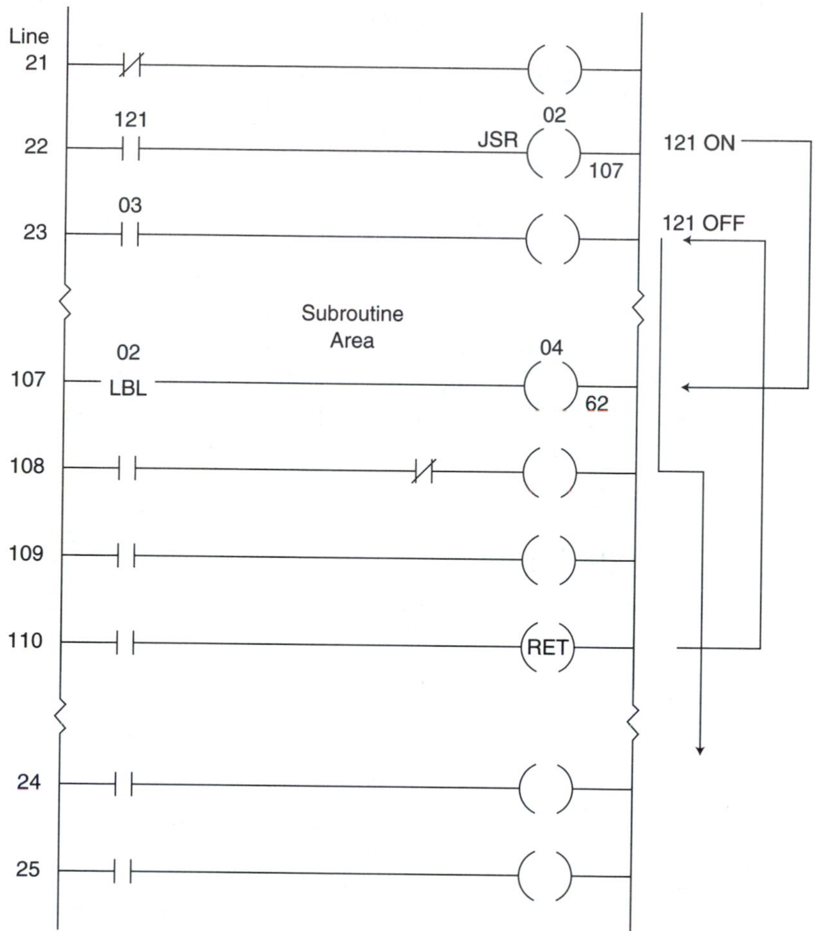

FIGURE 15–5
Jump-to-Subroutine Function

Referring again to figure 15–5, lines 21 through 23 are part of the main program. When input 121 is true, a jump to a subroutine at LBL, on line 107, takes place. Lines 107 through 110 contain the subroutine. When the return command (RET) is encountered, the subroutine returns to line 23 in the main program.

In figure 15–6 we see a water tank. Water enters at the top and is drained at the lower right. A water sensor, positioned near the top of the tank, detects when water reaches a dangerously high level. At that point an alarm light begins flashing and a relief valve opens.

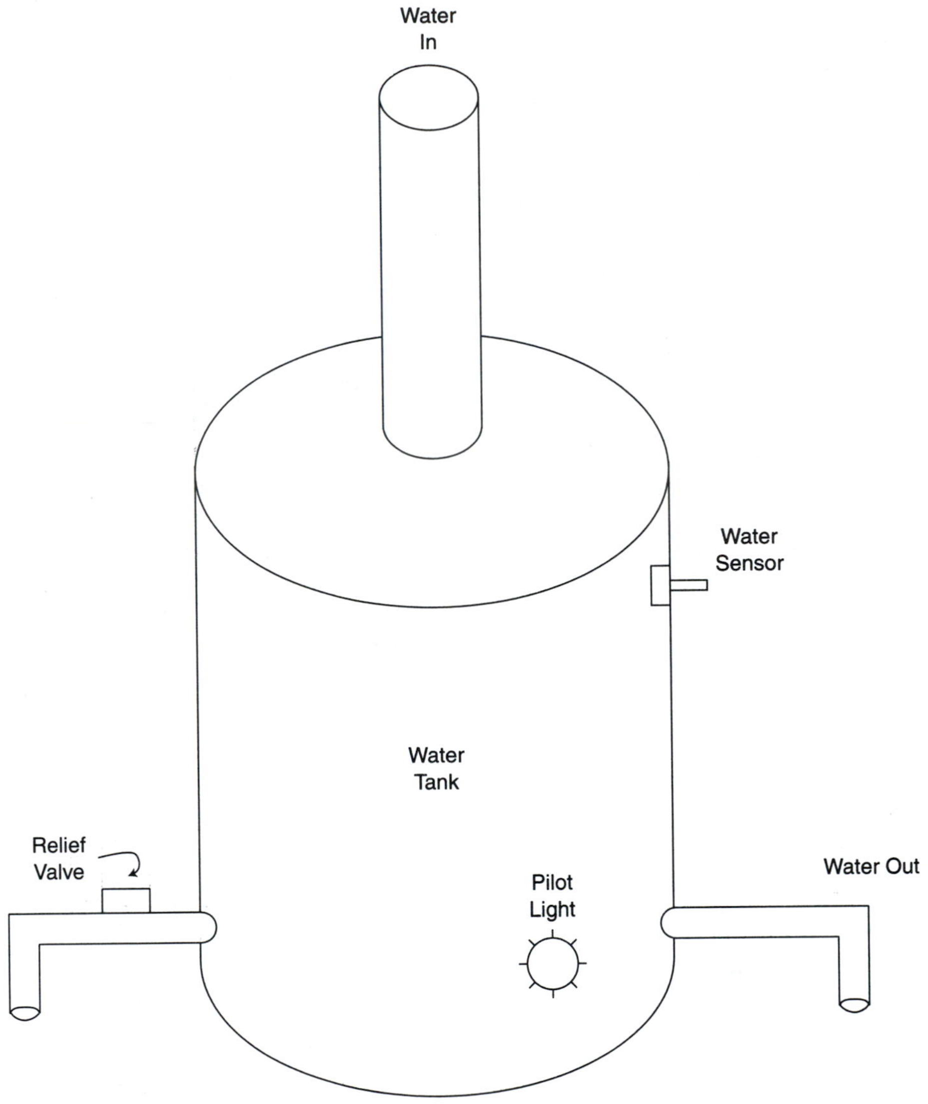

FIGURE 15–6
Flashing Pilot Light Subroutine

FIGURE 15–7
Ladder Logic Circuit with a Flashing Alarm as a Subroutine

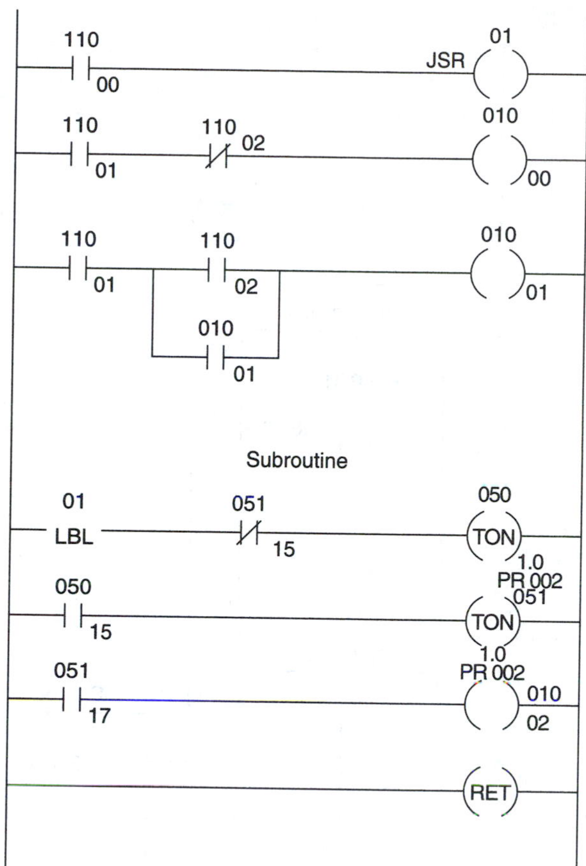

Figure 15–7 shows a ladder logic circuit with a flashing alarm as a subroutine. If the water sensor is triggered, contact 110 00 closes, the JSR is activated, and the program jumps to the subroutine area. The subroutine is continually scanned and the alarm output (010 02) flashes until contact 110 00 opens. When that happens, the subroutine is no longer called, and the processor continues scanning the main program.

Finally, subroutines are often *nested.* This means having a subroutine within a subroutine within a subroutine, etc. (see figure 15–8). Such subroutine nesting results in faster program operation since the program does not continually have to return from one subroutine to enter another.

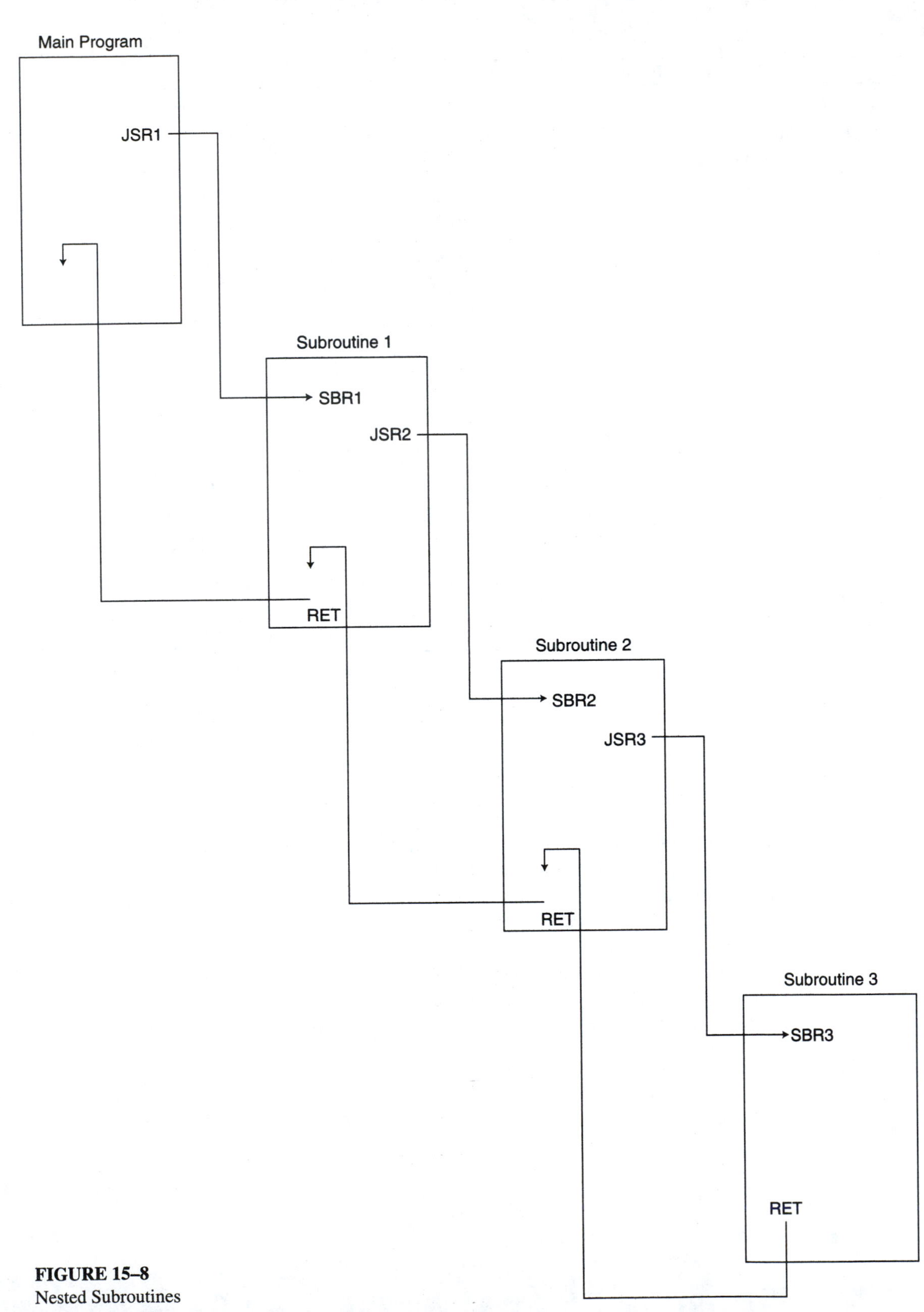

FIGURE 15–8
Nested Subroutines

FIGURE 15–9
Diagram for exercise 2

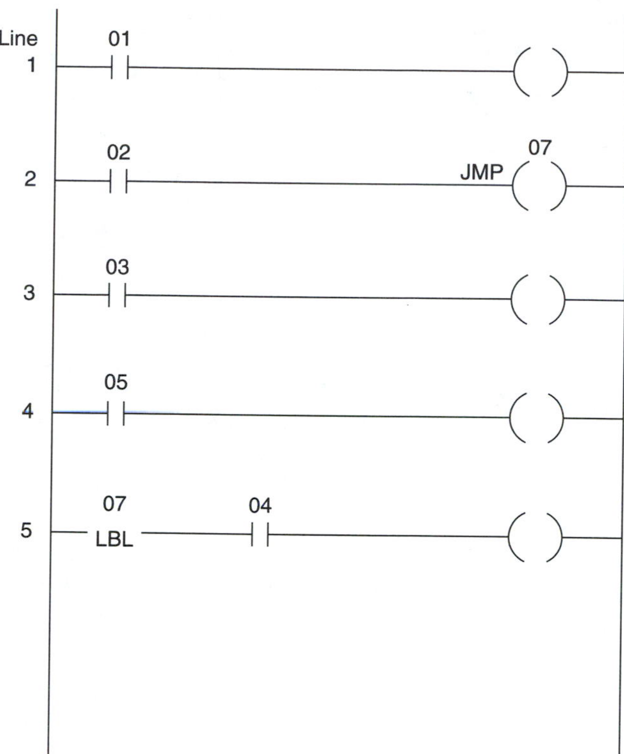

EXERCISES

1. Explain how JUMP differs from the SKIP and MCR functions.
2. Explain what happens in the ladder logic program shown in figure 15–9.
3. Explain what happens in the ladder logic program shown in figure 15–10.
4. Illustrate a process requiring a jump with nonreturn instruction and draw the ladder logic diagram.
5. Illustrate a process requiring a jump with return instruction and draw the ladder logic diagram.
6. Describe a process requiring the nesting of two subroutines.

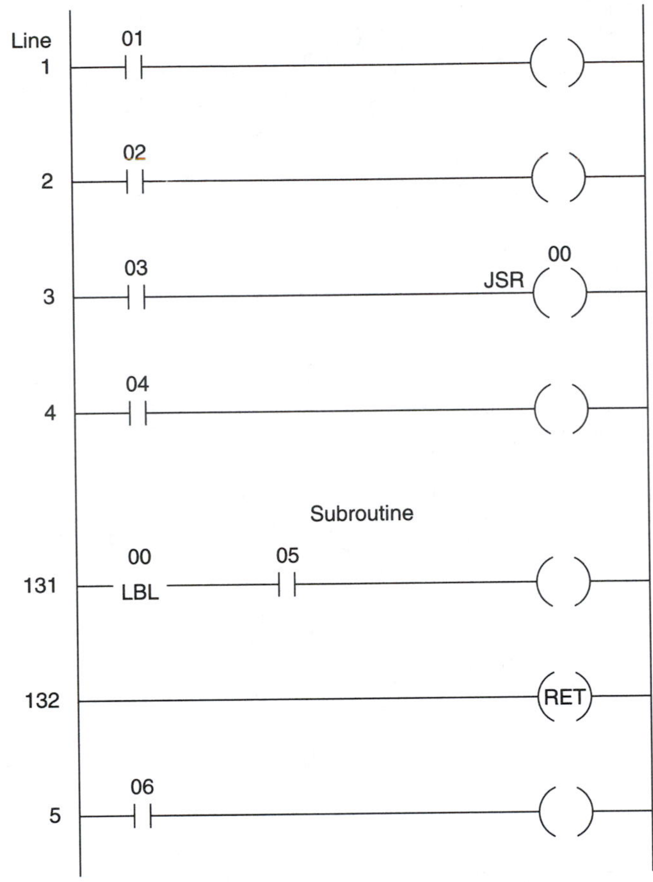

FIGURE 15–10
Diagram for exercise 3

16

PLC Data Move Systems

OUTLINE

OBJECTIVES

At the end of this chapter, you will be able to

- □ Describe the MOVE function.
- □ Apply the MOVE function to industrial problems in combination with other PLC functions.
- □ Describe the BLOCK MOVE function.
- □ Apply the BLOCK MOVE function to industrial problems in combination with other PLC functions.
- □ Describe the TABLE-TO-REGISTER MOVE (TR) and REGISTER-TO-TABLE MOVE (RT) functions.
- □ Apply the TR and RT functions.
- □ Describe other PLC MOVE functions.

16–1 INTRODUCTION

All computer systems, including PLCs, have the internal ability to move data, numbers, and bits from one location to another during computer operation. In smaller PLCs, the moving of data from one register to another internally is carried out automatically. The movement of data takes place, but we cannot monitor or control its operation.

In medium-size and larger PLCs, programming functions are available to control data movement. The PLC's programmable data moves data from one place and move it to another. This chapter explains three types of these data moving programming functions. First, the chapter covers the basic system of moving one register's contents into one other register. The basic MOVE function takes a word, byte, or group bit pattern from one place and moves it to another. Some PLC systems use a GET/PUT format instead of MOVE. We use MOVE.

The second type of PLC data move involves moving groups of data from two or more consecutive registers to two or more other consecutive registers. This second type is usually designated as BLOCK MOVE in PLCs. It moves a consecutive group of registers' data patterns to another consecutive group of registers. Some PLC formats call this FILE-TO-FILE. We use BLOCK MOVE.

The third type of data move involves two subtypes. One type sequentially moves data from designated group registers into a single register. This is called a TABLE-TO-REGISTER move. Some PLC formats call this FILE-TO-WORD. We use TABLE-TO-REGISTER. The other subtype takes the data value from a single register and moves its value (which is normally varying) sequentially to a portion of a table. The sequential moves are often at designated time intervals. This is called the REGISTER-TO-TABLE or FILE-TO-WORD function. We use REGISTER-TO-TABLE.

In all moves, the contents of the original source register are retained. You essentially then duplicate the source register's value in another register. Conversely, the destination register, which receives the duplicated new data, loses its previous value. In other words, the original value in the receiving register before the move is normally lost. If you wish to keep its original value for reference, additional programming is needed to duplicate and store it elsewhere before the move.

In section 16–5 a number of additional move functions are illustrated by manufacturer.

16–2 THE PLC MOVE FUNCTION AND APPLICATIONS

Figure 16–1 shows the elements of a MOVE function in block and in coil format. We discuss only the block format in this chapter. When the function is turned on through the enable circuit, the bit pattern from the specified source register is duplicated in the specified destination register. The source register is unchanged. The destination register pattern is replaced and lost when the new value is brought in. The function coil goes on when the MOVE function is completed. The coil operation can be used to interlock MOVE functions when there is more than one MOVE function in the program. Interlocking prevents energizing two or more contradictory moves when only one is desired. For example, two moves to the same register cannot take place at the same time.

FIGURE 16–1
The MOVE Function

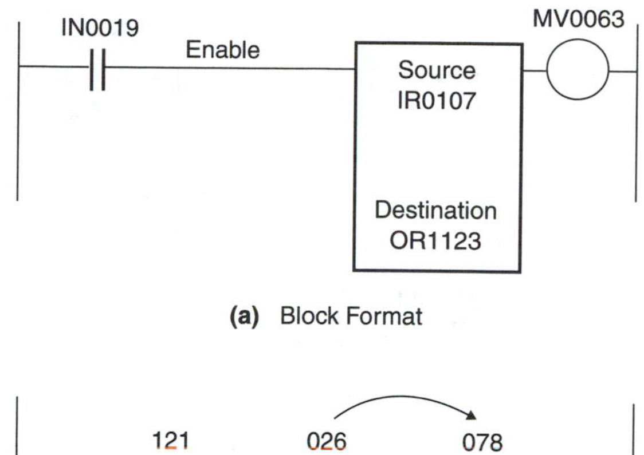

(a) Block Format

121 026 078
G PUT
10 16 16

When 121 is on, Value in location 026 (16) is duplicated (effectively transferred) to location 078.

(b) Coil Format

The types of registers that are accessible by moves vary among PLC models. Source locations are generally input registers, output registers, holding registers, or internal registers. Sources also include output and input group registers. Destination locations are the same, except that input and output register groups are usually not available as destinations.

In chapter 9 the timer program examples all use constant value numbers for time intervals. In PLCs it is possible to use a register's numerical content for the timing interval. For this situation, the numerical value of the holding register is specified as the time interval in the timer functional block.

It is often necessary to quickly change the time interval of a process timer. We can accomplish the change by varying the numerical value of the timer's time-interval holding register with the MOVE function.

Figure 16–2 shows an application of the MOVE function where either of two values of time may be moved into a PLC timer. The two different times are 7 and 15 seconds (shown in binary). Either time may be entered into the timer by closing input switch IN0007 or IN0009.

We will give a second example of single register moves, this time using an addition process. Suppose that the PLC ADD function can operate only with holding registers as operands. Suppose, also, that the numbers to be added enter the PLC into input registers, not holding registers. Furthermore, the result of the addition is to go out of the PLC into an output register. PLC data moves are therefore required to accomplish the total register configuration.

Figure 16–3 shows a register designation problem in block diagram form. The two numbers to be added enter the PLC in input registers. The data must be transferred to holding registers before addition can take place. After the addition is made, the result is located

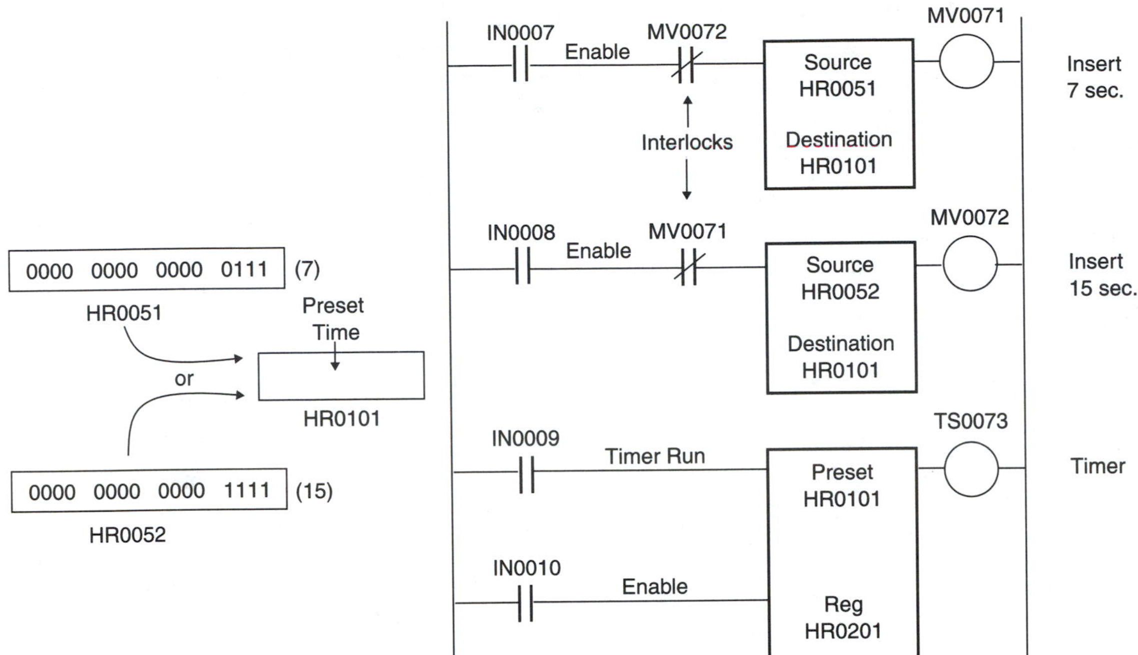

FIGURE 16–2
MOVE Timing Example

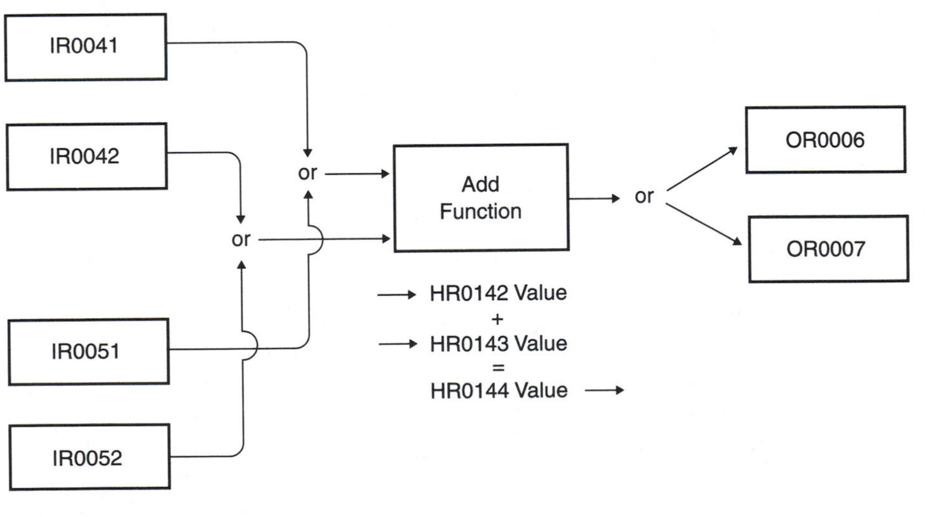

FIGURE 16–3
MOVE—Addition System

in a holding register. The output of the addition is required to appear in an output register. Another MOVE transfer is then required at the output.

Figure 16–4 illustrates how the PLC would be programmed to accomplish the total addition process of this example. IN0007 closure causes one group of numbers to be placed in the ADD function. IN0008 moves in other numbers. IN0010 causes the addition to take place. IN0017 or IN0018 causes the resulting number to be sent to either of two designated outputs.

FIGURE 16–4
MOVE—Addition Program

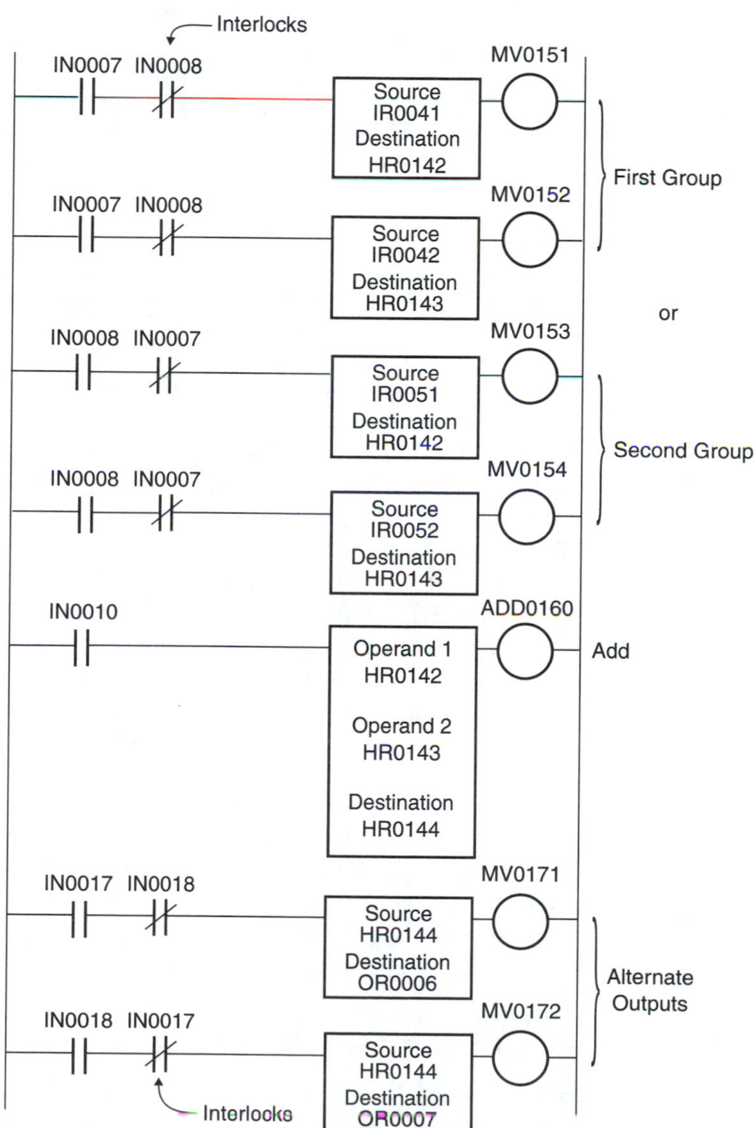

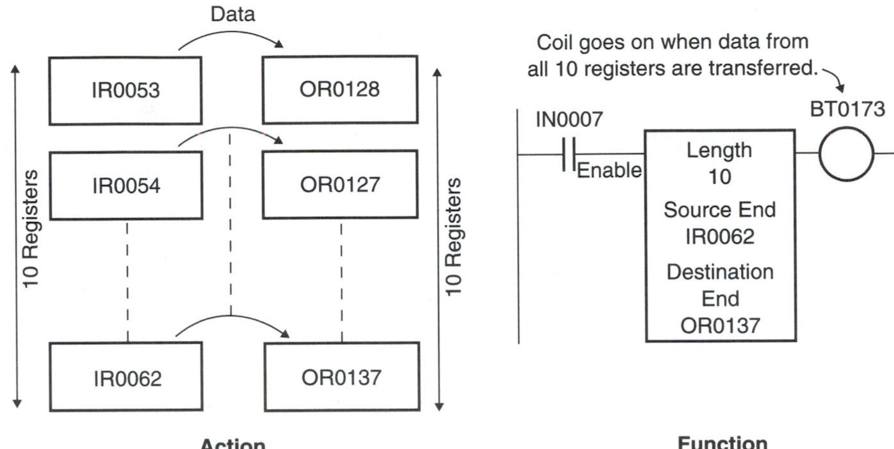

FIGURE 16–5
BLOCK TRANSFER Function

16–3 MOVING LARGE BLOCKS OF PLC DATA

It is sometimes necessary to move more data than the quantity that will fit into one address or register. One option is to use a number of individual MOVE functions. A better solution is to use one PLC function that will move many consecutive registers' data at once. These are called BLOCK TRANSFER (BT) functions. Suppose we need to move 147 bits from one location to another but have only 16-bit registers available. We would need to use nine full 16-bit registers (144 bits) plus part of another (3 bits). In this case, one BLOCK TRANSFER function does the work of ten MOVE functions.

Figure 16–5 illustrates the BLOCK TRANSFER function. Some PLCs call this function a TABLE-TO-TABLE function. In the functional block, specify the number of registers to be moved. Also specify the last register of the input sequence from which the data comes. Some PLC systems require the programmer to specify the first register instead of the last; this example uses the last-register system. Finally, specify the last register of the destination register sequence where the data is to be delivered. As usual, the enable input causes the data to be transferred.

The output coil comes on when the transfer of all registers is complete. The coil operation can be used to verify that the new data pattern in the output registers is complete. BLOCK TRANSFER uses up scan time in proportion to its size; a large amount of computer time can elapse for the BT operation. If the output data were to be utilized in the middle of the transfer, both old data and new data would be in the receiving registers, which could produce some operational problems.

16–4 PLC TABLE AND REGISTER MOVES

In the two-move functions described previously we first moved data from one register to another register. Then we moved data from one consecutive register group to another consecutive register group of equal length. A third type of move involves tables and a single register. The TABLE-TO-REGISTER (TR) function moves data sequentially from a specified portion of a large listing of data to a single register. Conversely, the REGISTER-TO-TABLE (RT) function moves data sequentially out of a single register into a specified portion of a table of registers.

Figure 16–6 shows in block diagram form how the TR function moves data. In a typical application, the receiving register operates a number of machines by bit picking, which is described completely in chapter 18. As different table register patterns are moved into the receiving register, the machines' on-off patterns will change.

A typical PLC function used to accomplish TR moves is shown in figure 16–7. It operates similarly to other MOVE functions. The TR function is first programmed for table length, which is the number of registers to be sequentially inputted. The second line in the TR function is the point at which the table transfer operation is to end. For example, the specified table length is 14 and the specified end register is IR 0058. The table of registers utilized runs from IR 0045 through IR 0058.

The third programmed input to the block is the pointer location, found in many, but not all, PLCs' RT and TR functions. The pointer is used to point to the register being moved at any given moment. The pointer location can be used for information only or for control

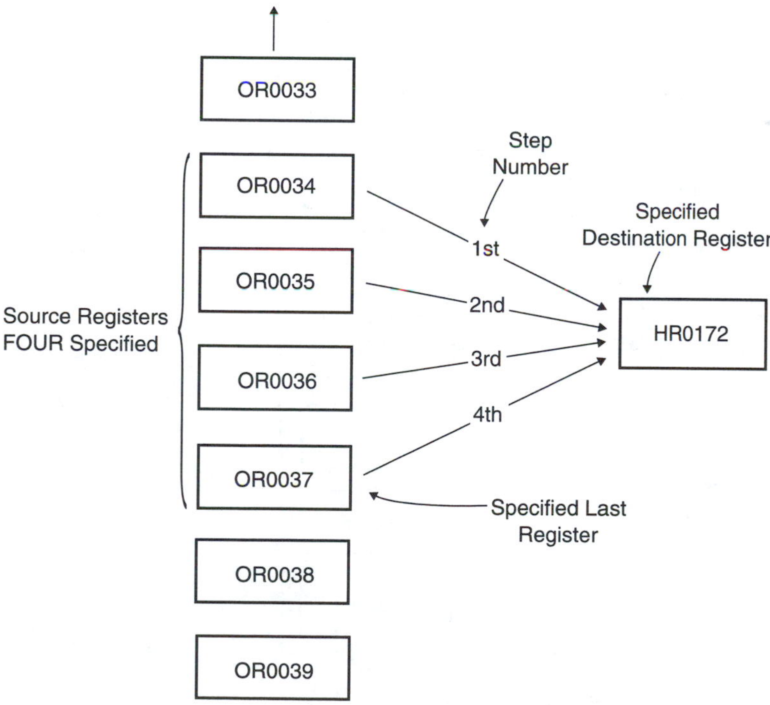

FIGURE 16–6
TABLE-TO-REGISTER MOVE System

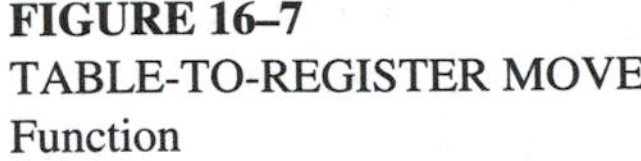

FIGURE 16–7
TABLE-TO-REGISTER MOVE Function

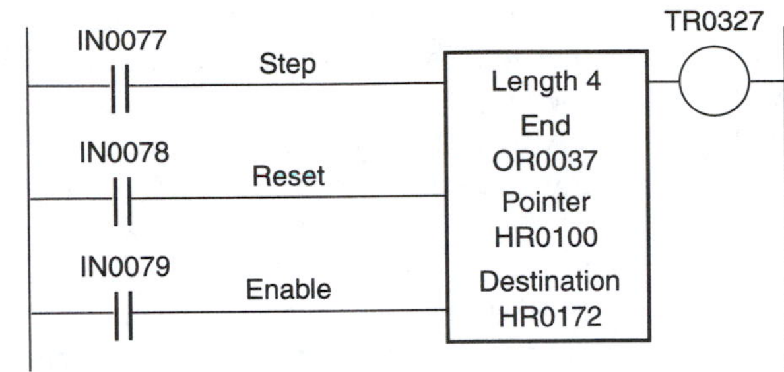

purposes. Finally, a single register destination for the successive data values is specified as a fourth input to the functional block.

The function is enabled when the lower input line is on. When the middle line, Reset, is off or turned off, the function is reset to the first register. The operational pointer, if included in the function, is also set at the first register when Reset is off. When reset is on, the function is operational and can be stepped. The top line is the step line. Whenever the step line is turned on, the function transfers data and moves down one register. To step again, the top step line must be turned off and back on. For time interval operation through

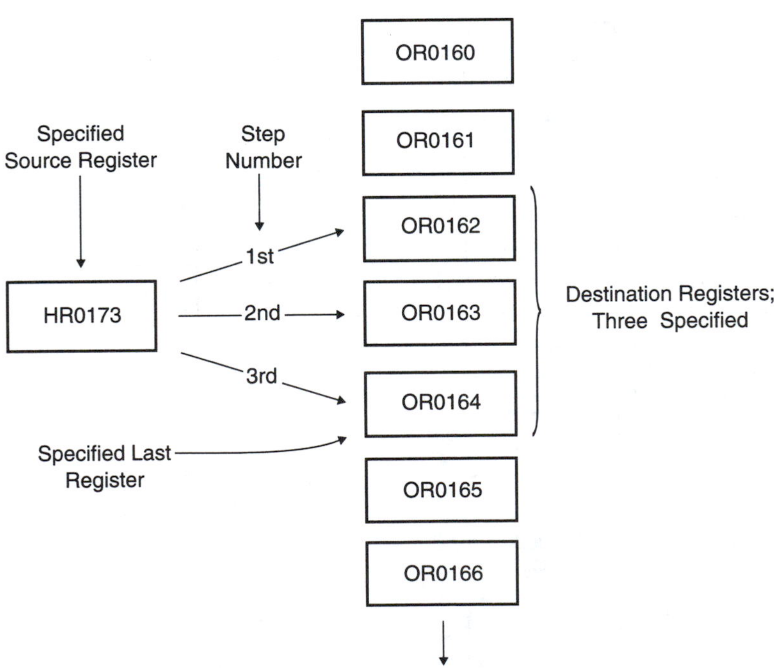

FIGURE 16–8
REGISTER-TO-TABLE MOVE System

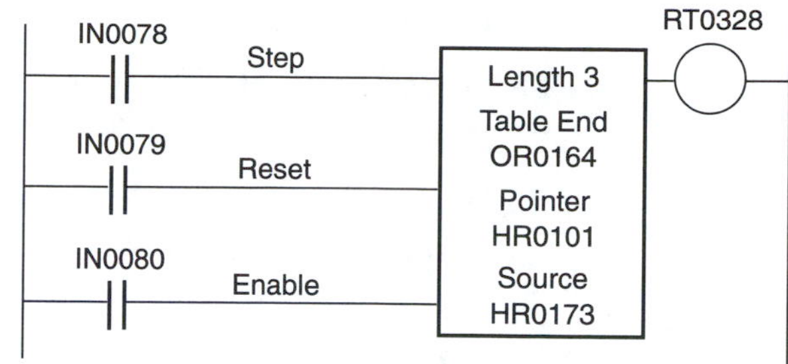

FIGURE 16–9
REGISTER-TO-TABLE MOVE Function

the table, a timer can be used to do the stepping. A timer contact would be used as the step contact for the function. In Chapter 19 we discuss a more advanced form of the TR function, the SEQUENCER function.

The REGISTER-TO-TABLE (RT) function is similar to the TR function. It moves data from a single register sequentially into a specified number of consecutive registers. A block diagram of an RT move is shown in figure 16–8.

Programming RT moves is similar to programming TR moves. The input lines operate in the same manner. The RT function is shown in figure 16–9. Table length denotes how many destination registers are to be used. The table end is the last destination register to be used. The pointer in the RT function operates similarly to the pointer in the TR function. The source in the RT function specifies the one register from which the data is to come.

An RT application might be the periodic recording of data. A single register could be programmed to indicate the value of a varying process parameter. The single register's value is constantly changing as the process changes. This register is used as the source. To record its value every 10 seconds for 5 minutes, we need 6 times 5, or 30, registers to record the required sequential readings. The function's step line is pulsed every 10 seconds. The table destination length needs to be 30 registers. The 10-second interval results are then recorded in order. The 30 sequential values appear in order in the specified series of 30 destination registers.

16–5 OTHER PLC MOVE FUNCTIONS

There are a number of other special-purpose MOVE functions in various PLC models. Some of these are described below, classified by manufacturer.

Omeron

MOVE NOT. Inverts a channel data or a four-digit constant and transfers it to a specified channel.

Cutler-Hammer

FMOV. Moves data on a repeat data basis to a series of destinations.

Allen-Bradley

BTR: Block Transfer Read. When the input rung goes true, the BTR instruction tells the processor to read data from the rack/group/module address and store it in the data file.

BTW: Block Transfer Write. When the input rung goes true, the BTW instruction tells the processor to write data stored in the data file to the specified rack/group/module address.

Modicon

SRCH. The SRCH instruction searches the registers in a source table for a specific bit pattern. The function will search the entire source table in a single scan until either a match is found or the end of table is reached.

BLKT. The BLKT (block-to-table) instruction combines the REGISTER-TO-TABLE and BKLM (block move) functions in a single instruction. In one scan it can copy data from a source block to a destination block in a table. The source block is of fixed length. The block within the table is of the same length, but the overall length of the table is limited only by the number of registers in your system configuration.

TBLK. The TBLK (table-to-block) instruction combines the TABLE-TO-REGISTER and BLKM (block move) functions in a single instruction. In one scan it can copy up to 100 contiguous 4× registers from a table to a destination block. The destination block is of a fixed length. The block of registers being copied from the source table is of the same length, but the overall length of the source table is limited only by the number of registers in your system configuration.

IBKR. The IBKR (indirect block read) instruction lets you access noncontiguous registers dispersed throughout your application and copy the contents into a destination block of contiguous registers. This instruction can be used with subroutines or for streamlining data access by host computers or other PLCs.

IBKW. The IBKW (indirect block write) instruction lets you copy the data from a table of contiguous registers into several noncontiguous registers dispersed throughout your application.

For details of the application of these functions, see the manuals listed in the bibliography.

TROUBLESHOOTING PROBLEMS

TS 16–1 Refer to figure 16–2. The timer always runs for 15 seconds. What could be wrong with the program or the process devices?

TS 16–2 Refer to figure 16–6. The incorrect values appear in HR0172. What could be wrong with the program?

TS 16–3 Refer to figure 16–8. The values from HR0173 appear in the wrong output registers. What could be wrong with the programming?

EXERCISES

1. A PLC counter is used for the process shown in figure 16–10. During its operation, three different count limits are used. The count chosen depends on the particular product being produced. The three count values are stored in IR0022, IR0023, and IR0024. The three limit counts are 5, 12, and 17, respectively. Inputs 0034, 0035, and 0036 are used to insert the three respective limit counts one at a time. The counter uses HR0345 for the preset time limit value. Design a PLC ladder program to accomplish the change in the process count limit using one counter and three MOVE functions.
2. A PLC subsystem has a SUBTRACT function, as shown in the block diagram in figure 16–11. The SUBTRACT function has three possible sets of two input numbers. The resulting SUBTRACT output is to be moved to one of three output registers. The output register chosen to receive the SUBTRACT result is determined by the position of a selector switch. Design a PLC ladder program to carry this out using six input MOVE functions and three output MOVE functions along with the SUBTRACT function.

FIGURE 16–10
Diagram for Exercise 1

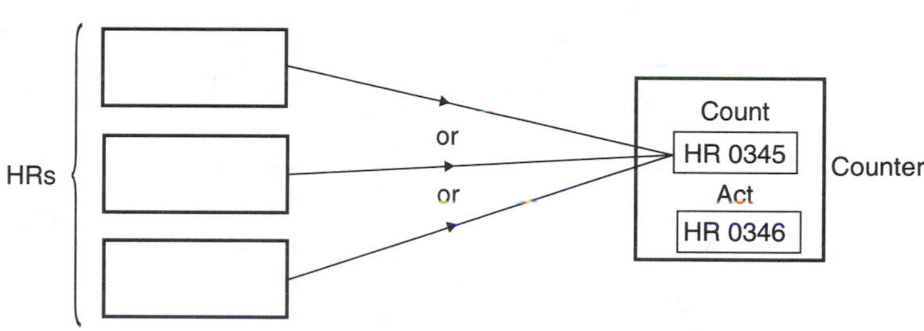

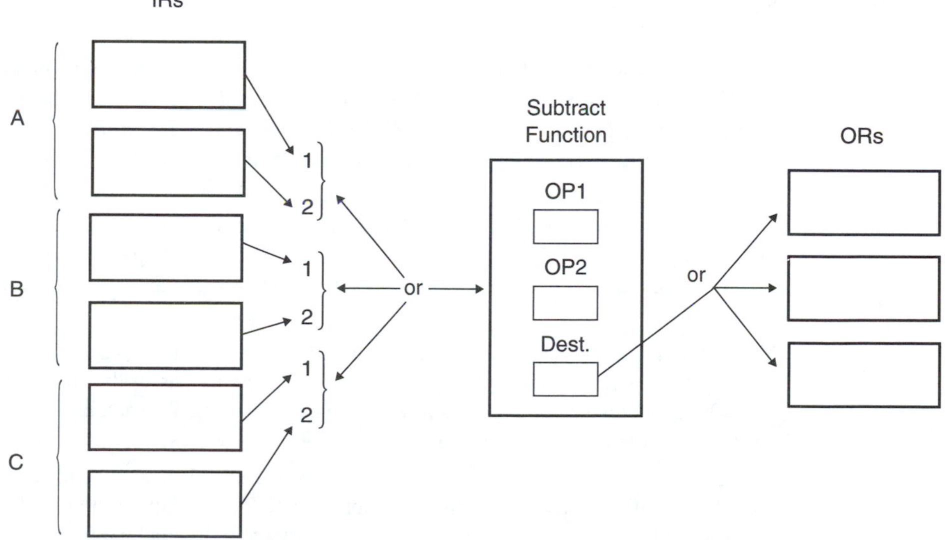

FIGURE 16–11
Diagram for Exercise 2

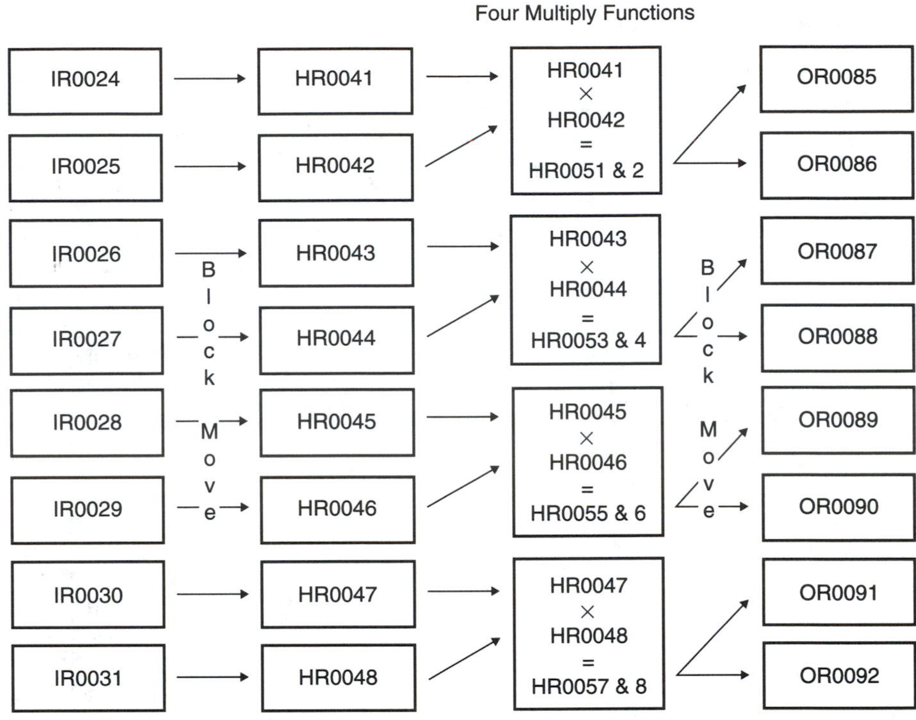

FIGURE 16–12
Diagram for Exercise 3

3. Use two BLOCK MOVE functions to accomplish the process shown in the block diagram in figure 16–12. The first BLOCK MOVE function moves the eight input registers' contents to the operand 1 and 2 positions in the four MULTIPLY functions. The resulting MULTIPLY values, which are two registers wide, are then to be moved by a second BLOCK MOVE function to eight output registers as shown.
4. There are 30 bit patterns of 27 bits each to be moved sequentially into OR0011, one every 7 seconds. Design a double TR function program with a timer to accomplish the data transfer. Two TR functions are required because there are more than 16 bits (the amount available in one register) to be transferred.
5. You are assigned the task of recording lap times at the Indianapolis 500. There are 200 laps. A laser sensor determines the exact moment your assigned car passes the start/finish line and turns on IN0042. You have 200 registers in which to place the time for each lap. Design a RT function system to accomplish the recording.
6. Develop a time system for exercise 5 that also records the elapsed time for each lap. Exercise 5 recorded total time at each lap's end. Add appropriate PLC circuitry with a SUBTRACT function and a second set of recording registers for recording the individual lap times.

17

Other PLC Data Handling Functions

OUTLINE

OBJECTIVES

At the end of this chapter, the student will be able to

- □ Describe the FIFO and other stacking functions.
- □ Apply the FIFO function.
- □ Describe the FAL function and its application.
- □ Describe the ONS function and its application.
- □ Describe the CLR function and its application.
- □ Describe the SWEEP function and its application.

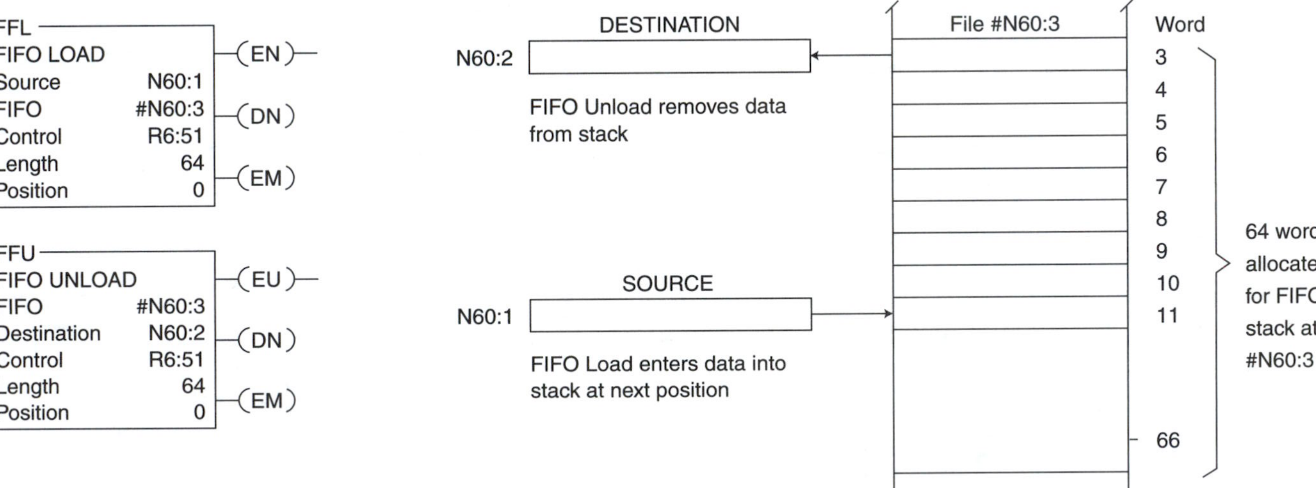

This Parameter	Tells the Processor:
Source (N60:1)	The location of the "next in" source word
FIFO (#N60:3)	The location of the stack (FIFO file)
Destination (N60:2)	The location of the "exit" word
Control (R6:51)	The instruction's address and control structure
Length (64)	The maximum number of words you can load
Position (0)	To start at the FIFO file address

a Allen-Bradley Format

The FIN instruction is used to produce a first-in queue. It copies the *source data* from the top node to the first register in a queue of holding registers. The *source* data is always copied to the register at the top of the queue. When a queue has been filled, no further *source* data can be copied to it.

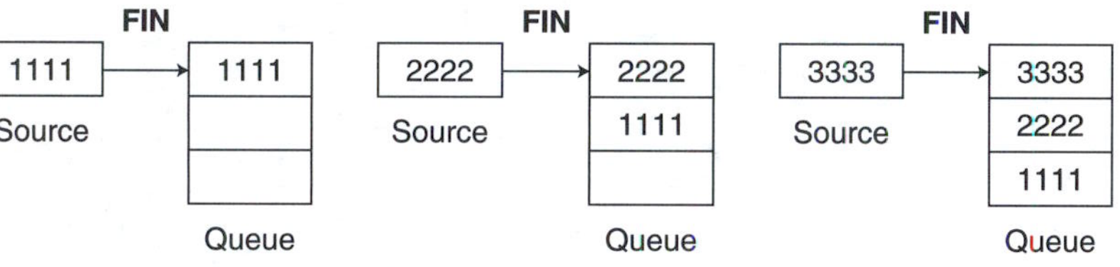

ON copies *source* bit pattern into queue

source data — Echoes state of the top input

queue pointer — Queue full

FIN *queue length* — Queue empty

The FOUT instruction works together with the FIN instruction to produce a first in-first out (FIFO) queue. It moves the bit pattern of the holding register at the bottom of a full queue to a *destination* register or to word that stores 16 discrete outputs.

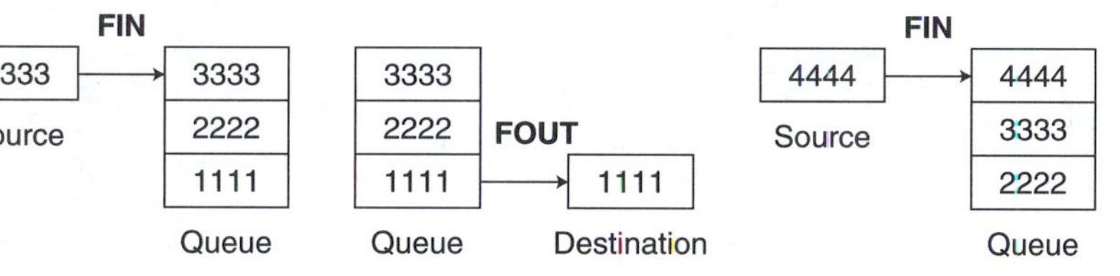

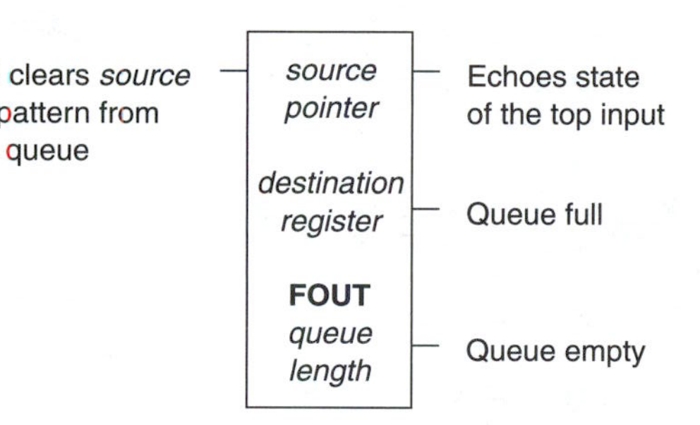

b - Modicon Format

FIGURE 17–1
FIFO Stacking Program Formats (Courtesy of Allen-Bradley and Modicon)

17–1 INTRODUCTION

In addition to the data handling functions covered in chapters, 14, 15, and 16, there are lots of other data handling functions. The quantity and nature of these functions vary from manufacturer to manufacturer and model to model. In this chapter we cover some of the other often-used data handling functions. In section 17–2 we discuss stacking functions and explain in detail the most used stacking function, FIFO. In section 17–3 the FILE ARITHMETIC AND LOGIC (FAL) function is covered. The FAL function is essentially a multiple COMPUTE (CPT) function, which was covered in section 11–6. In section 17–4, three other commonly used functions are explained: ONS, CLEAR, and SWEEP.

17–2 PLC FIFO FUNCTION

Commonly used PLC stacking systems include FIFO, first in–first out, and LIFO, last in–first out. The FIFO function retrieves data in the order stored and is found in most PLCs. The lesser-used LIFO function retrieves data in the reverse order stored and is normally an advanced option. We explain the use of the FIFO function in this section.

The FIFO function consists of two subfunctions. To put data into a stack, a function denoted FFL or FIN is used. To take data out of a stack, a function denoted FFU or FOUT is used. Figure 17–1 shows the FIFO programming formats for Allen-Bradley and Modicon. In both cases in figure 17–1, two functions are used in combination to accomplish FIFO. Information to programmed into the functions differs by manufacturer but includes:

- Source of data to be inserted into the stack
- Destination of data to be removed from the stack
- Length of the stack, n—typically, 64 maximum
- Beginning register (or word, or address) where the data is inserted
- Other special instructions, such as control system

Once the stack is full, the unload function must precede the input function. Otherwise, the input data will not be inserted. Example 17–1 shows the use of FIFO to obtain a running total for a production count.

EXAMPLE 17–1 You wish to keep a running total of both the last five working days' production and the past four weeks' production, 20 working days. How could this be done using FIFO?

Two FIFO programs would be set up. The first program would have n = 5. The source would be the figure for daily production and would be inserted at the beginning of the stack. An AD addition program would be made to add the five figures in the stack of five and display the resulting value as the five days' production. The second program would be set up for an n = 15. The destination for the first program would be the source of the second program. The destination of the second program is then not necessarily relevant. Then, to obtain the four-week total, another AD addition program would be made to total the values in the five registers from the first program and the 15 register values from the second program. The total would be the four-week production figure.

File Arithmetic and Logic (FAL)

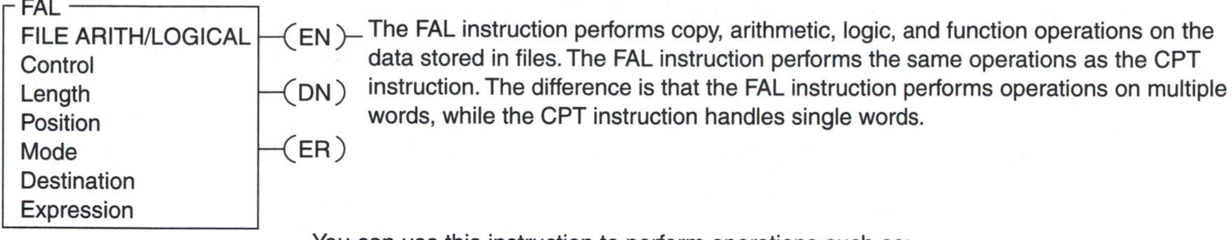

The FAL instruction performs copy, arithmetic, logic, and function operations on the data stored in files. The FAL instruction performs the same operations as the CPT instruction. The difference is that the FAL instruction performs operations on multiple words, while the CPT instruction handles single words.

You can use this instruction to perform operations such as:

- zero a file
- copy data from one file to another
- make arithmetic or logic computations on data stored in files
- unload a file of error codes one at a time for display

Table 9.B
FAL Operations

Type	Operator	Description	Example Operation
Copy	none	copy from A to B	enter source address in the expression enter destination address in destination
Clear	none	set a value to zero	0 (enter 0 for the expression)
Arithmetic	+	add	2 + 3 2 + 3 + 7 (Enhanced PLC-5 processors)
	−	subtract	12 − 5 (12 − 5) − 1 (Enhanced PLC-5 processors)
	*	multiply	5 * 2 6 * (5 * 2) (Enhanced PLC-5 processors)
	|	divide	24 | 6 (24 | 6) * 2 (Enhanced PLC-5 processors)
	−	negate	− N7:0
	SQR	square root	SQR N7:0
	**	exponential (x to the power of y)	10**3 (Enhanced PLC-5 processors only)
Bitwise	AND	bitwise AND	D9:3 AND D10:4
	OR	bitwise OR	D9:4 OR D9:5
	XOR	bitwise exclusive OR	D10:10 XOR D10:11
	NOT	bitwise complement	NOT D9:4
Conversion	FRD	convert from BCD to binary	FRD D14:0
	TOD	convert from binary to BCD	TOD N7:0

FIGURE 17–2
FILE ARITHMETIC AND LOGIC (FAL) Function (Courtesy of Allen-Bradley)

17–3 THE FAL FUNCTION

The FILE ARITHMETIC AND LOGIC (FAL) function is useful when a complex computation is performed on a series of data values. Instead of programming a COMPUTE (CPT) function repetitively for each group of data, you program only once for a FAL function. Then, for FAL, you specify the number of times the calculation should run (length) and where the calculation process should start in a stack of data values. The FAL function is essentially a multiple, sequential CPT function which saves programming time and program space. The Allen-Bradley FAL function and its operational capability are shown in figure 17–2.

17–4 THE ONE SHOT (ONS), CLEAR (CLR), AND SWEEP FUNCTIONS

The ONE SHOT (ONS) function is related to data handling so is included in this section. The ONS function is often used with a pushbutton input which is subject to contact bounce. If there is contact bounce, there are multiple input spikes of voltage. The PLC can misinterpret the input signal. When the input goes from false to true, ONS turns the rung on for one scan. The ONS output stays turned on for subsequent scans until the input goes from false to true. The ONS output going from true to false has no effect on the output, which eliminates the effect of spikes from bounce.

The CLEAR (CLR) instruction of function sets all the bits in a register or word to zero. This is useful when you wish to zero out a system before starting or restarting a process.

The SWEEP function is used when you wish to scan through a program or portion of a program at fixed intervals. The fixed interval needs to be longer than the scan time to be operational: for example, if scan time is 30 milliseconds (ms) and you set the sweep time to 20 ms. The scans will be made every 30 ms, the scan time. If you set the sweep time for 5 seconds, the program will be scanned every 5 seconds, which is longer than the scan time.

TROUBLESHOOTING PROBLEMS

TS 17–1 You have set up a FIFO program for a stack of 15. The source inputs data 15 times and then will not enter any more. What could be wrong with the program or setup?

TS 17–2 You have filled the stack of 15 initially in problem TS 17–1. The source data will not go into the stack. What could be wrong in this case?

EXERCISE

1. A hydroelectric dam PLC control program requires a running average for operational decisions. The running average is for the upstream water level, which is measured by a water height sensor and sent into the control panel. The running average is to be for the previous hour and measured every minute. Set up a FIFO system to accomplish this control parameter.

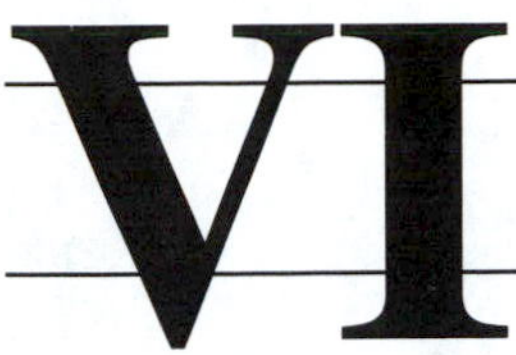

PLC Functions Working with Bits

18

PLC Digital Bit Functions and Applications

OUTLINE

OBJECTIVES

At the end of this chapter, you will be able to

- □ Describe the PLC digital bit control system.
- □ Describe the BIT PICK CONTACT function and its use.
- □ Use digital bits to turn outputs on and off.
- □ Modify and control digital bits in a register.
- □ Use shift registers to move digital bits within and through registers.
- □ Apply digital bit register systems to process control programs.

18–1 INTRODUCTION

Most PLCs are capable of working with digital bits. Instead of controlling output devices from individual contacts, these PLCs use register bits in groups. For example, if the on-off status of 16 machines must be controlled, just one of the 16 bits in a 16-bit register could control each of the 16 machines. If there are 157 machines to turn on and off, only 10 of these 16-bit registers are needed for on-off control (157/16 = 9.815, or 9 registers plus part of a 10th one). By contrast, a contact-coil ladder control would need 157 ladder lines in the program.

The PLC not only uses a fixed pattern of register bits, but can easily manipulate and change individual bits. The PLC can pick, set, latch, and manipulate the individual bits in chosen registers. It also can shift the register contents to the right or left. Register shifts can be set to move the bits one position per input pulse. Shifts may also be set for multiple position movement (two, three, or more). This multiple bit shift function is often designated the *N-bit shift*.

Functions discussed in other chapters also play a part in process control with digital bits. For example, MOVE enables you to replace the entire register contents in order to change the 16 output commands. If you want an on-off pattern changed, shift in an appropriate new register pattern. Moves of data into registers can be done for one register only, but data moves can be made for a consecutive series of many registers.

The digital bit system is the foundation of multiple machine control. The bit system is used extensively in all types of automation systems. One very powerful bit control system is the drum controller/sequencer discussed in Chapter 19.

18–2 BIT PATTERNS IN A REGISTER

In some PLCs, the internal slots for memory and operation are called addresses. In others, the slots are called registers. This chapter uses the word *register*.

Previous chapters dealt with using numerical values in registers for PLC function operation. This chapter is not concerned with a register's numerical value, but only its binary bit pattern status, that is, its pattern of 1's and 0's.

For example, start with the register bit setting shown in figure 18–1. A bit pattern has been inserted into the register by calling it up on the screen and then keyboard-inserting its desired bit values. For illustration purposes, the register bits in HR0207 have been arbitrarily given the values shown. HR0207 now has an equivalent binary code decimal (BCD) value of 7851 and an equivalent decimal value of 30,801. These BCD and decimal values are irrelevant at this point; only the binary pattern shown is useful. Binary bit patterns can be applied to (and from) any type of register, not just holding registers.

0111 1000 0101 0001 HR 0207

FIGURE 18–1
Register with a Binary Value

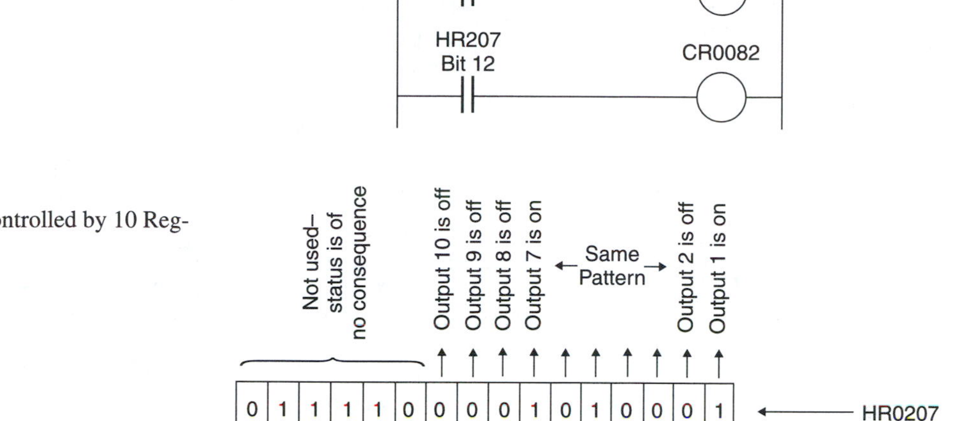

FIGURE 18–2
BIT-PICK CONTACT Control

FIGURE 18–3
Ten Outputs Controlled by 10 Register Bits

Suppose you wish to have outputs CR0081 and CR0082 controlled by a register bit status. To have CR0081 controlled by bit 11 in HR0207 and CR0082 controlled by bit 12, you would designate the contacts as shown in figure 18–2. A menu appears when you press the contact key on the keyboard. Instead of choosing CR or IN, as you have been doing, in this case you would choose BP.

Take the first 10 bits (from the right is standard) and use them to control 10 outputs, as shown in figure 18–3. The outputs with a feeder bit of 1 would be on, and those with 0 bit would be off. If you modify HR0207 to another pattern of bits, the outputs would change status accordingly. An appropriate BP contact system would be used.

18–3 CHANGING A REGISTER BIT STATUS

Suppose you wish to change bit 4 in HR0207 from 0 to 1. Call up register HR0207 on the screen and completely rewrite its bit pattern. Pushing Return would insert the pattern into the PLC CPU. Otherwise, move the cursor over bit 4 and change bit 4 only. This change process is very slow.

Bit status changes are more quickly accomplished by using one of three PLC functions: BIT SET (BS), BIT CLEAR (BC), and BIT FOLLOW (BF). We illustrate the first function using the fourth bit of holding register HR0207. When the BS function is enabled in figure 18–4, bit 4 of HR0207 is set to 1 (if it was not already a 1). Turning the function off would have no further effect on the bit—it would remain a 1.

The BIT CLEAR function, shown in figure 18–5, has the opposite effect of BIT SET. The example in figure 18–5 operates on bit 5 of HR0207. When enabled, the BC function would change bit 5 from 1 to 0. If you had applied BC to bit 6, nothing would have happened, because bit 6 is already a 0. When BC is turned off, nothing further happens.

FIGURE 18–4
BIT SET Function

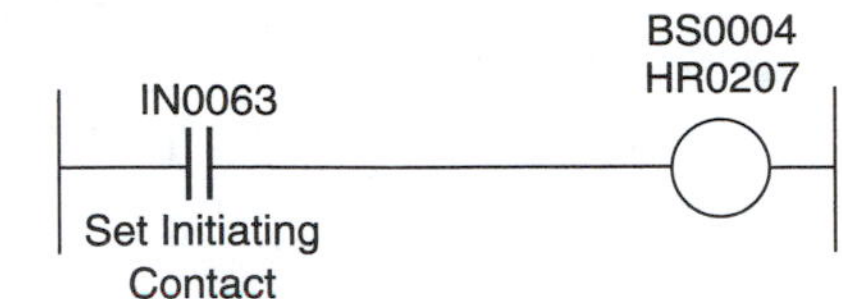

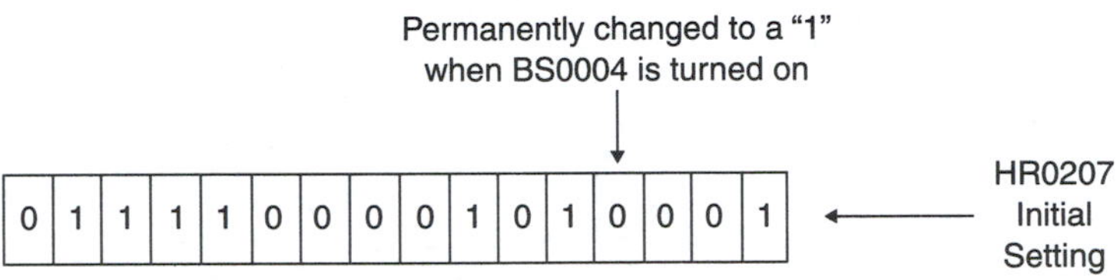

There is one more bit-operating function, the BF, or BIT FOLLOW, function. Go back to bit 4 of HR0207. Figure 18–6 shows the BF function as applied to this bit. When enabled, the function sets the bit to 1. When disabled, or off, the function sets the bit to 0. Notice how BF differs from BS and BC: on and off are both active and significant in the BIT FOLLOW function.

Figure 18–7 illustrates an application using bit status changes for BS, BC, and BF. The application involves a board-painting process that uses the bit modification functions. White, square boards are to be painted red in certain areas. There are 16 square sections on the board, as shown. There are 16 spray guns, one above each of the 16 sections, that spray perfect squares through a template.

When the spray guns (red) operate in the pattern of the original HR0207, the red/white pattern will be as shown. A bit of 1 would cause the corresponding spray to take

FIGURE 18–5
BIT CLEAR Function

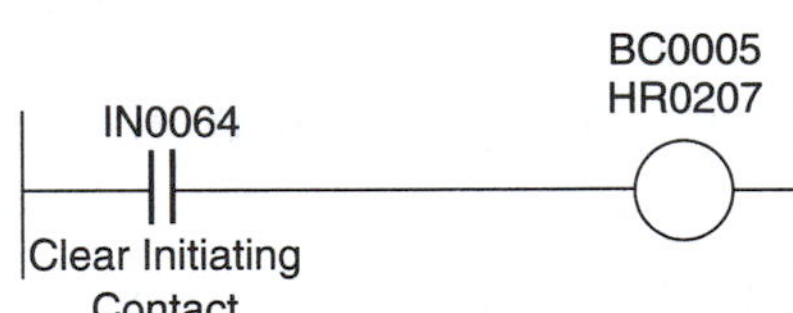

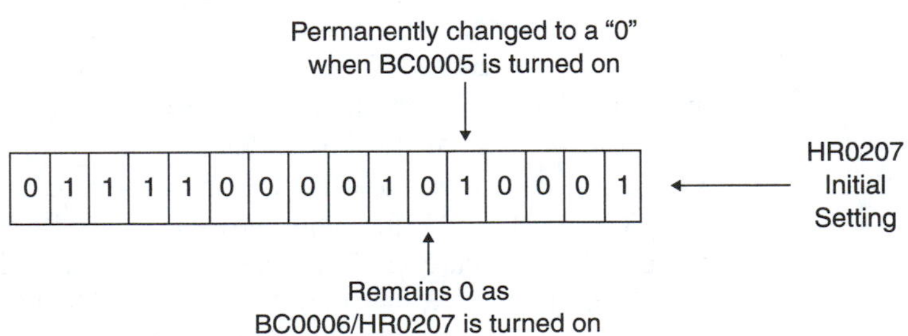

FIGURE 18–6
BIT FOLLOW Function

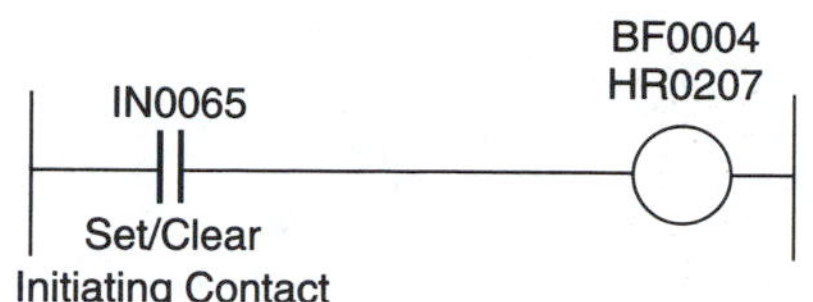

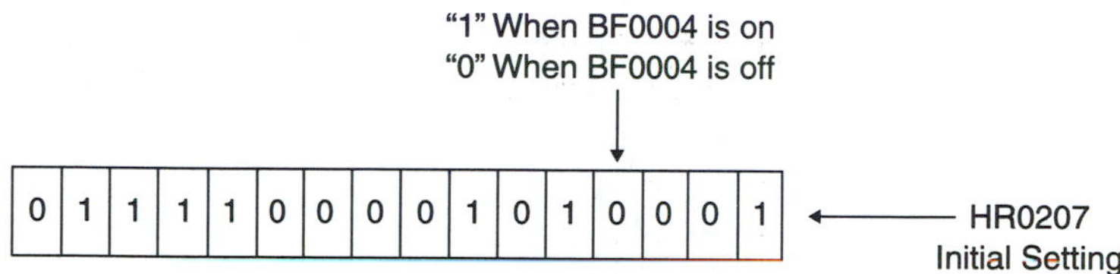

1 2 3 4
5 6 7 8
9 10 11 12
13 14 15 16

White Board Portions
Painted Red → ///

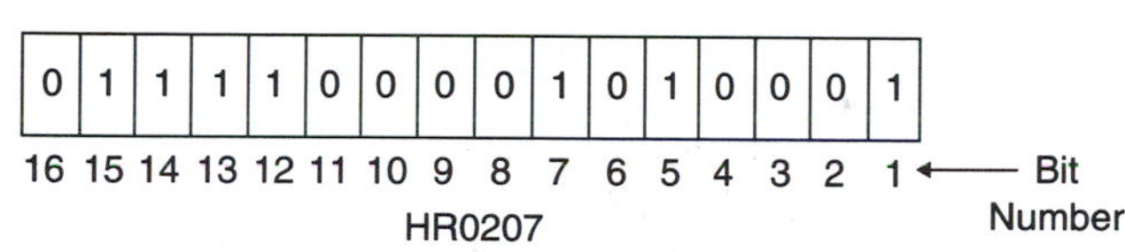

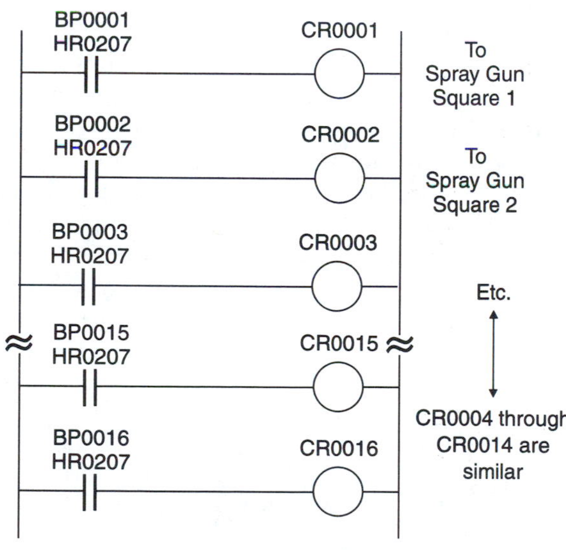

PLC Control

FIGURE 18–7
Spray-Paint Pattern and Program

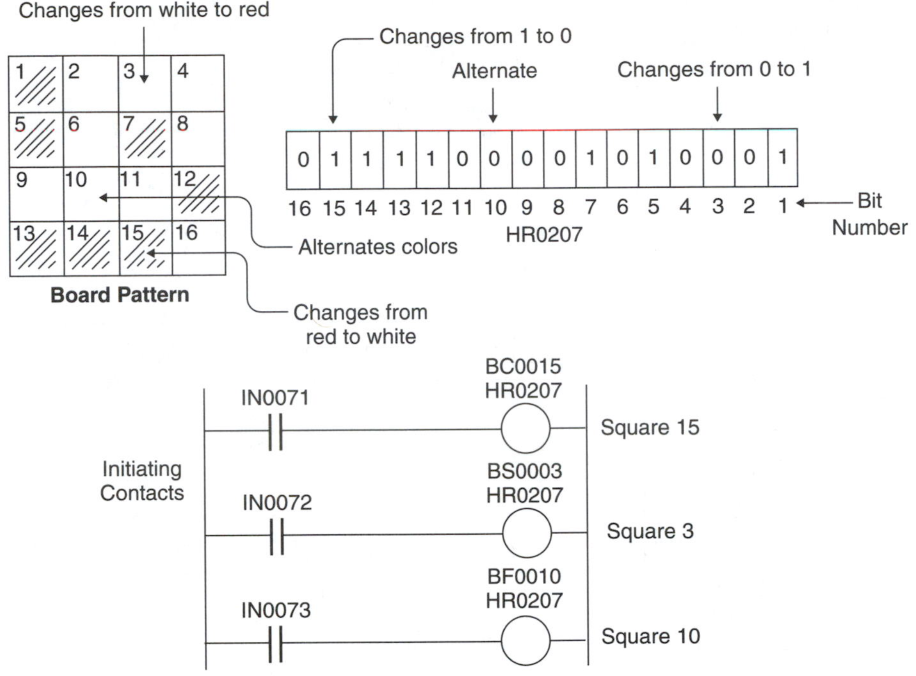

FIGURE 18–8
Revised Spray-Paint Pattern and Program

place. Each spray gun's operation is controlled by a corresponding PLC output coil, which is controlled by a PLC contact as shown. The input contact is described by two specified lines: a register number line and a bit number line. Note that you could use an output group register instead of 16 output coils. This illustration uses individual output coils.

As board model patterns are changed throughout the day, the red/white patterns will change. For example, square 15 is to be changed from red to white. This change may be made (permanently) by applying the BC function as shown in figure 18–8. To change square 3 from white to red, apply a BS function to bit 3. To change square 10 back and forth repeatedly between red and white, apply a BF function as shown in the figure.

A word of caution: Using BF on the same bit as BS or BC can cause problems; for example, if both BC and BF were working on bit 7 of HR0207, the BF would probably override the BC in most PLC models.

18–4 SHIFT REGISTER FUNCTIONS

The shift register functions enable the operator to move digital bits within and through the PLC registers. This is accomplished through SHIFT RIGHT, SHIFT LEFT, ROTATE, and MULTIPLE SHIFT. In this section we discuss each of those functions.

Figure 18–9 shows the operation of a SHIFT RIGHT (SR) function. This explanation uses only one register. Later sections show how to use more than one. There are normally three inputs to the functional block. The bottom input is normally the enabling input, as in previous PLC functions; the middle input determines whether a 1 or a 0 is inserted into the register when shifting; and when the top input is activated, the register shifts all bits one position to the right and a new bit is added on the left.

Whether the bit in the vacated register on the left becomes a 1 or a 0 depends on whether the middle input was on or off when the shift took place. If the middle data line is on, a 1 is entered and for off, a 0. The pattern in figure 18–10 illustrates how to use a shift register to produce the original values in HR 0207 shown in figure 18–1. A shift is made 16 times. This illustration starts with all 0's in the register. The register might have had any other pattern; it would not affect the final pattern, since all previous bit 1's and 0's are pushed out after 16 steps.

Another important part of this SR function is that the coil, or output, status may follow the status of the bit on the right. A 1 produces an output on, and 0 results in output off. If output status does not follow the bit leaving, you may bit pick the last bit of the shift register. A later application explains how output on-off is used. The bits are normally lost when they are pushed off to the right; however, the bit status can be saved and reused in the rotate function, which will be discussed shortly.

Now suppose that we need to control 45 machines or functions. Sixteen bits are not enough to control the process; we need to shift through three registers to cover the 45 outputs by placing the number 3 in the function block that asks for the number of registers. If we put HR0207 in as the starting register, we shift through HR0207, HR0208, and HR0209. Multiple register shifting is shown in figure 18–11.

SHIFT LEFT (SL) functions operate exactly like SHIFT RIGHT, except that bit status is inserted on the right. The bits shift to the left and leave on the left. The output coil status normally then follows the status of the last bit on the left.

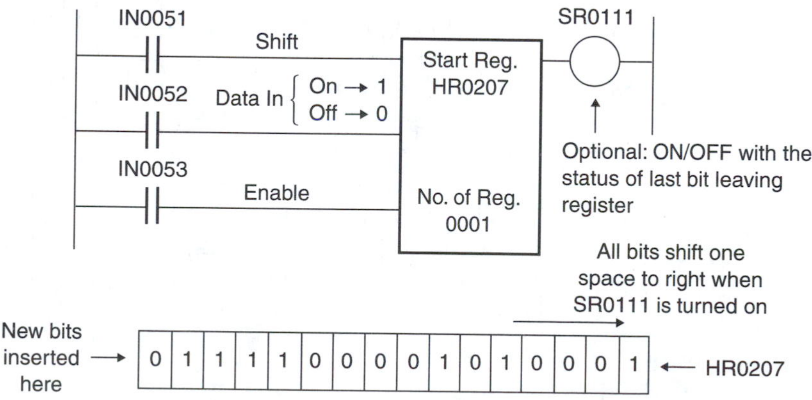

FIGURE 18–9
SHIFT RIGHT Function—One Register

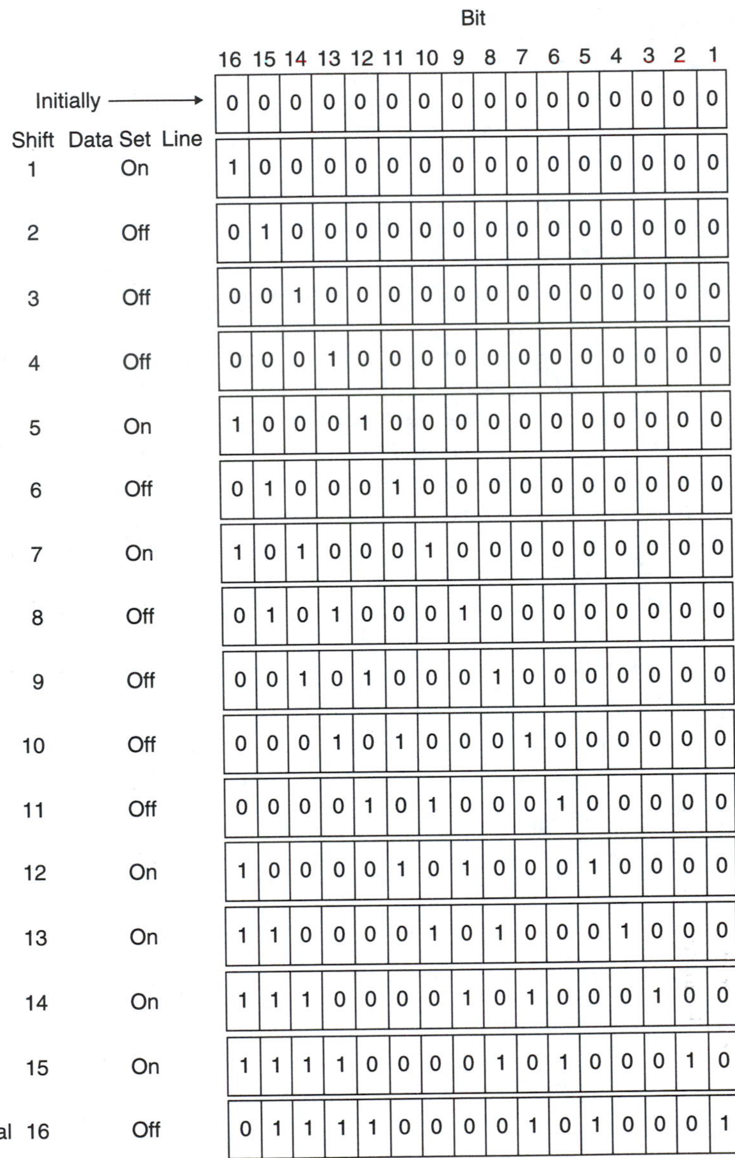

Shift	Data Set Line	Bit 16	15	14	13	12	11	10	9	8	7	6	5	4	3	2	1
Initially		0	0	0	0	0	0	0	0	0	0	0	0	0	0	0	0
1	On	1	0	0	0	0	0	0	0	0	0	0	0	0	0	0	0
2	Off	0	1	0	0	0	0	0	0	0	0	0	0	0	0	0	0
3	Off	0	0	1	0	0	0	0	0	0	0	0	0	0	0	0	0
4	Off	0	0	0	1	0	0	0	0	0	0	0	0	0	0	0	0
5	On	1	0	0	0	1	0	0	0	0	0	0	0	0	0	0	0
6	Off	0	1	0	0	0	1	0	0	0	0	0	0	0	0	0	0
7	On	1	0	1	0	0	0	1	0	0	0	0	0	0	0	0	0
8	Off	0	1	0	1	0	0	0	1	0	0	0	0	0	0	0	0
9	Off	0	0	1	0	1	0	0	0	1	0	0	0	0	0	0	0
10	Off	0	0	0	1	0	1	0	0	0	1	0	0	0	0	0	0
11	Off	0	0	0	0	1	0	1	0	0	0	1	0	0	0	0	0
12	On	1	0	0	0	0	1	0	1	0	0	0	1	0	0	0	0
13	On	1	1	0	0	0	0	1	0	1	0	0	0	1	0	0	0
14	On	1	1	1	0	0	0	0	1	0	1	0	0	0	1	0	0
15	On	1	1	1	1	0	0	0	0	1	0	1	0	0	0	1	0
Final 16	Off	0	1	1	1	1	0	0	0	0	1	0	1	0	0	0	1

FIGURE 18–10
Operation of the SR Function

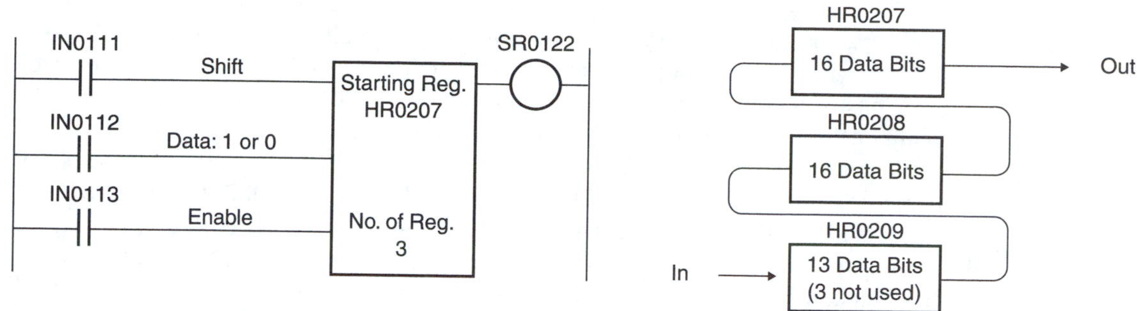

FIGURE 18–11
SHIFT RIGHT REGISTER—Multiple Registers

Suppose you wish to save the bit sequence status that leaves and is lost when using shift registers. You also may want to repeat a pattern again and again. This may be accomplished by using the REGISTER ROTATE function, which is found in some, but not all, PLCs. Its operation is shown in figure 18–12. This example uses the same pattern as the 45 functions in figure 18–11. Now the pattern movement is repeated again and again as a result of the automatic reentry system. For the previous shift registers, the pattern would have to be reentered manually or by MOVE for each time through; the ROTATE functions are automatically repetitive.

ROTATE functions are ROTATE RIGHT (RR) or ROTATE LEFT (RL). ROTATE systems can be of two other general types: full-register reentry or partial-register reentry.

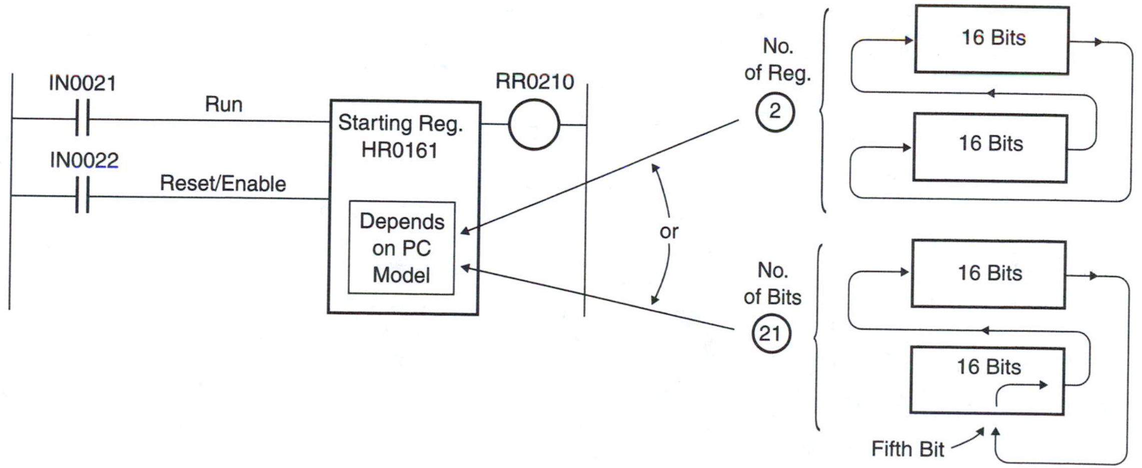

FIGURE 18–12
REGISTER ROTATE Function

The full-register system's reentry point can only be at the beginning of a register. With 16-bit registers, you must shift through 32, 48, 64, and other multiples of 16 bits. With the partial reentry type, you may choose the exact number of bits needed. For example, if you need only 27 bits, you could reenter the initial register at the 27-minus-16 point, which is the eleventh bit. The point would be specified by entering 27 bits into the block function.

Some advanced-function PLCs have SHIFT RIGHT and SHIFT LEFT functions that shift more than one bit at a time. These might be labeled MULTIPLE SHIFT RIGHT (MSR) and MULTIPLE SHIFT LEFT (MSL), or N-BIT RIGHT and N-BIT LEFT. Figure 18–13 shows a typical multiple-shift register. The MSR or MSL functions need one more piece of input information than the SR and SL functions; a specification of the number of shifts to be made at a time, N.

For example, assume that the number of shift steps, N, is set at 3. For an MSR function, it will put in three 1's or three 0's, depending on whether the serial-in switch is closed or open, respectively. The before and after register patterns are shown in figure 18–13, starting with the original register, HR0207.

Figure 18–14 gives a summary of the operation of eight types of shift registers discussed in this chapter. The bit identification numbering system in all of these can be of two types. For two registers, HR0207 and HR0208, the total bit numbers can go from 1 through 32; for three registers, it would be 1 through 48, and so on. The numbering system can also be 1 through 16 for each register only. See your PLC operational manual to determine your system's bit identification scheme.

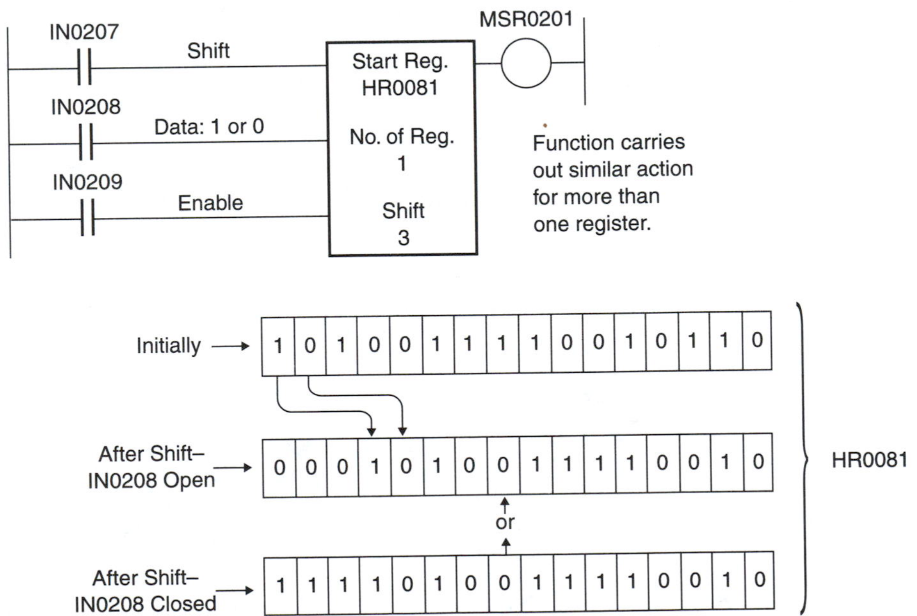

FIGURE 18–13
Typical MULTIPLE SHIFT RIGHT Function

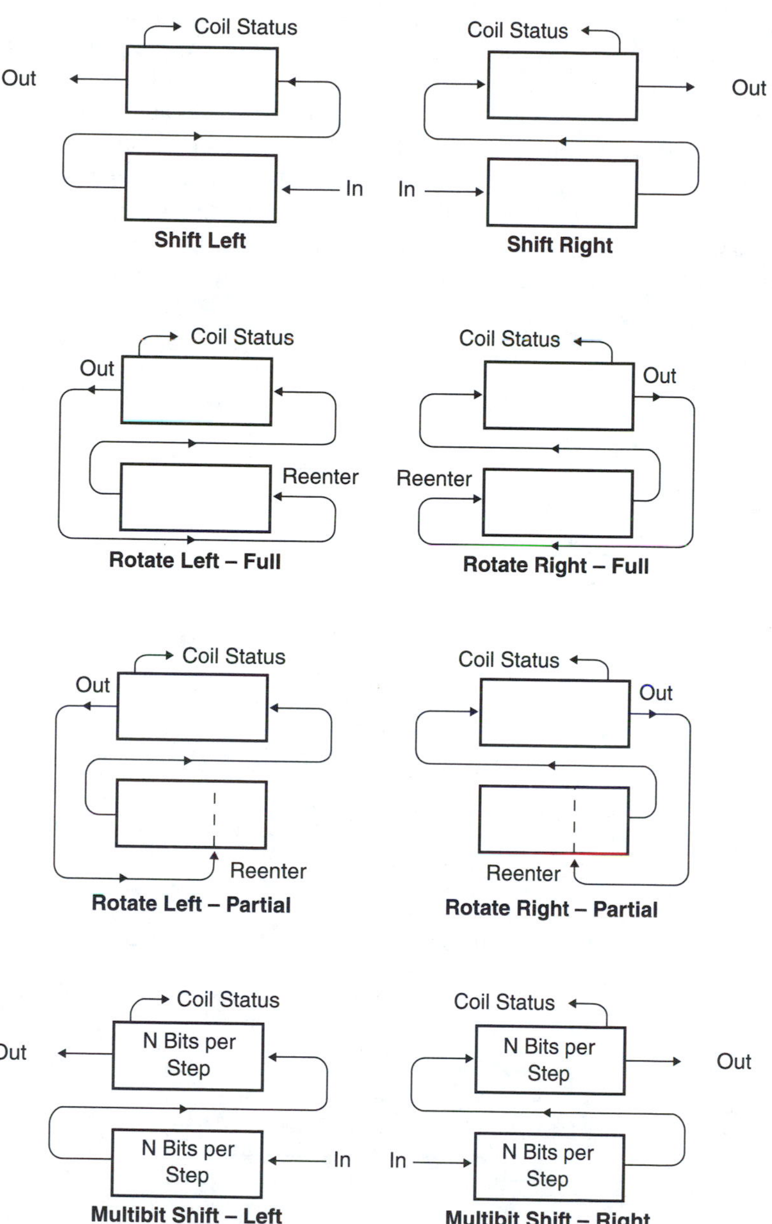

FIGURE 18–14
SHIFT REGISTER Operation Summary

18–5 SHIFT REGISTER APPLICATIONS

The first shift register application example involves controlling a light pattern. Figure 18–15 shows an arrangement of lights to give a flashing, moving arrow pattern that moves to the right. Each light is connected to a PLC input terminal. Each terminal is controlled by a bit location in the two registers shown. As the bit patterns are moved to the right, one step at a time, the light pattern moves to the right. The seven 1's in the registers are progressively moved to the right and then up to the next register; 0's are initially entered into the emptied slots. As the bits move, the lighted lamp pattern progressively moves to the right. In this system, a complete arrow (example 1 through 7) does not move all at once (to 8 through 14); the lighted lamps progress to the next arrow, one at a time, starting at the top (1, then 2, etc.).

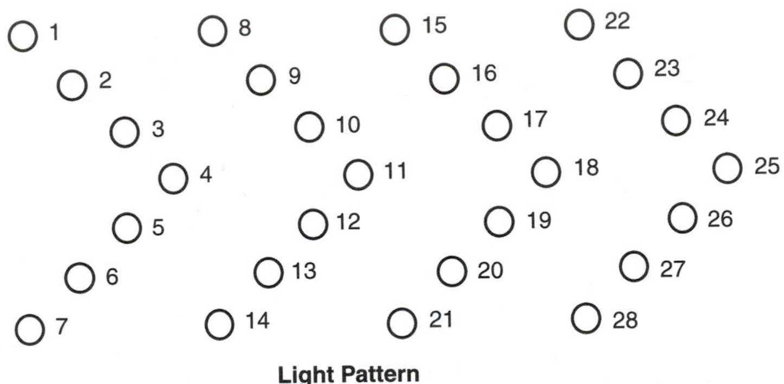

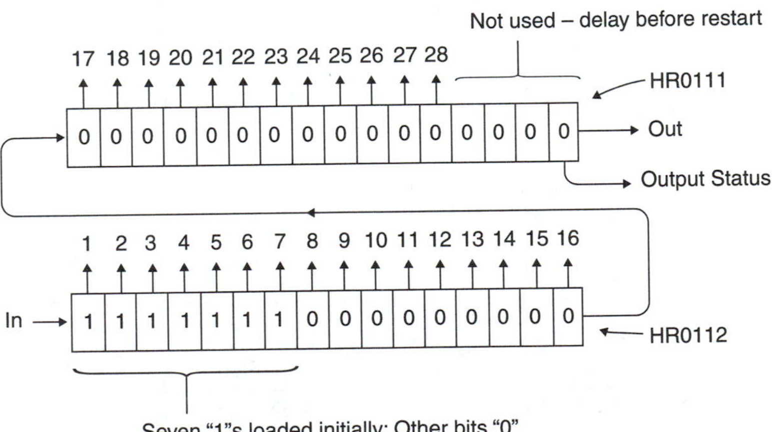

FIGURE 18–15
Flashing Arrow Pattern and Registers

Figure 18–16 illustrates the complete PLC function that accomplishes the movement. The speed of movement is controlled by the time set in TT0151, 5/10 of a second in this example. The SR register is therefore pulsed by the input shift line every 5/10 of a second by the timer. Enabling of the timer and the SR is accomplished by IN0050. As the 1's move through the registers, the corresponding outputs are progressively turned on by the 28 bit-pick functions shown at the bottom. The arrow then moves to the right.

To repeat the process, you could reload the 0's and 1's manually by opening or closing the data line input, or you could use the MOVE system to reload the two operating registers from two master registers. Another possible reload system is to reload from the output status. When the 1's reach the end of the registers, SR 0150 coil goes on for each step. This will cause a 1 to be reinserted into the initial slot. For a 0, the output is off and a 0 will be reloaded. At the end of 31 steps, we are reloaded as shown.

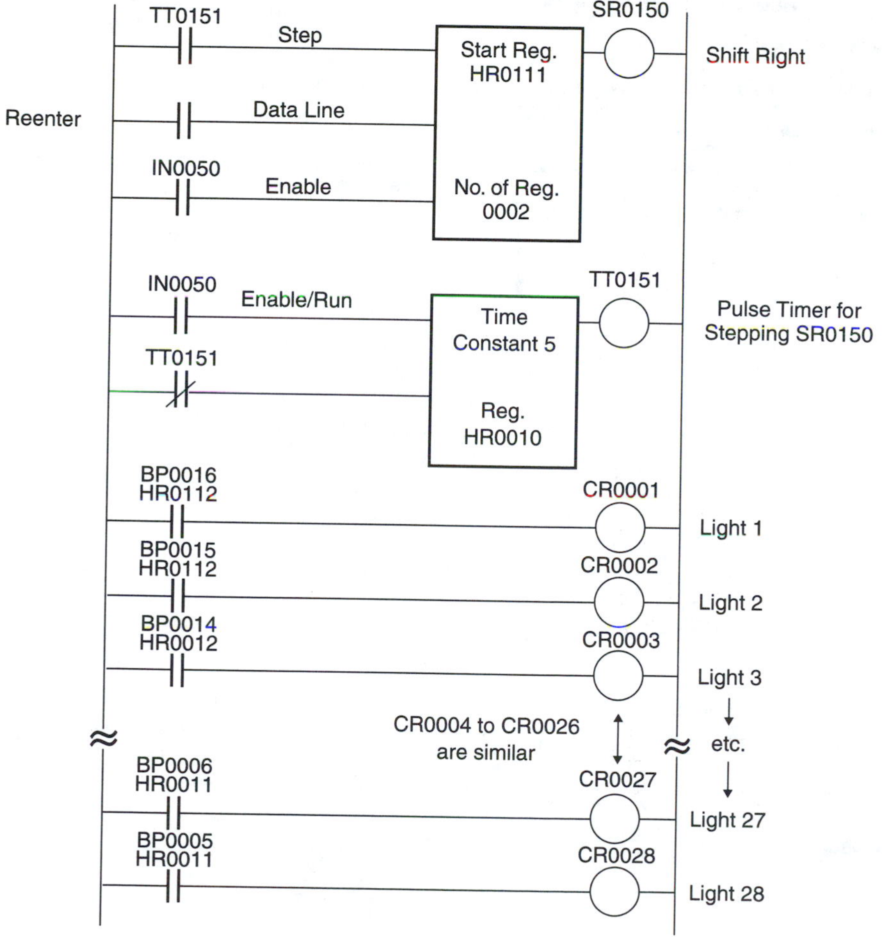

FIGURE 18–16
PLC Program for Flashing Arrow Movement

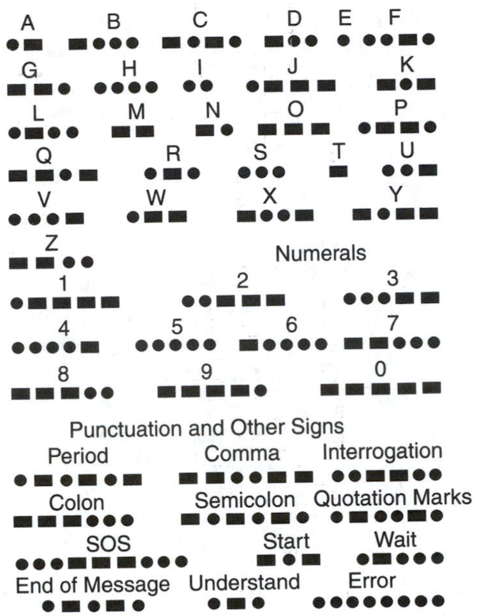

Morse Code System

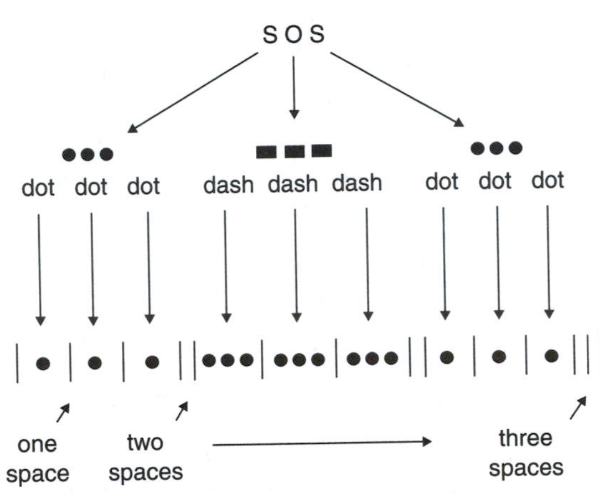

SOS Pattern

1. Dash (–) is three times as long as dot (•)
2. One space between dots and dashes
3. Two spaces between letters
4. Three spaces between words
5. Four spaces between sentences

Morse Code Rules

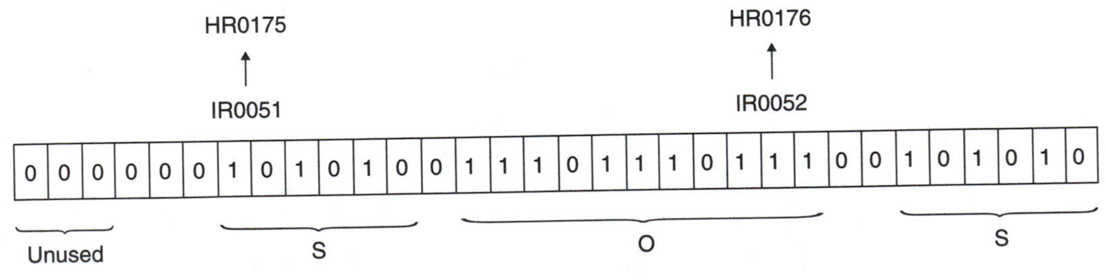

Register Pattern

FIGURE 18–17
Morse Code in SR Register Form

If you had used a ROTATE RIGHT function, the reloading would have been 32 automatic steps. An alternative output scheme for figure 18–16 could be one using output group registers.

The second example of the use of the shift register function involves controlling a code output. The application involves using the SR (or SL) register and the output coil instead of the individual bits for control and indicating. It uses the output as a Morse code indicator to a light or a buzzer. In the Morse code system, each letter is assigned a dot-dash configuration as shown in figure 18–17. Follow the Morse code rules to put dots, dashes, and intervals in registers. A dot is 1 on-bit wide (1) and a dash is 3 on-bits wide (111). Appropriate spacing is added between letters and words.

The resulting pattern for the international distress symbols SOS (Save Our Ship) is put in input registers HR0051 and HR0052 in the figure. The bit pattern is then moved into two HRs, 0175 and 0176. The register bits are then shifted to the right to the output. The resulting output on–off pattern represents the code pattern.

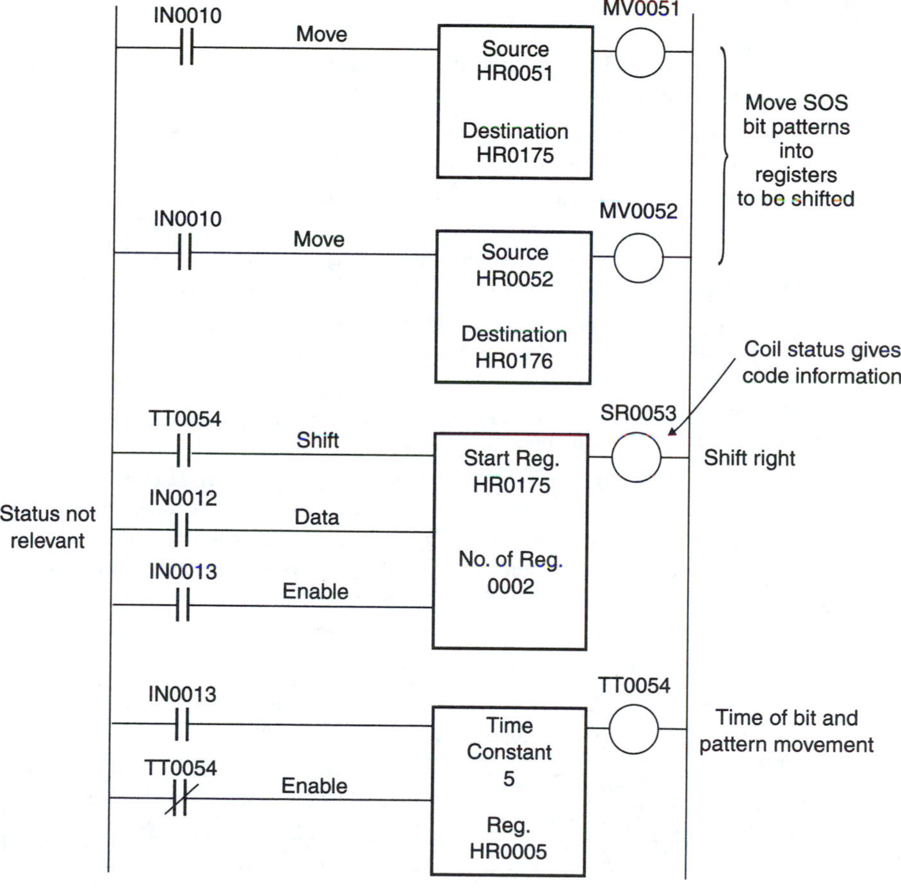

FIGURE 18–18
Morse Code Shift Register Program

Figure 18–18 shows the PLC program to produce the coded output. Turning on IN0010 will move the complete coded pattern from the IR registers into the two HR registers. Then, turning on IN0012 enables the SR function and starts the pulsing timer. The timer contact in the SR shift line causes a bit to be inserted into the SR every 0.5 second. Since the timer is set at 5/10 of a second, a dot (1 bit) will be 0.5 second long, and a dash (3 bits), 1.5 seconds long. Appropriate intervals are added between dots and dashes. To speed up or slow down the rate of code output, the timer's time interval may be changed.

Once the 32 bits have been shifted in, the message is over. We then have on or off signals for the output, depending on the SR data line setting during shifting. For SOS, we must reload the HR patterns into the IR registers again and start over. An alternate to the reload procedure would be programming in a ROTATE-RIGHT register system, which reloads continuously.

EXERCISES

The first three exercises are applications for register bit control programming. Include a sketch of the hardware needed for each exercise, as well as one for the program.

1. A product moves continuously down an assembly line that has 15 stations. Set up a register-controlled production line for the pattern for product A only. The A pattern is shown in figure 18–19.

2. Next, three products are sent down a 15-station production line. They are product A of exercise 1, and products B and C. The schedule of which stations are on or off for each of the three products is shown in figure 18–19. A selector switch is set according to which product is going down the production line. Design a PLC circuit to produce the three products to the required patterns.

3. There are two production lines. Either line may produce products A, B, or C. Expand your program from exercises 1 and 2 to control the two lines.

The final three exercises are applications of shift registers. Again, include a hardware sketch and a PLC program for the solution.

4. One assembly station produces two models of watches, E and F. Sixteen operations take place sequentially at the station. Stations are active or inactive, depending on the model being produced. The station pattern is shown in Figure 18–20. At the end of the day, all watches that have not passed the test are disassembled by running the watch assembly operations in reverse. Design a PLC system with SR for assembly and SL for disassembly. Use the same pattern register for both the SR and SL functions.

	Station Status														
Product	1	2	3	4	5	6	7	8	9	10	11	12	13	14	15
A	On	Off	Off	On	On	On	Off	On	Off	On	Off	On	Off	On	Off
B	On	On	Off	Off	On	Off	Off	Off	On	On	Off	On	On	Off	On
C	On	Off	On	On	On	Off	On	On	Off	On	On	On	Off	On	Off

FIGURE 18–19
Diagram for Exercises 1 and 2

	Station Status															
	1	2	3	4	5	6	7	8	9	10	11	12	13	14	15	16
Pattern E	On	Off	On	On	On	On	Off	Off	On	On	On	Off	Off	On	Off	On
Pattern F	On	On	Off	On	On	Off	On	On	Off	On	On	On	On	Off	Off	Off

FIGURE 18–20
Diagram for Exercise 4

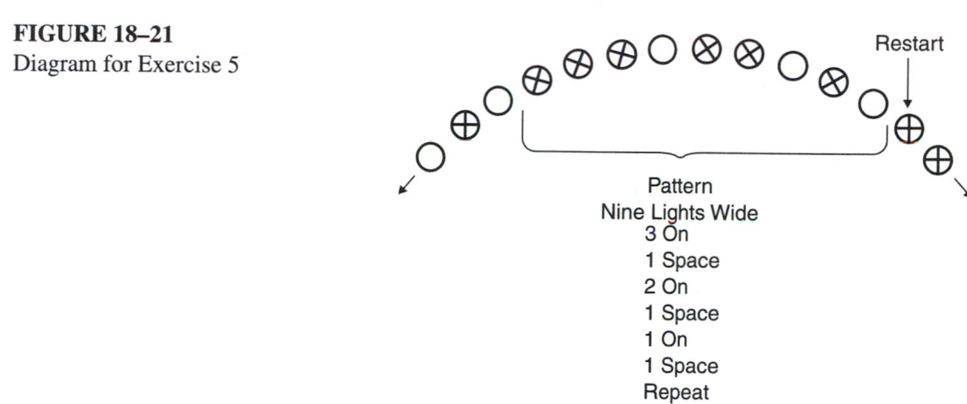

FIGURE 18–21
Diagram for Exercise 5

5. The lights in a circle are 8 degrees apart. The pattern of lighted lights is to rotate, as shown in figure 18–21, in either direction. The 8-degree steps take place every 3 seconds for clockwise rotation. When the pattern is rotating counterclockwise, steps are at 1-second intervals. Design a shift register system to accomplish this rotation.

6. Design a PLC system to put out the word ACME in Morse code to a buzzer. Select a speed of either 4/10 or 7/10 of a second per step. Additionally, select any other word of your choice for the output, and reprogram accordingly.

19

PLC Sequencer Functions

OUTLINE

OBJECTIVES

At the end of this chapter, you will be able to

- □ Compare the operation of a conventional drum switch with a PLC ladder program and a PLC SEQUENCER program.
- □ Describe the basic PLC SEQUENCER function.
- □ Apply the PLC SEQUENCER function to an operational program.
- □ Apply a timing sequence to a PLC SEQUENCER operation.
- □ Describe the other major PLC SEQUENCER functions and their application.
- □ Show how to cascade and chain sequencers.

19–1 INTRODUCTION

The PLC SEQUENCER function is often called the DRUM CONTROLLER function. We use the function designation DR, instead of SQ, which has already been designated for square root. The SEQUENCER concept has evolved from the mechanical drum switch, which is an important control device, but the PLC SEQUENCER function handles large sequencing control problems more easily than does the drum switch. Another advantage of the PLC is that its SEQUENCER programming is relatively straightforward and user friendly.

Traditional drum switches are manually operated. If a timing of the steps being controlled by the drum switch is required, manual operation timed by a clock is needed. The PLC SEQUENCER functions can operate between steps by programmed time sequences. This chapter explains how the PLC SEQUENCER function operates and can be applied to control problems. Some PLCs use the TABLE-TO-REGISTER or FILE-TO-WORD functions as an alternative to the SEQUENCER function. However, these functions are not as complete or versatile as the SEQUENCER function described in this chapter.

In addition to the basic SEQUENCER function, there are additional SEQUENCER functions which have specific additional uses in process control. These are described in this chapter. In some applications you need additional-size programs to accommodate large machines and long programs. These increased sizes are accommplished by cascading, sometimes designated as chaining, which is covered in the last part of the chapter.

19–2 ELECTROMECHANICAL SEQUENCING

Figure 19–1 shows a small, electromechanical drum controller. It is a three-position, six-electrical-terminal device. Its electrical internal connections are illustrated in figure 19–2 for each of its three positions. How is this drum switch used in process control? Four motor-reversing applications are illustrated in figure 19–3. Motor reversing is accomplished by reversing any two leads for three-phase AC, reversing the start leads with respect to the main leads for single-phase AC, or reversing the field leads with respect to the armature leads for DC.

Drum switches are limited to a maximum of seven positions and about 12 pairs of contacts. The switches cannot handle a process with 27 devices and 138 steps, for example,

FIGURE 19–1
Electromechanical Drum Switch

FIGURE 19–2
Internal Contact Switching for Figure 19–1

Left	Up Handle End	Right
Forward	Off	Reverse
1 O——O 2	1 O O 2	1 O O 2
3 O——O 4	3 O O 4	3 O O 4
5 O——O 6	5 O O 6	5 O——O 6

Internal Switching

unlike the PLC SEQUENCER function, which can easily handle the 27-by-138 control and more. The electromechanical drum switch in figure 19–1, however, has one major advantage. It is a good, economical control device for handling applications with a fixed sequence and a limited number of required contacts. It also handles high current.

To begin the discussion of PLC sequencers, a simple sequence of operation is shown in figure 19–4. Three lights are to be in the on or off state in five consecutive, different combinations. The five steps are to be in a given sequence, 1 through 5. A 1 indicates the light is to be on and a 0 off.

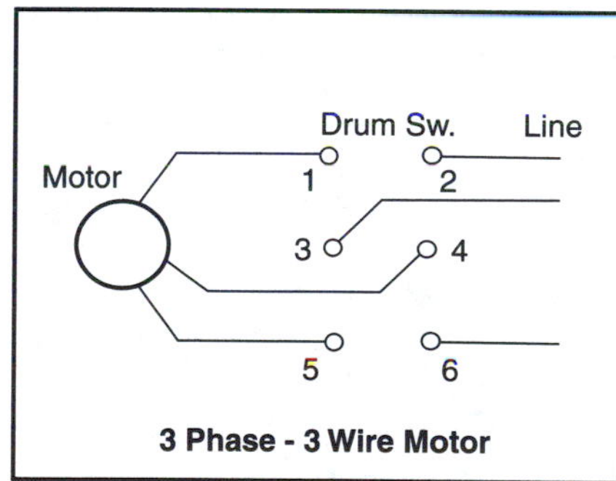

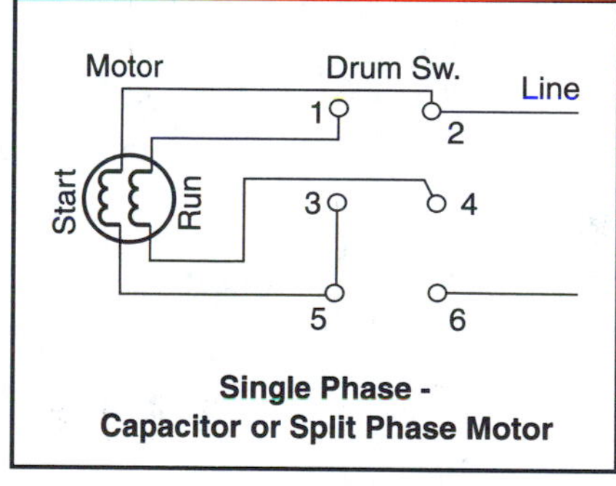

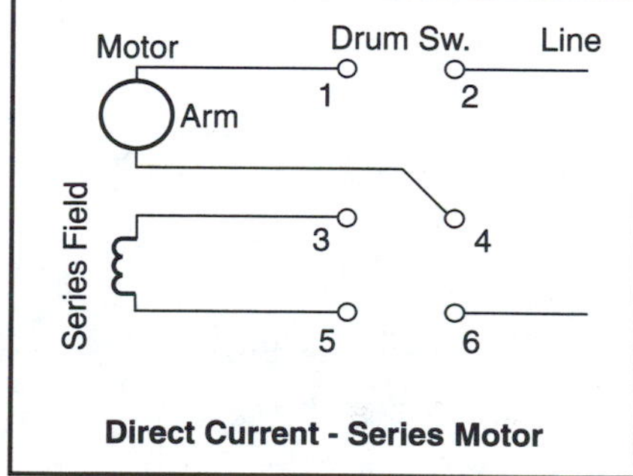

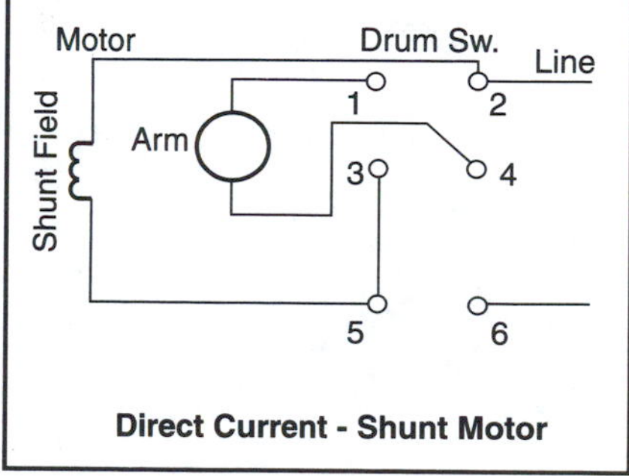

FIGURE 19–3
Motor-Reversing Applications for Figure 19–1

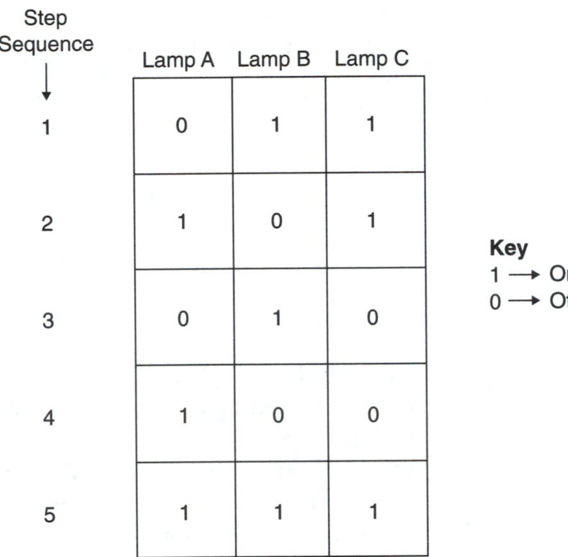

Step Sequence	Lamp A	Lamp B	Lamp C
1	0	1	1
2	1	0	1
3	0	1	0
4	1	0	0
5	1	1	1

FIGURE 19–4
Light Pattern Sequence

The required light pattern sequence of figure 19–4 can be accomplished by manually opening and closing three toggle switches as shown in figure 19–5. Figure 19–5 shows the open or closed status of the three toggle switches for each of the five required sequential patterns.

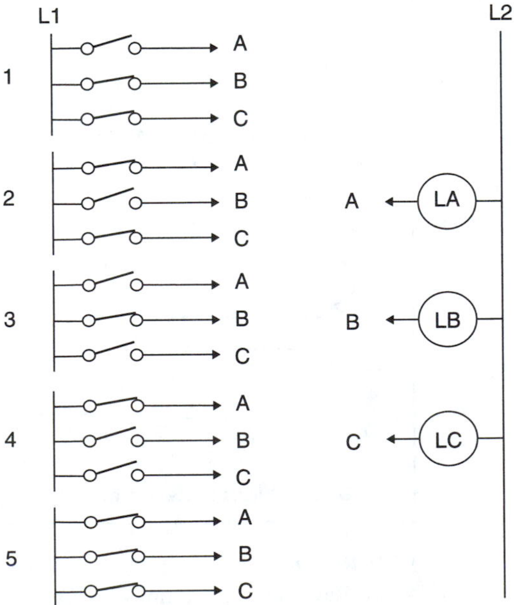

FIGURE 19–5
Toggle Switch Light Pattern Control

Sequence Number	*Close Switch # (Only)*
1	IN001
2	IN002
3	IN003
4	IN004
5	IN005

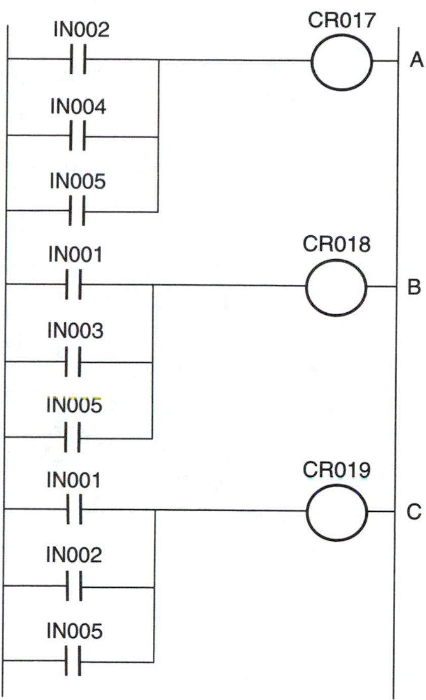

FIGURE 19–6
PLC Coil/Contact Light Pattern Control

As an alternative to using toggle switches, the three lights could be connected to the output of a PLC using the coil/contact system, as shown in figure 19–6. Inputs could be programmed as shown with five input switches, one for each pattern. Closing IN001 would result in sequence pattern 1, IN002, pattern 2 and so on, up through IN005, resulting in pattern 5.

19–3 THE BASIC PLC SEQUENCER FUNCTION

Now, suppose that instead of a three by five sequence, you have to program 14 outputs on–off through 47 steps. Using toggles as in Figure 19–5 or a PLC contact coil system for control would be a long and complicated system. However, programming the system with a PLC SEQUENCER function is relatively easy. A typical PLC function is shown in Figure 19–7. There are three inputs:

Step circuit	Each on pulse steps the function to the next HR pattern.
Reset circuit	Each on pulse resets back to the first HR.
Enable circuit	When on the function may be stepped, but not when off.

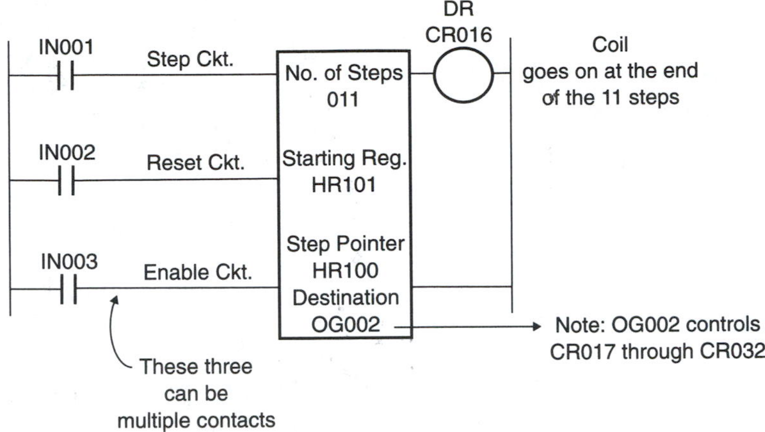

FIGURE 19–7
Typical PLC SEQUENCER Function

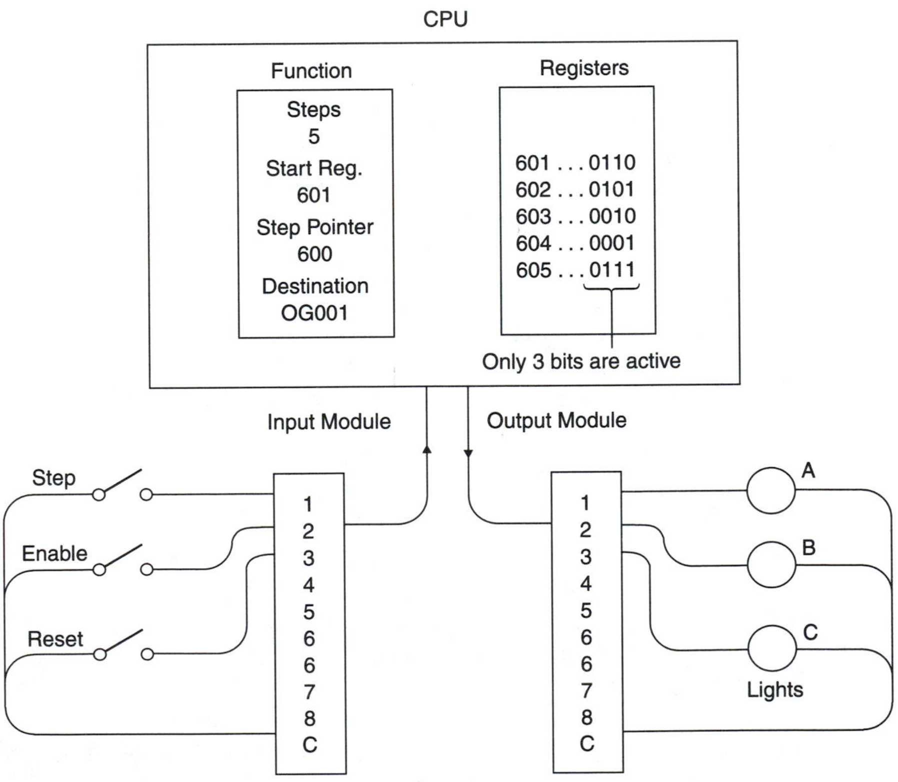

FIGURE 19–8
PLC SEQUENCER Function and Pattern for the Light Sequence

The SEQUENCER function block is programmed with four pieces of information:

Number of steps to be sequenced through
The starting register used for the sequence
Step pointer location, an HR that shows which step you are on
Destination is the OG register, a group of HRs that are to be controlled on and off by the SEQUENCER function.

Figure 19–8 shows how the PLC SEQUENCER function could be used for our three-light, five-pattern control. After the function is programmed, the on–off bit patterns are inserted into the registers used in the sequence. A series of required register patterns, along with the function block programming, is shown in Figure 19–8.

Programming for more steps and outputs would be done similarly. For example, a 47-step program for 14 outputs would have the first function in the block programmed with the number 47. The specified start register could be anywhere there are 46 more unused registers following. If 601 is the starting register, the series involves registers 601 through 647. Pointer and OG group lines in the block are programmed as required. Then, appropriate bit patterns have to be put into the 47 registers, HR601 through HR647.

As the SEQUENCER function operates, the output DRCR016 only comes on at the last step as an indicator. We show how this is used in a future example.

19–4 A BASIC PLC SEQUENCER APPLICATION WITH TIMING

We further illustrate the use of the SEQUENCER function by applying it to control of a dishwasher. We assume that the dishwasher has six functions that must be turned on and off periodically (an actual dishwasher has more). These are

- Soap release solenoid
- Input valve for hot water
- Wash-impeller operation
- Drain water valve
- Drain pump motor
- Heat element for drying cycle

The operational pattern for the dishwasher is shown in Figure 19–9. There are 11 steps. Therefore, we need 11 registers to control the pattern. On the right of the figure, the required register pattern for correct dishwasher sequencing is shown. The 11 HR bit patterns are programmed to match the required on-off patterns for each step. Only the first 6 bits of the 16 in each register are needed in our illustration.

The dishwasher sequence is controlled by the functional block previously shown in figure 19–7. The values in this figure are chosen so that they match the dishwasher example. In this example the stepping of the next interval would be accomplished by manually closing a switch to IN001, the step circuit.

A more complete program for our dishwasher control is shown in figure 19–10. We show a programmed time of 4 seconds. Actually, in the dishwasher's operation, the time would obviously be much longer. We use 4 seconds so that if you program this example,

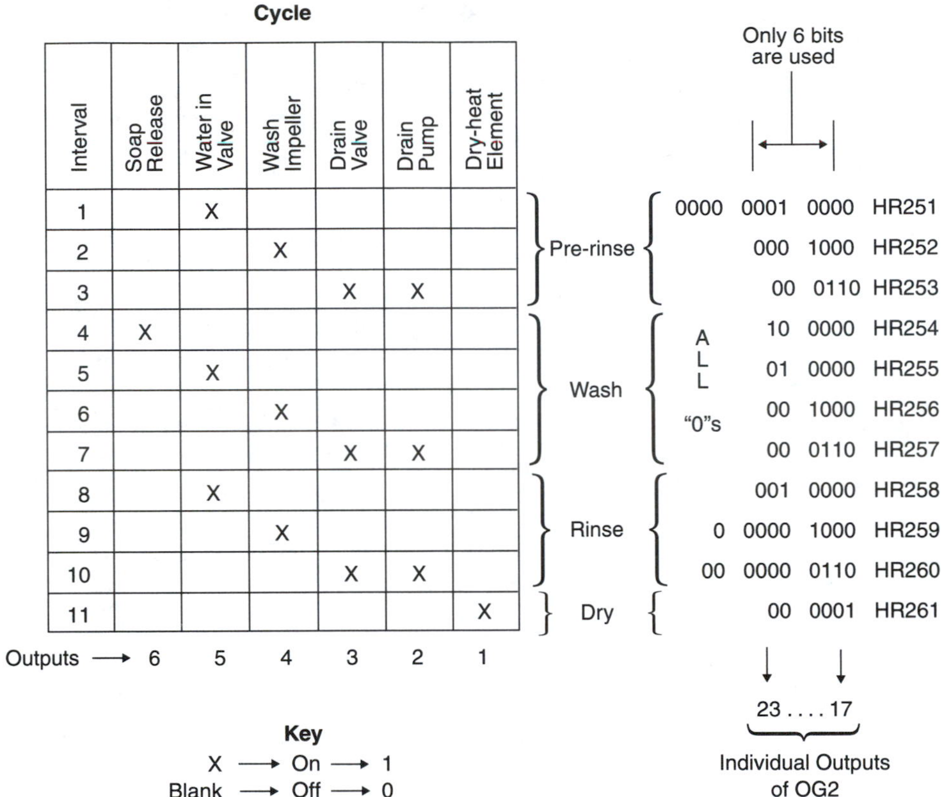

FIGURE 19–9
Dishwasher Function Matrix and Register Patterns

you won't have to wait long to run through the sequence. Additionally, there is a start–stop system to enable the sequence to stop at the end of the 11 steps. Start must be depressed to restart the sequencer at the beginning again. Note that 12 steps are programmed to complete 11 process steps. This is because the sequence stops by unsealing CR0039 when the last step is started. If you stop at the eleventh step, the unsealing never takes place. Stop takes place at the instant you get to step 12. The ladder sequence of figure 19–10 is

1. Push Start (momentary button); CR0039 goes on and seals.
2. Sequencer and timer enabled; process on step 1.
3. After 150 seconds, timer pulses on–off, stepping DR. (*Note:* If set up for demonstration, use 10 seconds to shorten demonstration time.)
4. Repeat for steps 2 through 11.
5. When DR steps to step 12, DR comes on. CR0039 is unsealed.
6. Ladder is reset and ready for step 1.

7. Whenever Stop is depressed, the system stops and resets. (This may or may not be safe. Other programming may be necessary to stop the process at the step it is on.)

You may also want each step to have different times. If the times are all multiples of 4 seconds, you can use repeat registers and lengthen the program. Assume that the steps have time interval values of 8, 4, 12, and so on, as shown in figure 19–11. The program could be lengthened as shown in the figure to accommodate these different times. The PLC program for this multiple interval is shown in figure 19–12.

If you want variable times instead of multiples, you would use two sequencers operating together. One sequencer would step through one group of registers for the output patterns, and the other would step through registers with the times. The times would be sequentially fed into the step timer's time interval register. A system for variable interval times is shown in figure 19–13. The system shown is one that could apply to the dishwasher problem of figure 19–8.

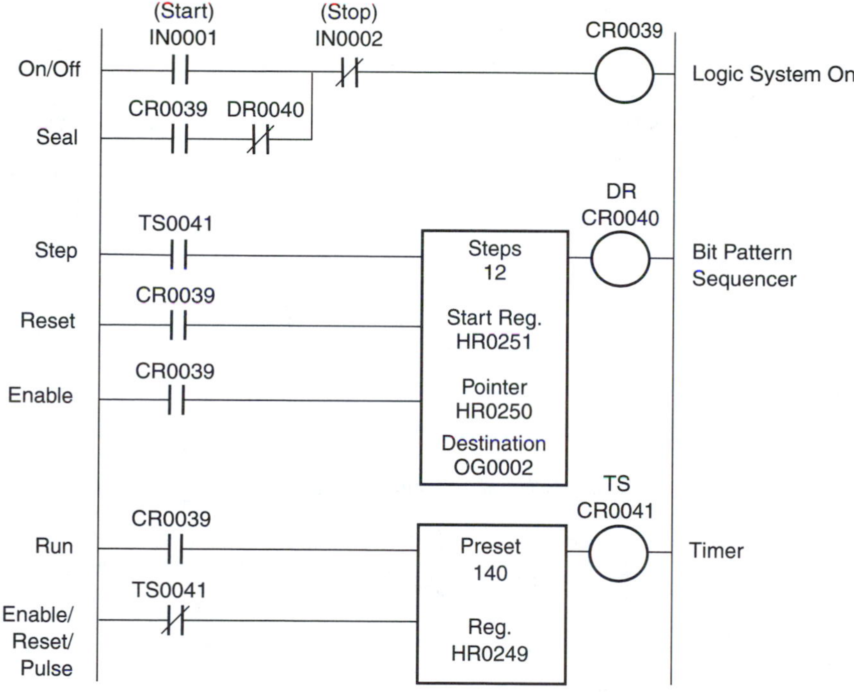

FIGURE 19–10
Sequencer with Timer Pulsing

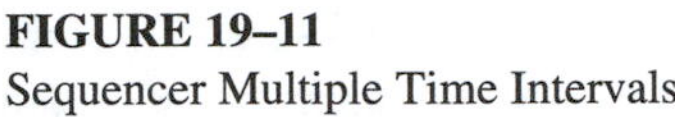

FIGURE 19–11
Sequencer Multiple Time Intervals

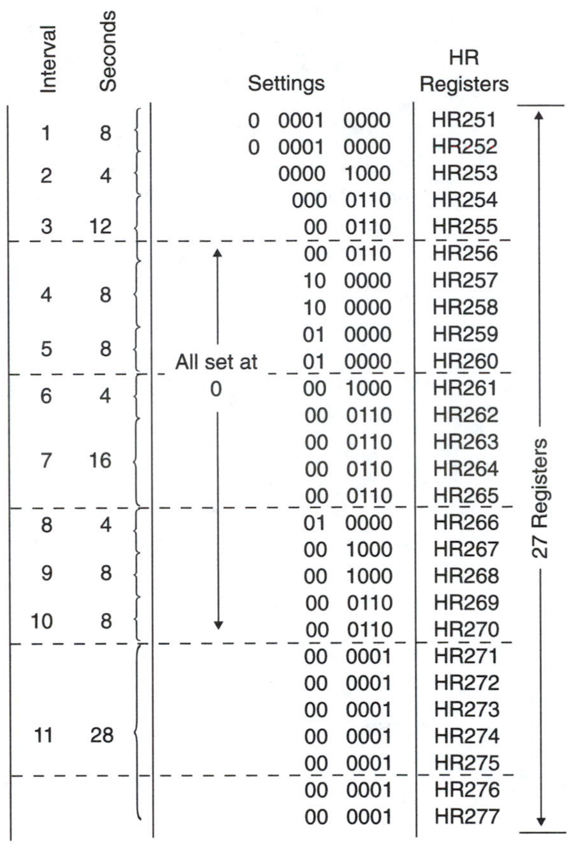

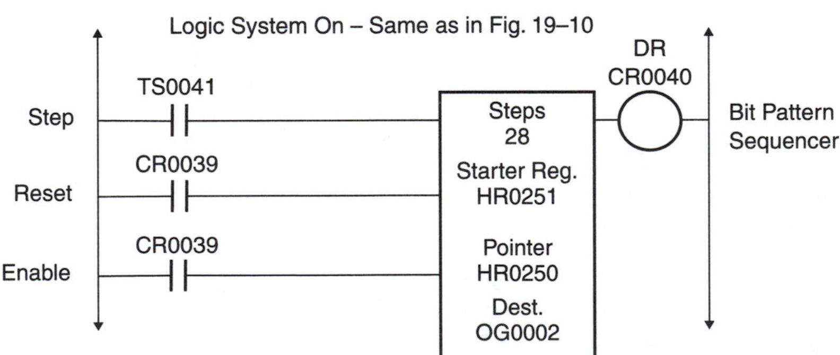

FIGURE 19–12
PLC Program for Figure 19–11

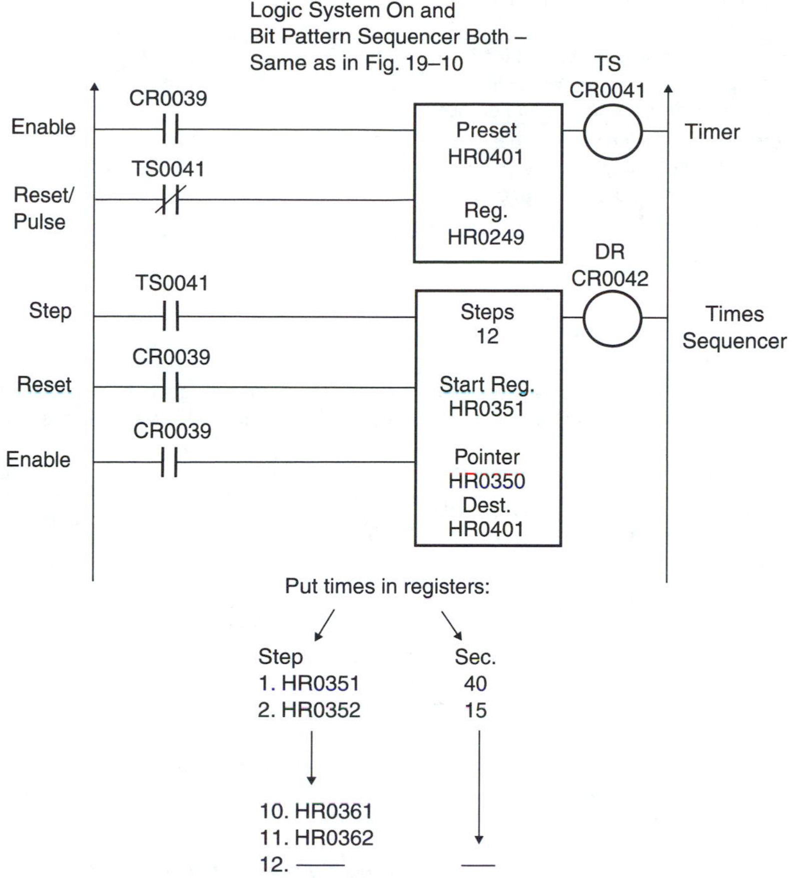

FIGURE 19–13
Sequencer Interval Timer Scheme

19–5 OTHER PLC SEQUENCER FUNCTIONS

Most medium-sized PLCs have the basic DRUM/SEQUENCER (DR) function that we described in sections 19–3 and 19–4. Some PLCs have DR functions which include a mask feature. Others have two additional SEQUENCER functions, one that carries out bit comparisons and another which transfers output data sequentially into a register or series of registers.

Figure 19–14 shows the three SEQUENCER functions of Allen-Bradley PLCs. The SQO function is similar to the DR function described previously but has a mask operation. Masking was described in section 12–4 and figure 12–12. The chapter 12 principle applies in the same way to the masked SEQUENCER functions.

The SQI function sequentially matches the preset input pattern with the output pattern through a mask. The SQI output goes on if all designated bits match. The SQL func-

If You Want to:	Use this instruction:
Control sequential machine operations by transferring 16-bit data through a mask to output image addresses	SQO
Monitor machine operating conditions for diagnostic purposes by comparing 16-bit image data (through a mask) with data in a reference file	SQI
Capture reference conditions by manually stepping the machine through its operating sequences and loading I/O or storage data into destination files	SQL

Description:

Use the SQI and SQO instructions in pairs to respectively monitor and control a sequential operation. Use the SQL instruction to load data in the sequencer file.

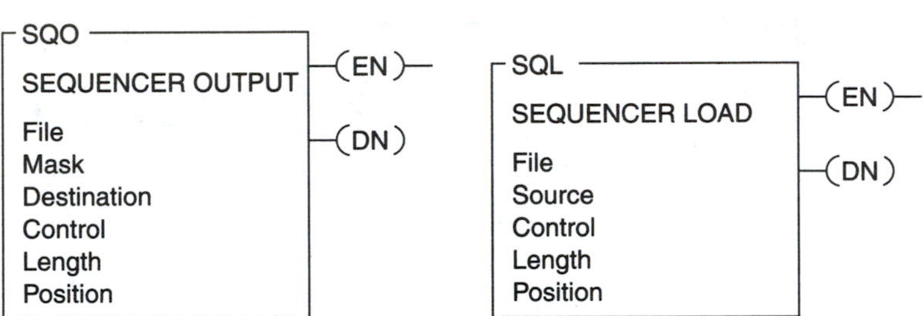

These instructions operate on multiples of 16 bits at a time. Place SQI instructions in series and SQO instructions in parallel in the same rung for 32-, 48-, 64-, or other bit operations.

Note

Each SQO instruction increments the control structure, so corresponding SQI instructions may miss parts of the source file.

Entering Parameters

When programming SQI and SQO instructions in pairs, use the same control address, length value, and position value in each instruction. The same applies when using multiple instructions in the same rung to double, triple or further increase the number of bits.

To program sequencer instructions, you need to provide the processor with the following information:

- **File** is the indexed address of the sequencer file to or from which the instruction transfers data. Its purpose depends on the instruction:

In this instruction:	The Sequencer File Stores Data for:
SQO	Controlling outputs
SQI	Reference to detect completion of a step or a fault condition
SQL	Creating the SQO or SQI file

FIGURE 19–14
Allen-Bradley SEQUENCER Functions

- **Mask** (for SQO and SQI) is a hexadecimal code or the address of the mask element or file through which the instruction moves data. Set (1) mask bits to pass data; reset (0) mask bits to prevent the instruction from operating on corresponding destination bits. Specify a hexadecimal value for a constant mask value. Store the mask in an element or file if you want to change the mask according to application requirements.

- **Source** (for SQI and SQL) is the address of the input element or file from which the instruction obtains data for its sequencer file.

- **Destination** (for SQO, only) is the destination address of the output word or file to which the instruction moves data from its sequencer file.

  If you use a file for the source, mask, or destination of a sequencer instruction, the instruction automatically determines the file length and moves through the file step-by-step as it moves through the sequencer file.

- **Control** is the address of the control structure in the control area (R) of memory (48 bits – three 16-bit words) that stores the instruction's status bits, the length of the sequencer file, and the instantaneous position in the file.

  Use the control address with mnemonic when you address the following parameters:

  – **Length (.LEN)** is the length of the sequencer file.

  – **Position (.POS)** is the current position of the word in the sequencer file that the processor is using.

For this instruction:	The Control Structure is Incremented:
SQO and SQL	By the instruction itself
SQI	Externally, either by the paired SQO with the same control address, or by another instruction

- **Length** is the number of steps of the sequencer file starting at position 1. Position 0 is the start-up position. The instruction resets to position 1 at each completion.

  The address assigned for a sequencer file is step zero. Sequencer instructions use (length + 1) words of data for each file referenced in the instruction. This also applies to the source, mask, and destination values if addressed as files.

- **Position** is the word location in the sequencer file. The position value is incremented internally by SQO and SQL instructions.

  Your ladder program can externally increment the position value of the SQI instruction. One way to do this is to pair it with the SQO instruction and assign the same control structure to both instructions.

FIGURE 19–14 (continued)

Block Structure

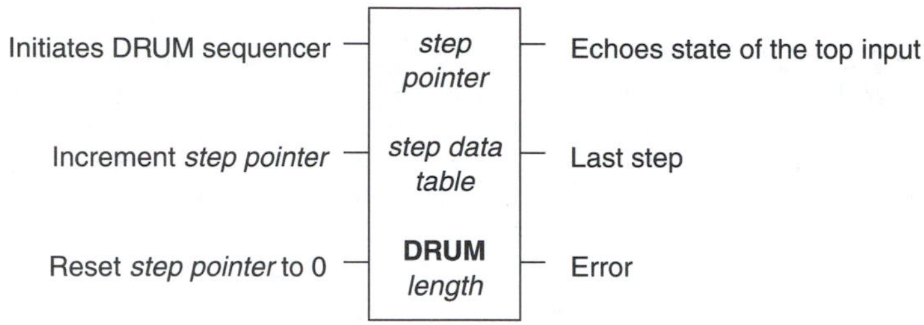

Inputs
DRUM has three control inputs. When the input to the top node is ON, the drum operation is initiated. When the input to the middle node is ON, the *step pointer* increments to the next step. When the input to the bottom node is ON, the *step pointer* is reset to 0.

Outputs
DRUM can produce three possible outputs. The output from the top node echoes the state of top input. The output from the middle node goes ON for the last step—i.e., when the *step pointer* value = *length*. The output from the bottom node goes ON if an error is detected.

a - DRUM

Block Structure

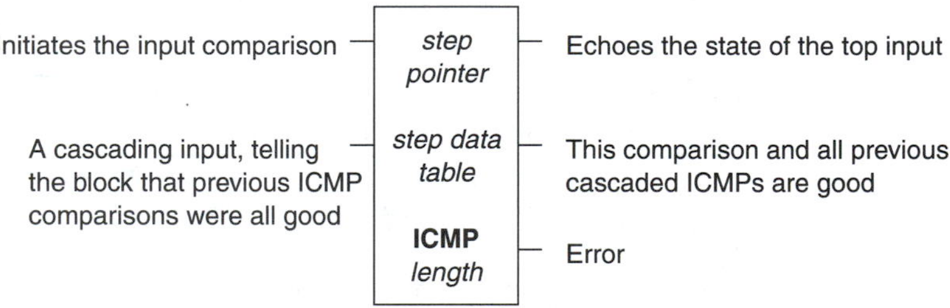

Inputs
ICMP has two control inputs (to the top and middle nodes). When the input to the top node is ON, the ICMP operation is initiated. When the input to the middle node is ON, the instruction passes the compare status to the middle output.

Outputs
ICMP can produce three possible outputs. The output from the top node echoes the state of top input. The output from the middle node goes ON to indicate a valid input comparison. The output from the bottom node goes ON if an error is detected.

b - ICMP

FIGURE 19–15
Modicon SEQUENCER Functions

Block Structure

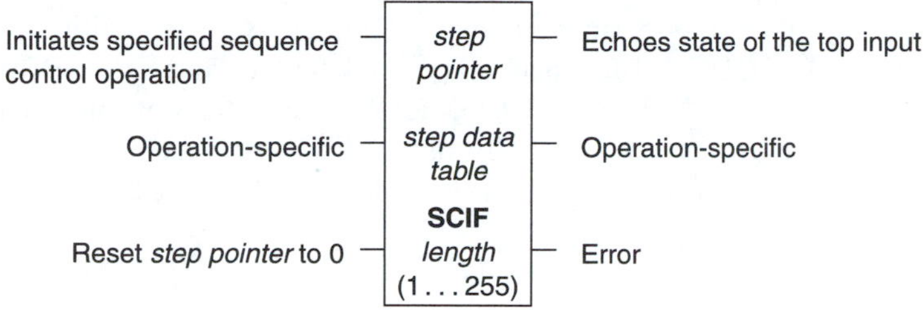

Inputs
SCIF has three control inputs. When the input to the top node is ON, the drum or ICMP operation is initiated.

When the input to the middle node is ON in drum mode, the *step pointer* increments to the next step. When this input is ON in ICMP mode, the instruction passes the compare status to the middle output.

When the input to the bottom node is ON in drum mode, the *step pointer* is reset to 0. The bottom input is not used in ICMP mode.

Outputs
SCIF can produce three possible outputs. The output from the top node echoes the state of top input.

In drum mode, the output from the middle node goes ON for the last step—i.e., when the *step pointer* = *length*. In ICMP mode, this output goes ON to indicate a valid input comparison.

The output from the bottom node goes ON if an error is detected.

c - SCIF

FIGURE 19–15 (continued)

tion captures reference (usually output) bit patterns in sequential registers or addresses each time the input is turned on.

Figure 19–15 shows some of SEQUENCER functions of Modicon PLCs. The DRUM SEQUENCER function is essentially the same as the DR function described previously. The ICMP (INPUT COMPARE) function is essentially the same as the Allen-Bradley SQI function. The SCIF function is a combined DRUM and ICMP function. The choice of operational mode is made by defining the value in the first register of the step data table.

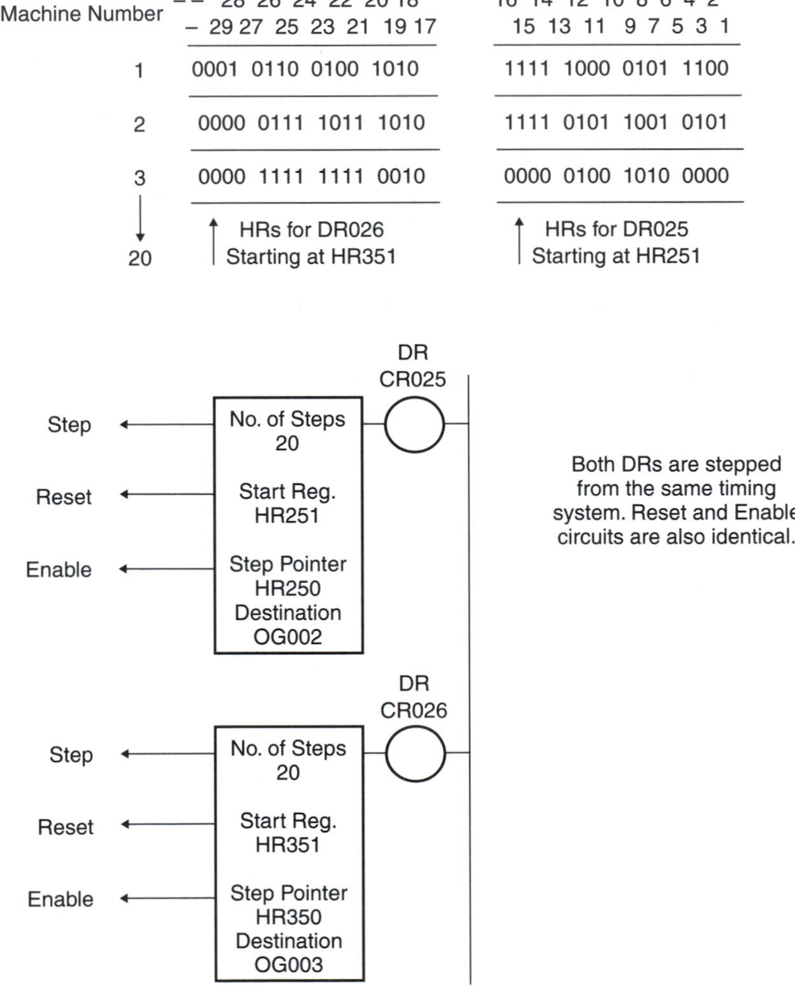

FIGURE 19–16
Expanded Sequencer Control for Many Outputs

19–6 CASCADING SEQUENCERS

Basic sequence functions have size limits. To obtain more steps, sequencers can be *cascading*. Some manufacturers designate this process as *chaining*. The single-function output capability may be 8 or 16 bits wide, and 8 or 16 output terminals can control only up to 8 or 16 machines. If there are 29 machines to turn on and off, even a single 16-bit program will not work. To add more output capability, you must chain, or expand, the number of bits across. Some formats have a place in the program or an address register to expand the bit capability. When a code key or number is inserted, it will automatically expand the number of outputs. Other formats, such as the one shown in figure 19–16, must have parallel functions programmed in. Figure 19–16 shows how 29 outputs can be controlled from a 16-bit function format. Two functions are run in parallel. The extra 13 bits are controlled by the second functional block.

The other sequencer dimension may need more steps for a process than one program block contains. Sequencers can have 64, 128, 256, and other numbers of steps per program block, depending on the format and PLC model. Suppose there is a 128-step function limit

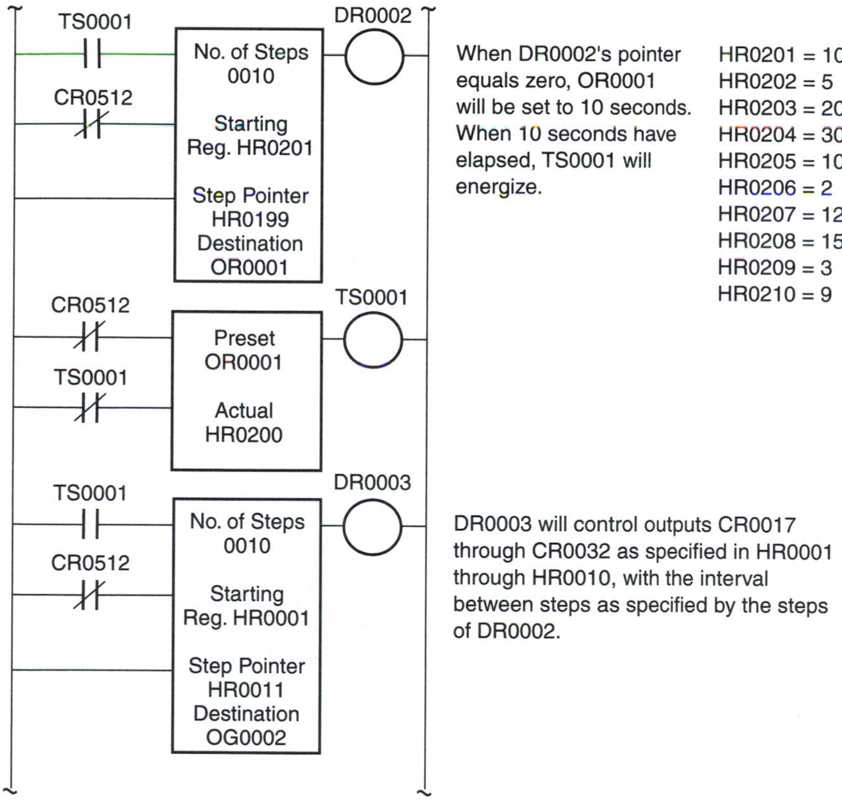

FIGURE 19–17
Expanded Sequencer Control for Many Steps

for one basic function. If you need 277 steps, you must chain three functions together to run the process: 128 plus 128 plus 21 steps of the third function. Chaining is accomplished by starting the next sequencer after the last step of the previous one. A way to do this for one format is shown in figure 19–17.

TROUBLESHOOTING PROBLEMS

You have programmed a DR function that has the malfunctions noted. What could be the cause?

TS 19–1 Refer to figure 19–10.

1. The program gets stuck at the end.
2. One of the registers produces an incorrect output.

TS 19–2 Refer to figures 19–12 and 19–13.

1. The time value in HR0375 keeps changing.
2. The program stops halfway through.

EXERCISES

1. Place the machine matrix shown in figure 19–18 in a PLC sequencer program. Program for manual, event-actuated operation.
2. Add individual times to the machine problem in exercise 1 for each step, using the same time interval for each step. Then reprogram the problem using varying times, multiples, or variables, depending on your particular PLC program format.
3. Create an operations scheme for a washer/dryer combination, using figure 19–8 for reference. Program a sequencer to run the wash/dry process. Make sure the program stops at the end of the cycle. It should not repeat until a reset switch is actuated. (*Note:* Do not use actual times in minutes for the program. Use seconds. Otherwise, program checkout takes half an hour or longer.)
4. Cascade a PLC for chained outputs for four or five intervals. Exceed the number of output bits for one program block. For example, if the limit is 8, use 11. Choose your own pattern arrangement.

FIGURE 19–18
Diagram for Exercise 1

Step Number ↓	Machine Number 7	6	5	4	3	2	1
1	—	On	—	—	—	—	—
2	—	—	On	On	—	—	—
3	On	—	—	—	On	On	On
4	On	On	On	On	On	On	On
5	—	—	—	—	—	—	—
6	On	—	On	—	On	—	On
7	—	On	—	On	—	On	—
8	—	—	On	—	—	—	On

Key — = Off

5. Cascade a PLC for chained steps for 4 or 5 output bits. Exceed the number of output steps for one program block. For example, if the limit is 128, use 135. Again, choose your own bit patterns.
6. *Extra credit:* Combine the system of exercises 4 and 5 and chain both output bits across and both output steps down.
7. Compare the SEQUENCER function of this chapter to the TABLE-TO-REGISTER function of Chapter 16 on data move systems. What are the advantages of the SEQUENCER function over the TABLE-TO-REGISTER function?

20

Controlling a Robot with a PLC

OUTLINE

OBJECTIVES

At the end of this chapter, you will be able to

- □ Develop a "coil and contact" (input/output) control system to operate a basic robot.
- □ Develop a drum controller/sequencer control system to operate a basic robot.
- □ Describe the electrical connection and interfacing system required to connect a PLC to an industrial robot.
- □ Develop a drum controller/sequencer control system to operate an industrial robot.

20–1 INTRODUCTION

This chapter illustrates how a PLC may be used to run a robot. The robots to be used for illustration are *pick-and-place* robots, which have various discrete positions for their gripper assembly. The positions are determined by discrete signals—"on" causes the robot's axes and manipulators to move to one extreme position, and "off" moves them to the other extreme position. The other type of robot, the *continuous path* type, is not discussed in this chapter. This type must be controlled by analog computers, or PID (see chapter 23) control devices, which are much more complicated and are not discussed in detail in this book.

The chapter progresses through various levels of complexity of pick-and-place robot control. First, on–off switches are used to manipulate a basic robot. Second, a drum controller/sequencer is used for control of the same robot. Then PLC control systems are used for industrial-type pick-and-place robots.

20–2 BASIC TWO-AXIS ROBOT WITH PLC SEQUENCER CONTROL

We progress through two schemes of controlling the basic pick-and-place robot shown in figure 20–1. We use switches first, and then a drum controller/sequencer. The robot used for illustration starts operating from the position shown, which is the "at-rest," lower-left,

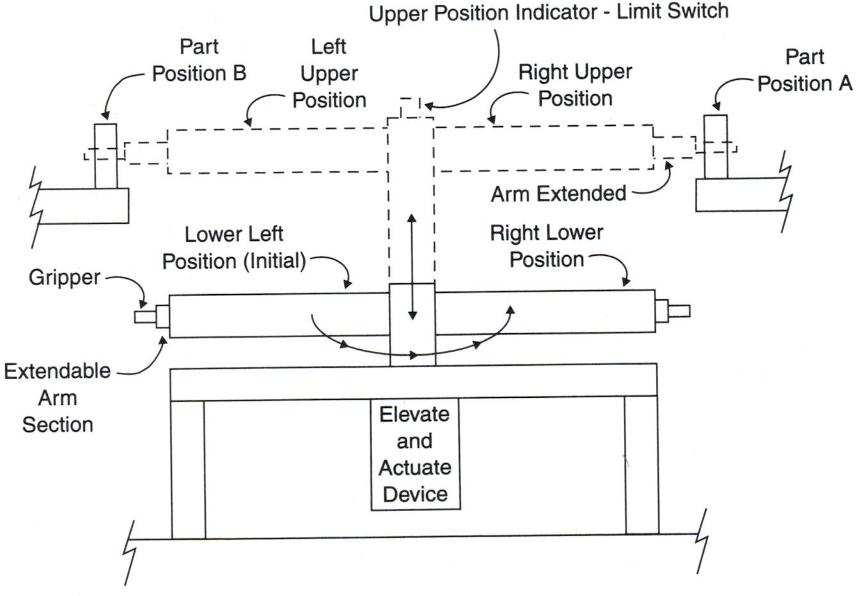

FIGURE 20–1
Basic Pick-and-Place Robot

initialized position. The step-by-step sequence of operation to move a part from position A to position B is as follows:

1. Arm is initially in the down-left position as shown. Gripper is open and not extended.
2. Arm moves to upper position.
3. Arm rotates to right.
4. Hand extends to position A.
5. Gripper closes, gripping part.
6. Arm swings back to the left to position B.
7. Gripper opens, releasing part.
8. Hand retracts.
9. Arm lowers to the initial position.

For illustration, assume that the robot has four powered pneumatic solenoids. If all solenoids are off, no air is applied to the robot's actuators. In this initial position, the robot is in the lowered, left position with the hand retracted and the gripper open. Energizing each of the four solenoids causes the following action to occur:

1. ROTATE: arm rotates full right.
2. RAISE: arm rises to the upper position.
3. EXTEND: hand extends from the arm.
4. GRIP: the gripper closes.

More than one solenoid can be energized in combination to facilitate operation. If a solenoid is not energized, the function is in the other extreme initial position, opposite those listed.

An operational matrix for the robot to move a part from position A to position B is shown in figure 20–2. An O indicates the opposite position; down, left, in, or open.

A simple control system for the robot shown in figure 20–1 could consist of four switches, one for each motion. One disadvantage of the four-switch control is that someone would have to do the controlling continuously. In addition, turning off a switch would not immediately stop the arm; it would spring-return to its initial position, which would be hazardous to anyone expecting it to stop immediately.

FIGURE 20–2
Part Movement Robot Operational Matrix

Step	Up	Rotate Right	Hand Out	Grip Close
Initialized	O	O	O	O
1	X	O	O	O
2	X	X	O	O
3	X	X	X	O
4	X	X	X	X
5	X	O	X	X
6	X	O	X	O
7	X	O	O	O
8	O	O	O	O

There could also be problems in mechanical interferences during operation. In the upper position, with the arm extended, moving the arm down could break off the arm on the conveyor below it. Also, if the gripper opened up while the arm was making a swing, the part would be dropped or thrown outward.

For these and other reasons, a ladder diagram with interlocks and sensors included for the robot's control should be developed. We do not develop the complete ladder con-

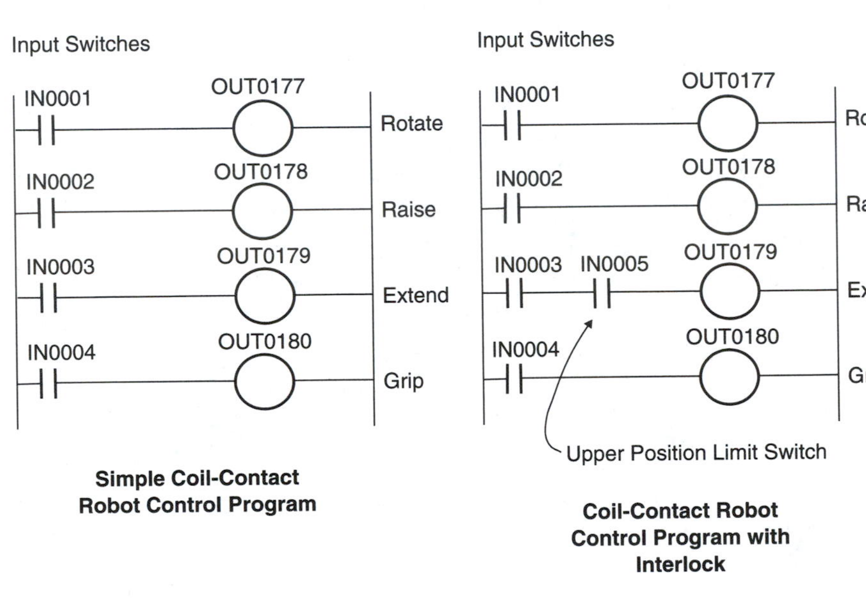

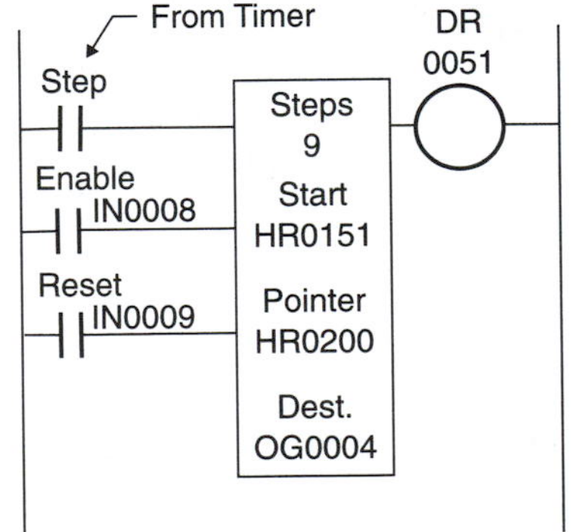

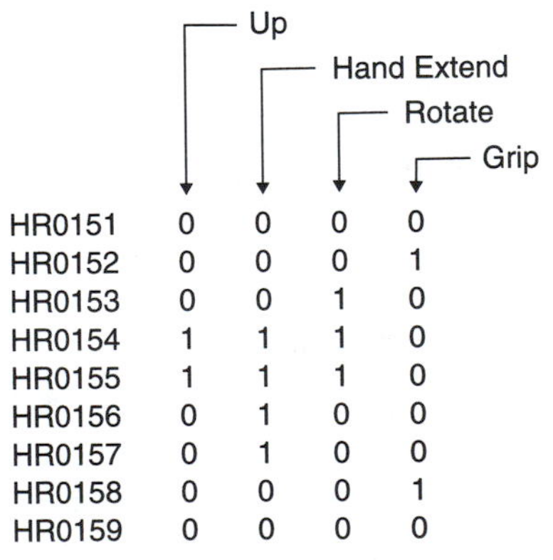

	Up	Hand Extend	Rotate	Grip
HR0151	0	0	0	0
HR0152	0	0	0	1
HR0153	0	0	1	0
HR0154	1	1	1	0
HR0155	1	1	1	0
HR0156	0	1	0	0
HR0157	0	1	0	0
HR0158	0	0	0	1
HR0159	0	0	0	0

Register Pattern - 4 Bits of 8 or 16

DR/SQ Robot Control Register Patterns

FIGURE 20–3
PLC Program for Robot Control

trol system in this chapter, but we do develop a drum controller/sequencer program of the type commonly used for robotic control.

Two basic programs for controlling the robot of figure 20–1 are shown in figure 20–3. The first program is a PLC version of the switch/relay system. The second is a DR function and registers. The DR is step-pulsed at intervals by a timer. The timer's preset times are determined by the interval or intervals required for each operation to be completed. As an example, if the arm swing time is four seconds, the time interval before starting the next step should be five seconds or more.

Again for this program, there are some possible operational problems. If a pneumatic cylinder failed or something became jammed, the PLC program would continue unabated. There could be equipment damage or even personnel injury. Additional programming would be necessary to include interlocks, sensors, positive emergency stops, and the like. The procedure for developing additional programming is covered in chapters 5 and 7.

The robot previously described is more accurately called *fixed automation*. Several different MOVE sequences would have made it a true robot. If, for example, you occasionally moved the part from B to A, instead of A to B, you would need two programs with different on-off patterns. The ability to change quickly from one program to another makes a system robotic. The insertion of different register patterns could be accomplished by MOVE functions. Note that a robot is loosely defined as a *reprogrammable manipulator*.

Next, consider the robotic control for the work cell shown in figure 20–4. The parts may come in on conveyor A or conveyor B and go out on either A or B. There are three possible process operations in the work area: drill 1/2 or 3/4 inch, countersink, and counterbore. A drilling operation for one of the two sizes is always done; countersink and counterbore may or may not be included for any particular part.

For this process, then, there are 32 possible combinations of individual moves for a given part (five alternatives of two possible operations is 2 raised to the fifth power, or 32). This requires a program catalog of up to 32 different programs. The various possible com-

FIGURE 20–4
Work Cell with Drilling/Boring Operations

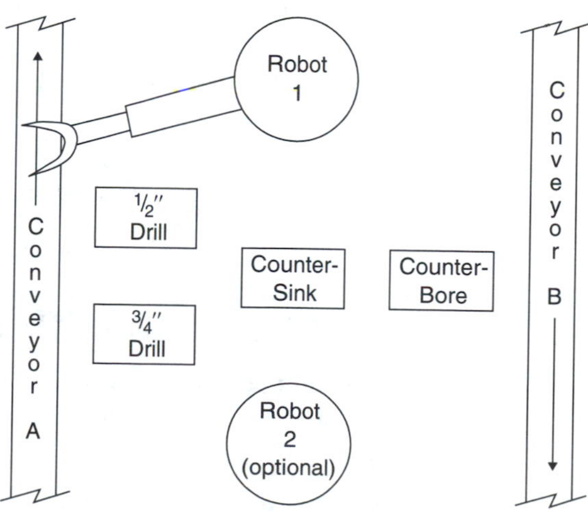

In on Conveyor	Drill Size	Counter-Sink	Counter-Drill	Out on Conveyor	
A	1/4″	Yes	Yes	A	
or	or	or	or	or	
B	1/2″	No	No	B	
	× 2 =	× 2 =	× 2 =	× 2 =	Combinational
2	4	8	16	32	Possibilities

FIGURE 20–5
Combinations of Operations for Figure 20–4

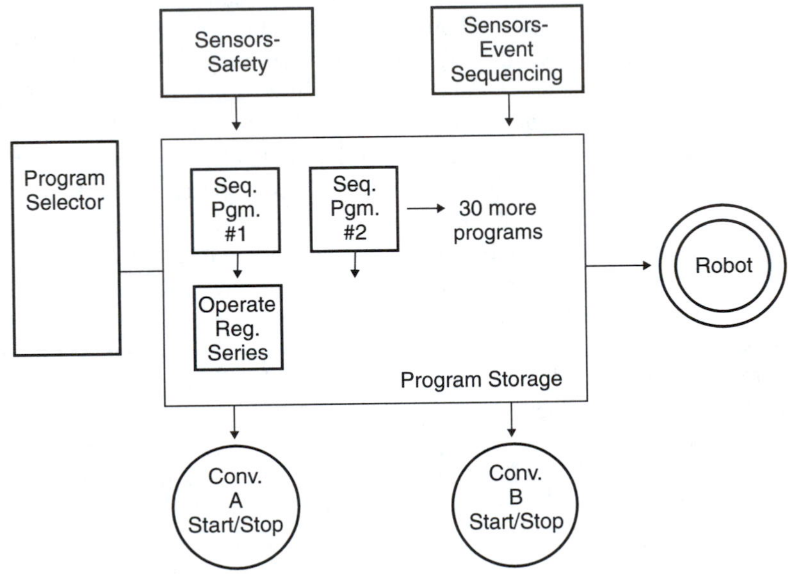

FIGURE 20–6
Programming Scheme for the Work Cell

binations are shown in figure 20–5. We do not write detailed PLC programs for the work-cell operation here, but we do show the PLC programming system in block form, as in figure 20–6.

20–3 INDUSTRIAL THREE-AXIS ROBOT WITH PLC CONTROL

An industrial-type robot is shown in figure 20–7. It has various motion and gripping capabilities similar to those in figure 20–1, namely:

1. Arm moves up or down, elevate.
2. Arm rotates 180 degrees.

3. Gripper rotates 180 degrees.
4. Gripper opens and closes.
5. Gripper extends and retracts.
6. Slide left or right, 5 stations—2 ends, 3 intermediate.

The robot shown in figure 20–7 is operated by applying 110 volts AC to various pneumatic solenoids. Each solenoid, when energized, lets its air valve supply air for operating the various functional motions. When off, no air is supplied, and the function is in its initial position. As in the robots discussed previously, the positions used are the extreme ones at one end of travel or the other. The only exception to end positions is motion 6: for slide right or left, there are three intermediate stopping points in addition to the two at the ends of travel.

This industrial robot differs from the previous one in that motions cease when a function is deenergized. It also differs in that some opposite motions require two inputs, one for each direction.

Various control methods are used to operate the robot or to program the robot to operate automatically. These include manual control, mechanical drum control, EPROM integrated circuit chip control, and computer control. Computer control may be further subdivided. One computer type has a teach pendant, which is used in conjunction with a

FIGURE 20–7
Industrial Pick-and-Place Robot (Courtesy of TII Robotics)

computer memory. It can record, remember, and repeat a series of steps performed manually. The other type is a straight computer program.

Also, many robotic devices can be controlled from a small computer (a master program is entered from a disk or computer memory). However, an increasing number of robots, such as the one in figure 20–7, are being controlled from programmable controllers.

The illustrated robot could be supplied with a PLC control package built in by the robot manufacturer. The next section illustrates how to use your own PLC to interface with the robot shown.

Before programming the PLC to control the robot, you must develop a scheme to connect and interface the PLC with the robot. Figure 20–8 shows the pin connections to the

Below is a listing of the 1/0 numbers and the letter that corresponds to the 1/0 on the cable between the 1/0 rack and control panel designed for the PLC. Also listed are the robot pin # and the corresponding function.

PLC Input #	Cable Letter	Robot Pin #	Robot Function
1	A	—	—
2	B	—	—
3	C	—	—
4	D	—	—
5	E	—	—
6	F	—	—
7	G	—	—
8	H	22	Aux. Input
9	J	23	Aux. Input
10	K	24	Aux. Input
11	L	25	Aux. Input
12	M	2	Station 1
13	N	3	Station 2
14	P	4	Station 3
15	Q	5	Station 4
16	R	6	Station 5
	S	7	Common Input

PLC Output #	Cable Letter	Robot Pin #	Robot Function
17	T	8	Grip
18	U	10	Elevate
19	V	15	Extend
20	W	11	Rotate CW
21	X	12	Rotate CCW
22	Y	13	Slide Right
23	Z	14	Slide Left
24	a	9	Rotate Grip
25	b	19	Aux. Output
26	c	20	Aux. Output
27	d	21	Aux. Output
28	e	—	—
29	f	—	—
30	g	—	—
31	h	—	—
32	j	—	—
	k	16	Common Output

FIGURE 20–8
Robot Control Cable Pin # Scheme

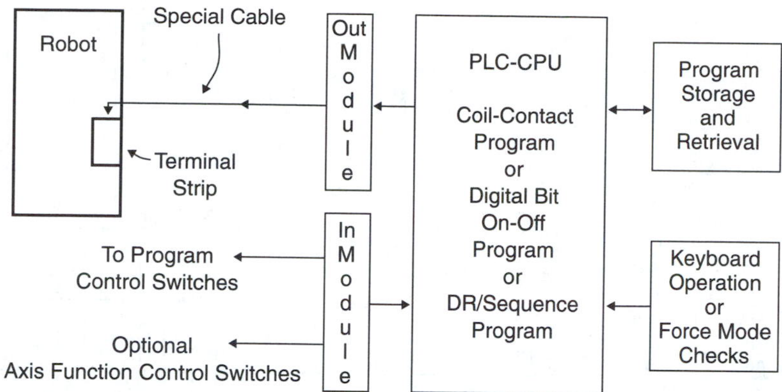

FIGURE 20–9
Robot-PLC Control System Block Diagram

robot and the necessary color code/wire numbers of the connecting cable. Since the robot uses 110 volts, you need a 110-volt interface I/O for PLC inputs and outputs. If you connect the ground and group common connections by direct wiring, the PLC needs only 13 output ports. You therefore would choose a 16-output PLC output module. Figure 20–9 shows a block diagram of this control scheme. Note that interlocks are not considered in this discussion, but would be included in advanced programs.

Now that the PLC and the robot are connected, you can choose a programming format. You may choose a contact and coil control format (explained earlier in this chapter), or you could use a digital bit program similar to those in chapter 18. This illustration uses the drum/sequencer discussed in chapter 19, which is the most commonly used format.

Figure 20–10 is a drum/sequencer program constructed on a program coding sheet. Blank coding sheets are furnished by the robot manufacturer for ease of program formulation. The sequential steps of operation go from top to bottom. A step description is written on the left of the sheet under *Sequence of Events*. The off-on pattern is indicated by X's for on and a blank space (or sometimes O's) for off. A listing of functional status for each motion is listed along the top (Rotate CCW, Rotate CW, etc.).

The sequence of events to accomplish the operational objectives is determined and listed. Then, the X's are filled in for the functions that will be actuated. Note that in many cases, more than one function must be on at a given step. The final part of the procedure is to record the assigned PLC operation numbers for each step (28, 27, etc.). Our program uses the output group 2 register, OG2. Therefore, the 16 individual outputs are 17 through 32. A cross-reference of cable numbers and letters corresponding to operational numbers is shown at the bottom of figure 20–8.

Figure 20–11 shows a typical PLC program that could be used for the robot sequencer program. The programming follows the system of chapter 19. It is operated on a 5-second time interval, as set in the timer program. Note that the pattern of regular bits cor-

Switch No. / Step No. / Time Interval / Sequence of Events						ROTATE CCW	ROTATE CW	UP		SLIDE RIGHT		GRIP	SLIDE LEFT	EXTEND			GRIP ROTATE
Station						28	27	26	25	24	23	22	21	20	19	18	17
HOME POSITION (INITIALIZE)	H																
ROTATE CW SLIDE RIGHT	1					X				X							
SLIDE LEFT, MAN. UP	2							X					X				
MAN. UP, EXTEND	3							X						X			
MAN. UP, EXTEND, GRIP	4							X				X		X			
MAN. UP, EX., GRIP, ROT. CW	5						X	X				X		X			
GRIP, EX., GRIP ROTATE	6											X		X			X
EXTEND, GRIP ROTATE	7													X			X
GRIP ROTATE, MAN. UP	8							X									X
SLIDE RIGHT, MAN. UP	9							X		X							
EXTEND, MAN. UP	10							X						X			
GRIP, EXTEND, MAN. UP	11							X				X		X			
GRIP, EX., MAN. UP, ROTATE CCW	12					X		X				X		X			
GRIP, EXTEND	13											X		X			
EXTEND	14													X			
NEUTRAL POSITION	15																
SLIDE LEFT, GRIP ROTATE	16												X				X
EXTEND, GRIP ROTATE	17													X			X
GRIP, EXTEND, GRIP ROTATE	18											X		X			X
MAN. UP, SLIDE RIGHT, GRIP, EX.	19							X		X		X		X			
MAN. UP, EXTEND	20							X						X			
MAN. UP	21							X									
ROTATE CW, GRIP ROTATE	22						X										X
EXTEND, GRIP ROTATE	23													X			X
GRIP, EXTEND, GRIP ROTATE	24											X		X			X
MAN. UP, GRIP, SLIDE LEFT, EX.	25							X				X	X	X			
	26																
	27																
	28																
	29																
	30																

FIGURE 20–10
Program Code Sheet for Programmable Logic Controller (Courtesy of TII Robotics)

responds to the coding sheet of figure 20–10. Only the first 12 bits are needed, because there are only 12 functions to be controlled.

The program shown in figure 20–10 is time based. In actual operation, for equipment and personnel safety, the program could be entirely or partially event based. There would be limit switches or sensors for position or part-in-place indication. An ensuing operational step would not be able to start until the preceding step is completed successfully. For example, if there were no part to be picked up, the sequence (not shown) would stop or the arm would return to home position.

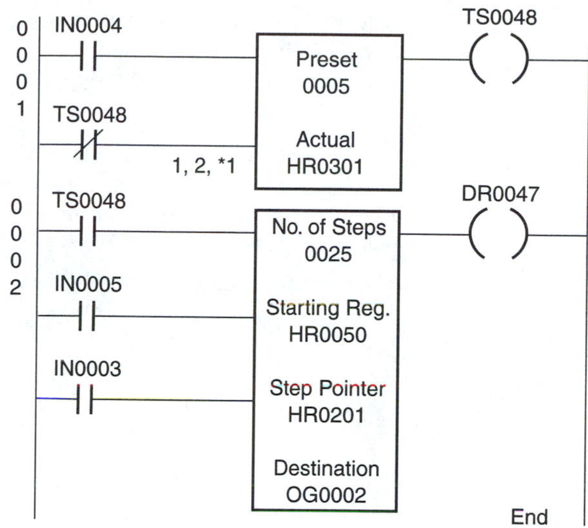

FIGURE 20–11
PLC Robot Control Program and Register Pattern

HR0050	0000	0000	0000	0000
HR0051	0000	1000	1000	0000
HR0052	0000	0010	0001	0000
HR0053	0000	0010	0000	1000
HR0054	0000	0010	0010	1000
HR0055	0000	0110	0010	1000
HR0056	0000	0000	0010	1001
HR0057	0000	0000	0000	1001
HR0058	0000	0010	0000	0001
HR0059	0000	0010	1000	0000
HR0060	0000	0010	0000	1000
HR0061	0000	0010	0010	1000
HR0062	0000	1010	0010	1000
HR0063	0000	0000	0010	1000
HR0064	0000	0000	0000	1000
HR0065	0000	0000	0000	0000
HR0066	0000	0000	0001	0001
HR0067	0000	0000	0000	1001
HR0068	0000	0000	0010	1001
HR0069	0000	0010	1010	1000
HR0070	0000	0010	0000	1000
HR0071	0000	0010	0000	0000
HR0072	0000	0100	0000	0001
HR0073	0000	0000	0000	1001
HR0074	0000	0000	0010	1001
HR0075	0000	0010	0011	1000

EXERCISES

1. Refer to the robot in figure 20–1. Develop a pattern similar to that in figure 20–2 for a different sequence of operation. You must move a part from position B to position A. The robot starts at the lower-left, initialized position.
2. For exercise 1, develop a drum sequencer program to move the part from B to A.
3. The workstation/robot in figure 20–12 is similar to the one in figure 20–1. It differs in that it has six active positions, not four. It has a positioning solenoid for each of the six positions shown. It also has arm-extend and gripper-close actuators, as in figure 20–1. Develop a coil-and-contact PLC program to accomplish moving a part from LM to UR. Also develop a program to move a part from UR to LL.
4. The robot shown in figure 20–13 is similar to the robot in figure 20–7. Develop a timed sequence

FIGURE 20–12
Diagram for Exercise 3

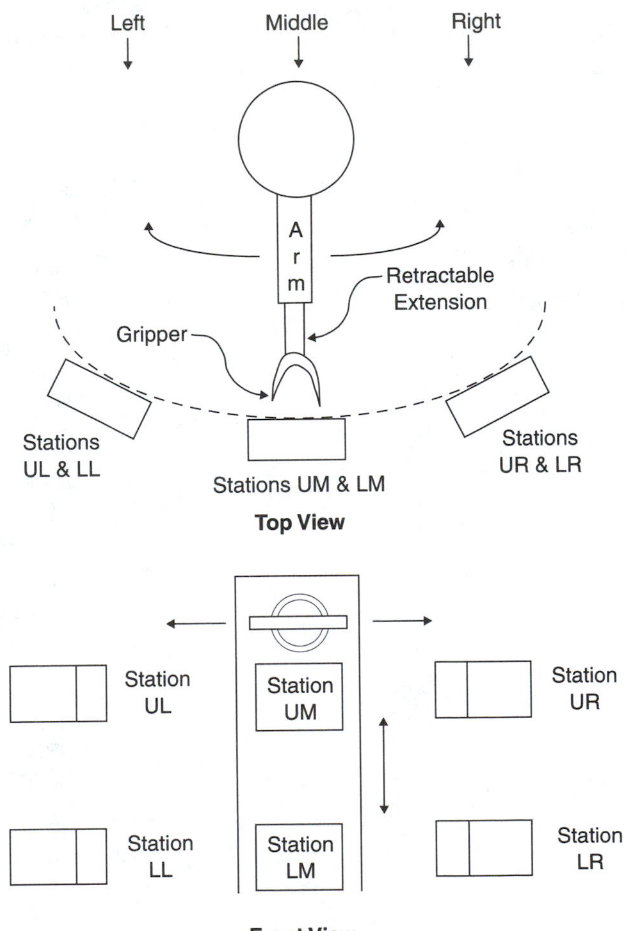

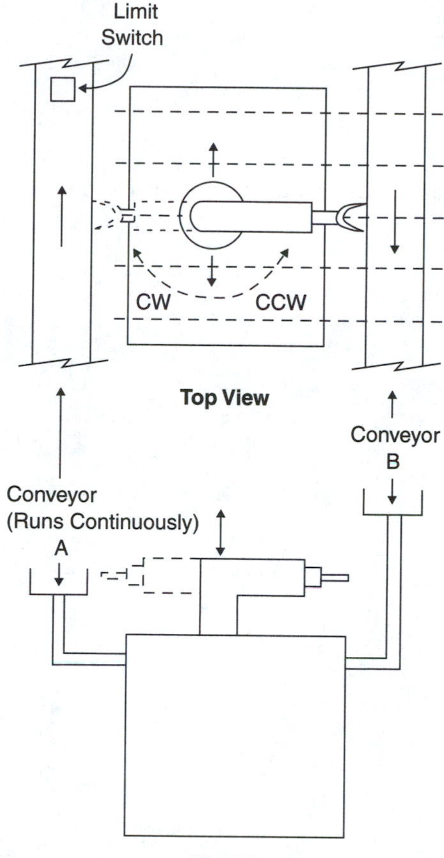

Motions are:
1. Along center axis;
   5 possible stopping points
2. Arm up and down
3. Arm rotate 180°
4. Extend
5. Grip

FIGURE 20–13
Diagram for Exercise 4

program (times of your choice) to accomplish the sequence below. An example program was illustrated in Figure 20–10.

1. Initial position is as shown.
2. Pick up part at A5.
3. Move part to B4.
4. Return to initial position.

5. Develop a timed sequence program to accomplish this longer program in the same manner as in exercise 3.

1. Initial position is as shown.
2. Pick up part at B1.
3. Move part to A4.
4. Conveyor moves part along.

5. Part reaches position A1.
6. Limit switch actuated by part.
7. Robot picks up part at A1.
8. Part is moved to B5.
9. Return to initial position.

For this exercise, two programs may be needed, separated by an LS1 actuation ladder logic line.

21

PLC Matrix Functions

OUTLINE

OBJECTIVES

At the end of this chapter, you will be able to

- □ Define the concepts of the COMPLEMENT and COMPARE functions.
- □ Describe the PLC matrix construction system in register form.
- □ Describe and program the following matrix functions: AND, OR, XOR, COMPLEMENT, COMPARE.
- □ Use matrix functions in combination to simulate combination gates such as NAND and NOR.

21–1 INTRODUCTION

The word *matrix* can bring to mind a complicated and tedious mathematical procedure involving determinates, comparisons, cross multiplication, and other time-consuming operations. The PLC matrix system eliminates the complication by enabling you to do a large number of comparisons or logic operations in a concise and orderly manner. The numbers involved in regular matrix algebra can be any decimal value: 13, -28, 45.782, 134567.2, and so on; the PLC matrix system involves only 1's and 0's. Furthermore, the PLC matrix system does not involve cross multiplication. It is a special method for handling bulk data manipulations.

The PLC matrix works with bits in one or two matrices and produces one resulting matrix. Chapter 6 covers digital gates. This chapter applies these various digital gates in large groups.

21–2 APPLYING MATRIX FUNCTIONS TO REDUCE PROGRAM LENGTH

Suppose that you had 207 pilot lights, each of which is to go on only if both of two contacts are closed. This means that you have to program 207 lines with two contacts and a coil for a pilot light for each line. See figure 21–1a for a conventional PLC program for the 207 lines. An alternative to the 207 lines of programming is one line of a matrix function. In this case we would program an AND matrix, which is explained in the next section. Energizing the matrix function effectively scans all 207 lines for Examine on and turns the pilot lights on or off. There is, of course, a catch. You must move each contact's on–off status (1 or 0) into appropriate registers by a MOVE or BLOCK MOVE operation, and then move the resulting status out to the pilot lights in a similar manner.

Next, suppose that you had 207 pilot lights, each of which is to go on if either or both of two contacts is on. (See figure 21–1b for this conventional programming.) Alternatively, you can use only one line of programming with the same process as before, using an OR matrix. Other comparable matrix functions are available as well. This enables the programmer to save a large number of lines of programming, if data movement into and out of registers is feasible.

21–3 THE PLC AND AND/OR MATRIX FUNCTIONS

Four coils, each of which can be energized by two inputs in series, result in four AND situations. The coils are programmed on the PLC in the usual manner shown in figure 21–2.

The upper section of figure 21–3 shows the original four coils and eight inputs arranged in a 2-by-2 matrix. A 1 represents on and a 0, off, in the conventional manner. Each bit of matrix A is used with the corresponding bit of matrix B. The bits, when used in a PLC AND matrix, are analyzed for an AND situation. The result of the analysis is put in the corresponding bit location of matrix C.

In actual operation, the input data, or status, is contained in two series of registers. The AND operation for the two series then takes place, and the results are put in another series of registers. This data is then moved to output registers. The equivalent register operation is shown in the lower section of figure 21–3.

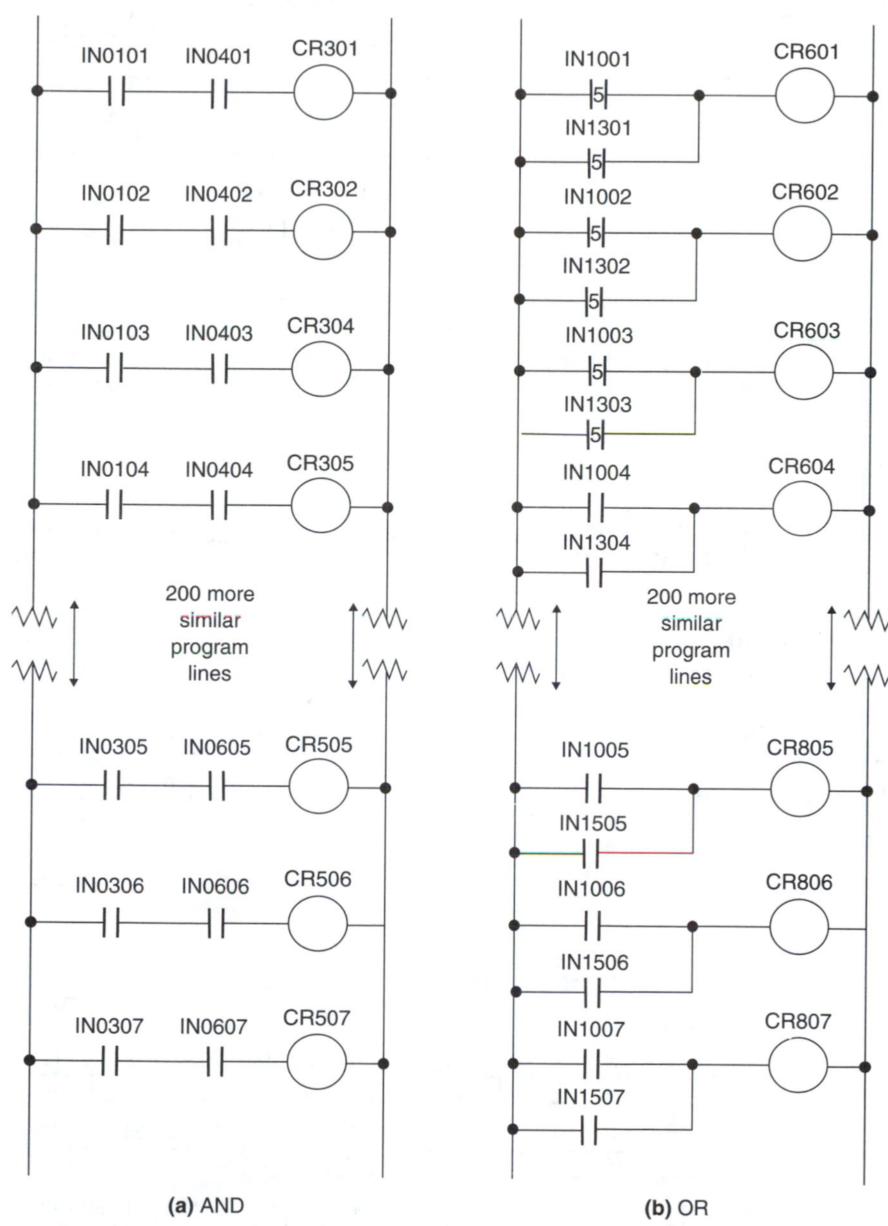

FIGURE 21–1
Long Repetitive Programs

FIGURE 21–2
Four Outputs with Two Series Inputs Each

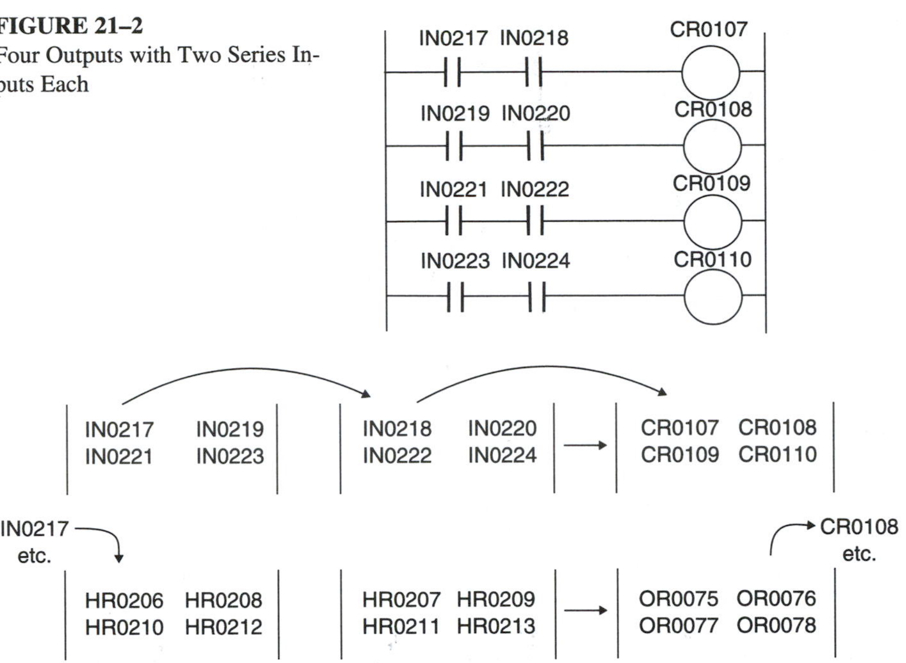

FIGURE 21–3
Matrix Arrangement for Figure 21–2

FIGURE 21–4
Two-by-Two AND Analysis for Figure 21–3

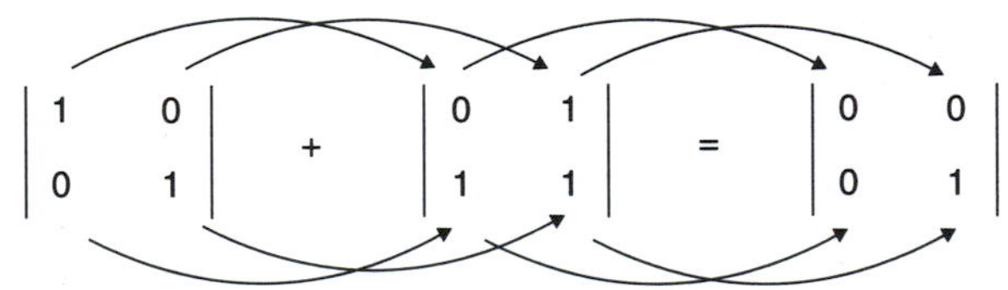

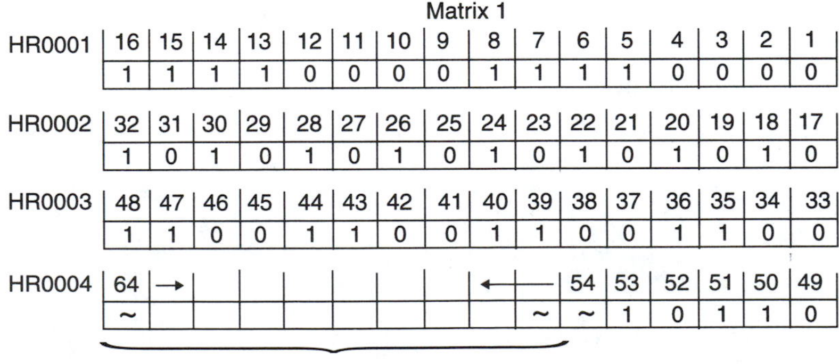

Matrix 1

HR0001	16	15	14	13	12	11	10	9	8	7	6	5	4	3	2	1
	1	1	1	1	0	0	0	0	1	1	1	1	0	0	0	0
HR0002	32	31	30	29	28	27	26	25	24	23	22	21	20	19	18	17
	1	0	1	0	1	0	1	0	1	0	1	0	1	0	1	0
HR0003	48	47	46	45	44	43	42	41	40	39	38	37	36	35	34	33
	1	1	0	0	1	1	0	0	1	1	0	0	1	1	0	0
HR0004	64	→							←		54	53	52	51	50	49
	~									~	~	1	0	1	1	0

Not used—
Status irrelevant

FIGURE 21–5
PLC Matrix for 53 Functions

Figure 21–4, a matrix AND operation, assumes that some of the inputs are on and some are off for figure 21–3. The resulting outputs shown are determined by the PLC multiple AND analysis.

You have used a 2-by-2 matrix for four AND functions. Next, suppose you had 53 coils, each with two series inputs for actuation. It would take a long time and a lot of PLC memory to program the 106 contacts and the 53 coils. Using the AND matrix system makes programming a lot more straightforward. A typical PLC uses sixteen 16-bit registers to give a 256-bit matrix. Using two input register matrices and one output matrix does up to 256 AND functions at once. This illustration uses only 53 of the 256 available register bits and four registers.

Figure 21–5 shows how to use four of the 16 registers available to perform the 53

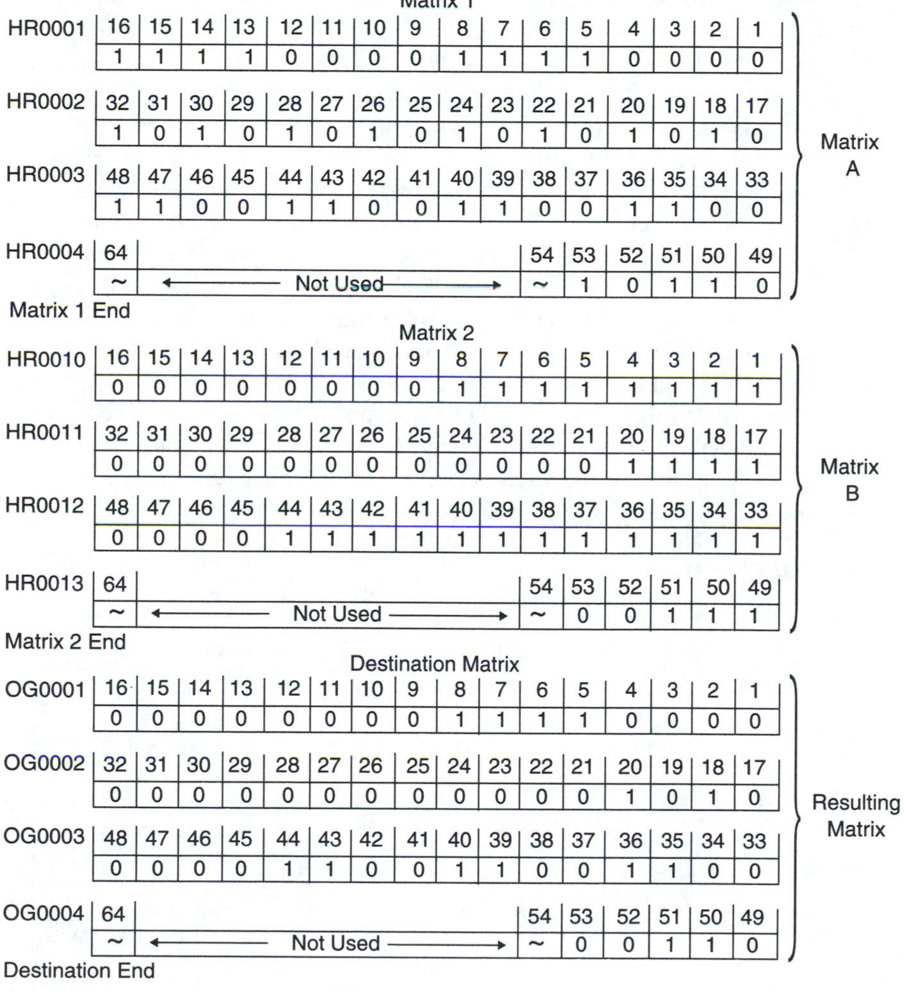

FIGURE 21–6
AND Matrix Results

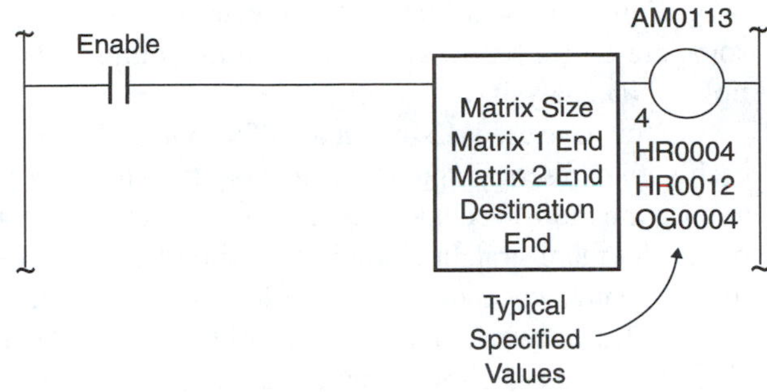

FIGURE 21–7
Typical PLC AND Matrix Function

AND functions. Tell the PLC how many registers will be used in the operation; in this case, four. We use three full registers for the first 3 × 16, or 48, bits. The last 5 bits go in the first part of the next (fourth) register. This leaves 11 unused bits in the fourth register. This way, the other 13 registers are not involved, thus saving memory.

Assume some on and some off statuses for the 53 input AND matrices. The results of the matrix operation appear in another matrix, as shown in figure 21–6.

How is the PLC programmed to do the PLC AND operation? Figure 21–7 shows a typical PLC AND function. The coil is assigned a number in the usual manner. The AND function is carried out when the input is turned on, as in most other functions. The illustration shows a general block configuration and a typical, specific, programmed AND function. In figure 21–7, you must also tell the PLC which registers to use by specifying the last register of each group of inputs and the output. Some program systems specify the first register, not the last. In this case, the registers used for each part are the one specified plus the previous three. The coil of all matrix functions goes on when the functional operation is completed.

Matrix size varies among manufacturers. Maximum allowable size also varies, depending on the type of register used. Figure 21–8 shows one manufacturer's permissible variations.

The OR matrix operates like the AND matrix, except that the bits in two matrices are compared on an OR logic basis instead of by an AND analysis. Figure 21–9 uses the same

FIGURE 21–8
Typical Allowable PLC Matrix Size

Type	Limit
HR	≤ 1792
IR	≤ 32 (PLC-700) ≤ 8 (PLC-900A) ≤ 16 (PLC-900B)
OR	≤ 32 (PLC-700) ≤ 8 (PLC-900A) ≤ 16 (PLC-900B)
IG	≤ 16 (PLC-700) ≤ 8 (PLC-900A/B)
OG	≤ 32 (PLC-700) ≤ 8 (PLC-900A) ≤ 16 (PLC-900B)

Various PLC Model Numbers

53 on and off patterns that were used for A and B in the AND example in figure 21–6. The results in matrix C are now determined on an OR basis instead of an AND basis.

Figure 21–10 is a typical OR function layout. The size limitation of the matrix varies in the same manner as the AND, as shown in figure 21–8. Operational procedure is essentially the same.

The EXCLUSIVE OR gate (XOR) is somewhat like the OR function, except that the output is not on when both inputs are on. The output is off when neither input is on, as in the OR function, and the output is on when either input is on, as with the OR function. However, when both inputs are on, the output is off. Figure 21–11 uses the same 53 inputs as in previous examples. The output matrix in this XOR example shows how the XOR is applied.

The PLC function for XOR is shown in figure 21–12. Size limits and operation are

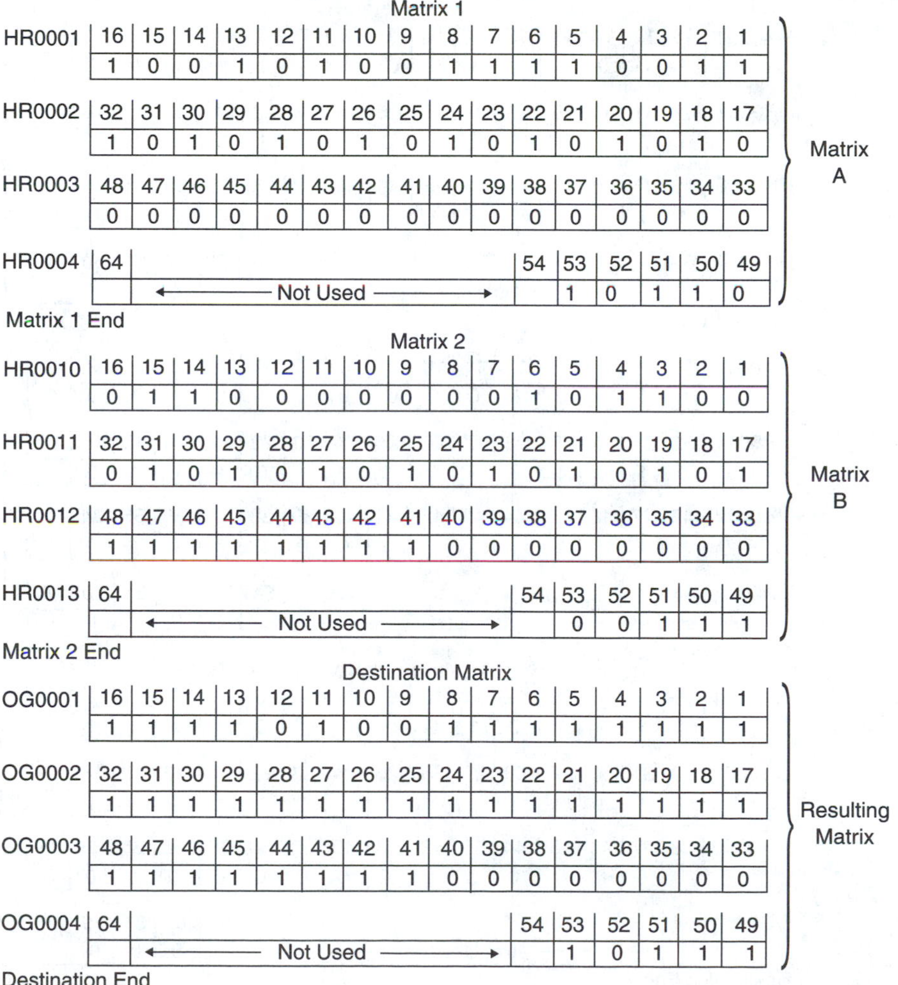

Matrix 1 (Matrix A)

HR0001	16	15	14	13	12	11	10	9	8	7	6	5	4	3	2	1
	1	0	0	1	0	1	0	0	1	1	1	1	0	0	1	1
HR0002	32	31	30	29	28	27	26	25	24	23	22	21	20	19	18	17
	1	0	1	0	1	0	1	0	1	0	1	0	1	0	1	0
HR0003	48	47	46	45	44	43	42	41	40	39	38	37	36	35	34	33
	0	0	0	0	0	0	0	0	0	0	0	0	0	0	0	0
HR0004	64	← Not Used →									54	53	52	51	50	49
												1	0	1	1	0

Matrix 1 End

Matrix 2 (Matrix B)

HR0010	16	15	14	13	12	11	10	9	8	7	6	5	4	3	2	1
	0	1	1	0	0	0	0	0	0	0	1	0	1	1	0	0
HR0011	32	31	30	29	28	27	26	25	24	23	22	21	20	19	18	17
	0	1	0	1	0	1	0	1	0	1	0	1	0	1	0	1
HR0012	48	47	46	45	44	43	42	41	40	39	38	37	36	35	34	33
	1	1	1	1	1	1	1	1	0	0	0	0	0	0	0	0
HR0013	64	← Not Used →									54	53	52	51	50	49
												0	0	1	1	1

Matrix 2 End

Destination Matrix (Resulting Matrix)

OG0001	16	15	14	13	12	11	10	9	8	7	6	5	4	3	2	1
	1	1	1	1	0	1	0	0	1	1	1	1	1	1	1	1
OG0002	32	31	30	29	28	27	26	25	24	23	22	21	20	19	18	17
	1	1	1	1	1	1	1	1	1	1	1	1	1	1	1	1
OG0003	48	47	46	45	44	43	42	41	40	39	38	37	36	35	34	33
	1	1	1	1	1	1	1	1	0	0	0	0	0	0	0	0
OG0004	64	← Not Used →									54	53	52	51	50	49
												1	0	1	1	1

Destination End

FIGURE 21–9
OR Matrix Results

FIGURE 21–10
OR Matrix Function

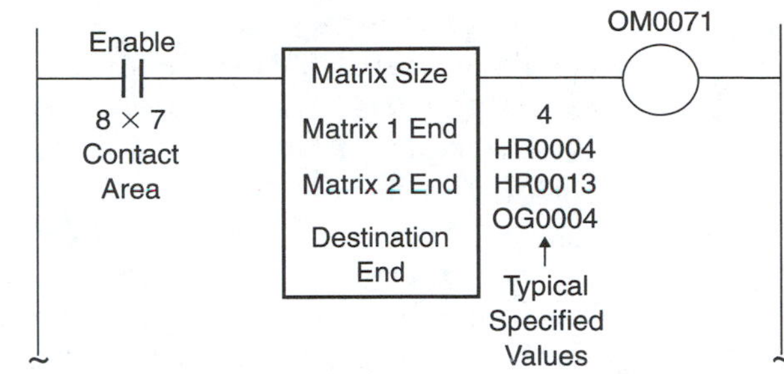

Matrix 1 (Matrix A)

HR0001	16	15	14	13	12	11	10	9	8	7	6	5	4	3	2	1
	1	0	0	1	0	1	0	0	1	1	1	1	0	0	1	1
HR0002	32	31	30	29	28	27	26	25	24	23	22	21	20	19	18	17
	1	0	1	0	1	0	1	0	1	0	1	0	1	0	1	0
HR0003	48	47	46	45	44	43	42	41	40	39	38	37	36	35	34	33
	1	1	1	1	1	1	1	1	0	0	0	0	0	0	0	0
HR0004	64										54	53	52	51	50	49
		← Not Used →										1	0	1	1	0

Matrix 1 End

Matrix 2 (Matrix B)

HR0010	16	15	14	13	12	11	10	9	8	7	6	5	4	3	2	1
	0	1	1	0	0	0	0	0	0	0	1	0	1	1	0	0
HR0011	32	31	30	29	28	27	26	25	24	23	22	21	20	19	18	17
	0	1	0	1	0	1	0	1	0	1	0	1	0	1	0	1
HR0012	48	47	46	45	44	43	42	41	40	39	38	37	36	35	34	33
	1	1	1	1	1	1	1	1	0	0	0	0	0	0	0	0
HR0013	64										54	53	52	51	50	49
		← Not Used →										0	0	1	1	1

Matrix 2 End

Destination Matrix (Resulting Matrix)

OG0001	16	15	14	13	12	11	10	9	8	7	6	5	4	3	2	1
	1	1	1	1	0	1	0	0	1	1	0	1	1	1	1	1
OG0002	32	31	30	29	28	27	26	25	24	23	22	21	20	19	18	17
	1	1	1	1	1	1	1	1	1	1	1	1	1	1	1	1
OG0003	48	47	46	45	44	43	42	41	40	39	38	37	36	35	34	33
	0	0	0	0	0	0	0	0	0	0	0	0	0	0	0	0
OG0004	64										54	53	52	51	50	49
		← Not Used →										1	0	0	0	1

Destination End

FIGURE 21–11
XOR Matrix Results

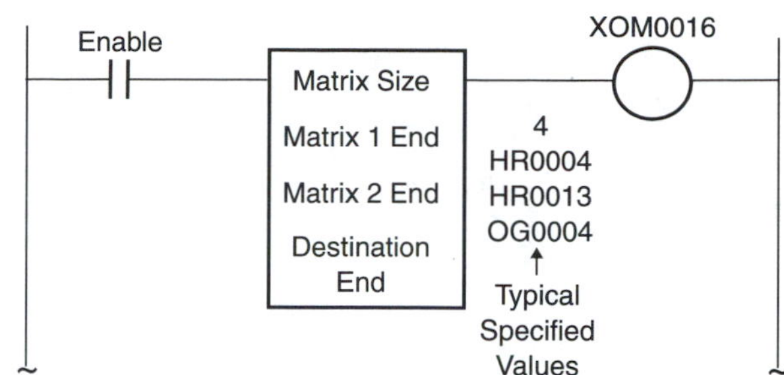

FIGURE 21–12
XOR Matrix Function

the same as for the AND and OR functions. One possible function of the XOR matrix function is setting all bits in a matrix to 0. This is accomplished by doing an XOR of a given matrix twice. Any 1 becomes 0 and any 0 becomes 0 also.

21–4 THE PLC COMPLEMENT AND COMPARE MATRIX FUNCTIONS

In some cases in a long PLC program, you may wish to turn a number of devices to their opposite state. The COMPLEMENT function allows you to do so. All devices that are on can be turned off, and vice versa. Effectively, this function changes all 1's in a matrix of applicable registers to 0's and all 0's to 1's. Figure 21–13 shows the result of complementing register A, from our previous example, to register C.

Origin Matrix

HR0001	16	15	14	13	12	11	10	9	8	7	6	5	4	3	2	1
	1	0	0	1	0	1	0	0	1	1	1	1	0	0	1	1
HR0002	32	31	30	29	28	27	26	25	24	23	22	21	20	19	18	17
	1	0	1	0	1	0	1	0	1	0	1	0	1	0	1	0
HR0003	48	47	46	45	44	43	42	41	40	39	38	37	36	35	34	33
	1	1	1	1	1	1	1	1	0	0	0	0	0	0	0	0
HR0004	64										54	53	52	51	50	49
		← Not Used →										1	0	1	1	0

Destination Matrix

Complemented Matrix

OG0001	16	15	14	13	12	11	10	9	8	7	6	5	4	3	2	1
	0	1	1	0	1	0	1	1	0	0	0	0	1	1	0	0
OG0002	32	31	30	29	28	27	26	25	24	23	22	21	20	19	18	17
	0	1	0	1	0	1	0	1	0	1	0	1	0	1	0	1
OG0003	48	47	46	45	44	43	42	41	40	39	38	37	36	35	34	33
	0	0	0	0	0	0	0	0	1	1	1	1	1	1	1	1
OG0004	64										54	53	52	51	50	49
		← Not Used →										0	1	0	0	1

FIGURE 21–13
COMPLEMENT Matrix Results

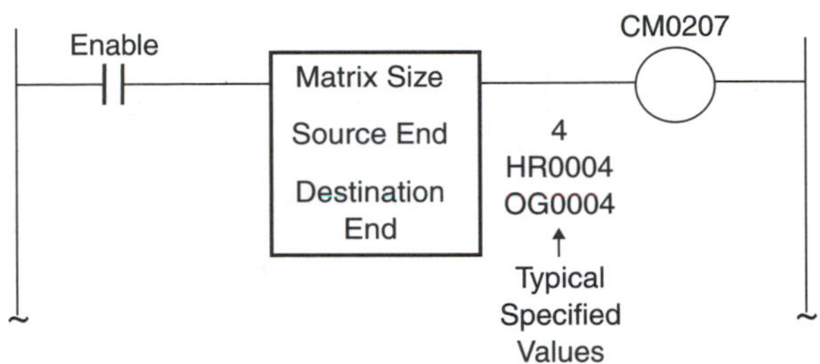

FIGURE 21–14
COMPLEMENT Function

Inputs		Output	Same?
A	B	C	↓
0	0	1	Yes
0	1	0	No
1	0	0	No
1	1	1	Yes

FIGURE 21–15
COMPARE Function Truth Table

Matrix 1 (Matrix A)

HR0001	16	15	14	13	12	11	10	9	8	7	6	5	4	3	2	1
	1	0	0	1	0	1	0	0	1	1	1	1	0	0	1	1
HR0002	32	31	30	29	28	27	26	25	24	23	22	21	20	19	18	17
	1	0	1	0	1	0	1	0	1	0	1	0	1	0	1	0
HR0003	48	47	46	45	44	43	42	41	40	39	38	37	36	35	34	33
	0	0	0	0	0	0	0	0	0	0	0	0	0	0	0	0
HR0004	64	← Not Used →									54	53	52	51	50	49
												1	0	1	1	0

Matrix 1 End

Matrix 2 (Matrix B)

HR0010	16	15	14	13	12	11	10	9	8	7	6	5	4	3	2	1
	0	1	1	0	0	0	0	0	0	0	1	0	1	1	0	0
HR0011	32	31	30	29	28	27	26	25	24	23	22	21	20	19	18	17
	0	1	0	1	0	1	0	1	0	1	0	1	0	1	0	1
HR0012	48	47	46	45	44	43	42	41	40	39	38	37	36	35	34	33
	1	1	1	1	1	1	1	1	0	0	0	0	0	0	0	0
HR0013	64	← Not Used →									54	53	52	51	50	49
												0	0	1	1	1

Matrix 2 End

Destination Matrix (Compared Matrix)

OG0001	16	15	14	13	12	11	10	9	8	7	6	5	4	3	2	1
	0	0	0	0	1	0	0	1	0	0	1	0	0	0	0	0
OG0002	32	31	30	29	28	27	26	25	24	23	22	21	20	19	18	17
	0	0	0	0	0	0	0	0	0	0	0	0	0	0	0	0
OG0003	48	47	46	45	44	43	42	41	40	39	38	37	36	35	34	33
	0	0	0	0	0	0	0	0	1	1	1	1	1	1	1	1
OG0004	64	← Not Used →									54	53	52	51	50	49
												0	1	1	1	0

Destination End

FIGURE 21–16
COMPARE Matrix Results

FIGURE 21–17
COMPARE Function

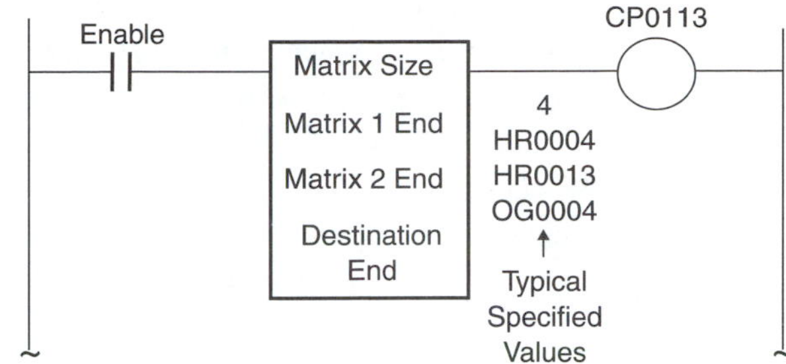

The PLC function for COMPLEMENT is shown in figure 21–14. Limits and operation are again the same.

Another form of matrix logic is the COMPARE function, which compares two bits. If they are the same, the function outputs a 1. If the original bits are different, it outputs a 0. Figure 21–15 is a truth table for this function.

Again using the 53 bits in the matrix from the previous example, compare matrix A bits with those in B, as shown in figure 21–16. Matrix C shows which are the same and which are different by putting out 1 or 0, respectively.

The PLC function for COMPARE is shown in figure 21–17. The usual size and operation description again applies. A frequently used form of the COMPARE function is the SEARCH matrix, which operates similarly to COMPARE. Your PLC might have a different name for this type of operation; check your operations manual.

21-5 COMBINATION PLC MATRIX OPERATIONS

Several matrix operations can be combined to perform special functions.

The NAND gate is an AND gate with an inverted output. To perform a matrix NAND, program two matrix operations in series, as shown in figure 21–18. First the AND matrix operation is performed from two input matrices. The result is placed in an output matrix. Next, the AND output matrix is complemented by the COMPLEMENT matrix function. The final output from the COMPLEMENT function is the NAND result for the original two matrices.

The NOR matrix operation is also a combination of two consecutive matrix operations. The results of an OR matrix are run through a COMPLEMENT matrix. The complemented resulting values are the NOR result of the original two matrices. Figure 21–19 shows how this function is programmed. The process is similar to the previously described NAND function.

Inversions of inputs are accomplished by using a COMPLEMENT function on the front end of the matrix operation. If only a portion of the inputs must be inverted before use,

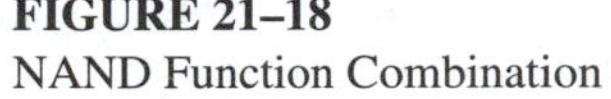
FIGURE 21–18
NAND Function Combination

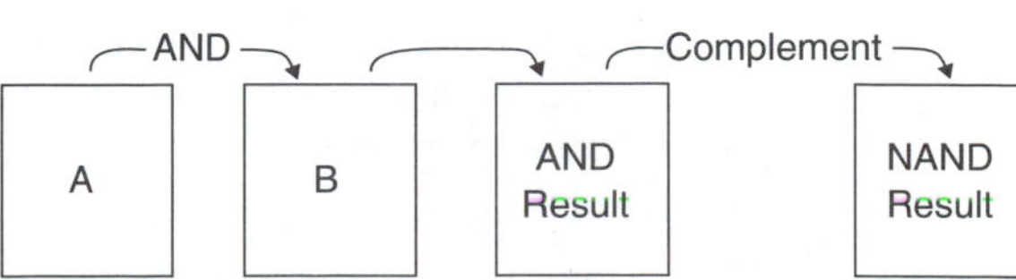

FIGURE 21–19
NOR Function Combination

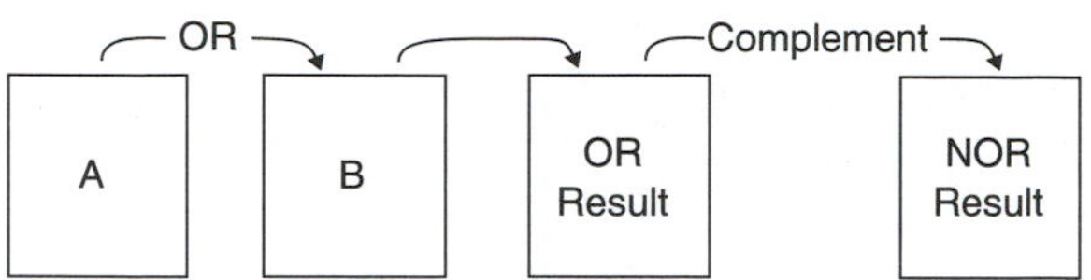

other detailed operations are needed. These might be a number of MOVE functions or other data move systems available in the PLC function list. The EXCLUSIVE OR function may be used to perform a selective complementation.

EXERCISES

The following two matrices are to be used for exercises 1 through 7:

Matrix A								Matrix B							
1	0	0	1	1	1	1	0	0	1	1	0	1	0	0	1
0	0	1	1	0	0	0	1	1	1	1	1	1	1	1	1
1	1	1	0	0	1	1	1	1	0	1	0	1	0	1	1
0	1	0	1	0	0	0	1	0	0	1	1	1	1	0	0
1	1	0	1	1	1	0	0	1	0	1	0	1	0	1	1
0	0	0	1	0	0	1	0	1	1	0	1	0	0	1	0
0	0	0	0	0	0	0	1	1	0	1	1	0	0	0	1
1	0	1	0	1	0	1	0	0	1	0	0	0	1	1	1

1. AND A and B to determine matrix C.
2. OR A and B to determine matrix D.
3. XOR A and B to determine matrix E.
4. COMPARE A and B to determine matrix F.
5. COMPLEMENT matrices A, C, and E to determine matrices G, H, and J.
6. NAND B and C to determine matrix K.
7. NOR A and C to determine matrix L.

Examine the following examples by programming them on a PLC:

8. Figure 21-6, AND.
9. Figure 21-9, OR.
10. Figure 21-11, XOR.
11. Figure 21-13, COMPLEMENT.
12. Figure 21-16, COMPARE.
13. Figure 21-18, NAND.
14. Figure 21-19, NOR.
15. Create two matrices of your own design with 43 active bits. In a 16-bit PLC, matrix analysis will require two full registers plus part of a third. Find the solutions to functions of your choice manually and by the PLC. Compare the results for exact correspondence.
16. How could you use matrix functions to set all bits to a 1 in a matrix?

VII

Advanced PLC Functions

22

Analog PLC Operation

OUTLINE

22–1 Introduction □ **22–2** Types of PLC Analog Modules and Systems □ **22–3** PLC Analog Signal Processing □ **22–4** BCD or Multibit Data Processing □ **22–5** PLC Analog Output Application Examples

OBJECTIVES

At the end of this chapter, you will be able to

- □ Differentiate between discrete and analog operation of a PLC.
- □ List and define the various major types of PLC analog inputs and outputs.
- □ Describe the data flow and number conversions involved in PLC analog operation.
- □ Convert input signals to a form usable by input modules.
- □ Convert output module signals to usable values for output devices.
- □ Describe the internal PLC operation for analog I/O operation.
- □ Program a PLC for use with both BCD and binary analog systems.
- □ Apply the analog PLC function's operation to industrial problems.

22–1 INTRODUCTION

So far we have dealt with discrete PLC operation; input and output statuses have been on or off. In this chapter we consider analog PLC operation. Analog PLC control can be used to control any process with variables as a control consideration. Many medium-size and large PLCs are able to deal with analog signals as having discrete functions. For analog operation, the level of a PLC input signal is sensed by an analog input module. In addition, the level of the output can be a variable value sent to the process from an analog output module. The PLC analog input capability enables you to monitor such devices as thermal indicators, pressure transducers, electrical potentiometers, and many other data input devices with varying signal values. Output PLC analog control devices can be positioned at many intermediate positions. This output control contrasts with discrete control, which operates only at its two extremes.

BCD PLC analog input and output value ranges are divided into a number of steps. BCD analog input devices include thumbwheels, encoders, and the like. Analog output devices control such devices as digital numbers, seven-segment displays, and stepper motors.

PLC analog capabilities allow many different actions for one single input, depending on the input's value. For example, a process in which 20 lights are used to indicate how full a tank is in 5 percent increments needs only one analog input and one sensor; a discrete system needs 20 on–off sensors and 20 inputs. Analog output programs have similar advantages; for example, a single analog output can position a valve in many different positions.

Analog capability enables you to control continuous processes in such industries as chemical and petroleum. Any number of variable input signals can be received by a PLC module and then processed mathematically by the CPU. The resulting analog value or values are then sent to an output module. The analog output module signal then controls a variable process or processes.

22–2 TYPES OF PLC ANALOG MODULES AND SYSTEMS

Analog PLC systems are of two general types: BCD and straight numerical. The BCD analog PLC system is sometimes called the multibit type. (Chapter 13 covers the BCD numbering system in detail.) Figure 22–1 shows the operation of a thumbwheel input to an input BCD module. BCD codes are fed into the PLC input module from the thumbwheel output. Other possible BCD-type inputs are bar-code readers and encoders. A BCD output module is also shown in figure 22–1. In this case, BCD codes are fed from the output module to a numerical indicating device. BCD output devices include such things as digital number displays, variable position actuators, and stepper motors.

The other general PLC analog system is the straight numerical type. Some typical ranges of the modules available for these systems are shown in figure 22–2. The PLC numerical type of module is used for a large variety of input devices, the most common of which is the electrical potentiometer. The potentiometer is used to input a linear, varying, electrical value to the input module. The potentiometer can be a type that reads temperature, pressure, distance, position, or electrical values. Other inputs include thermocouplers,

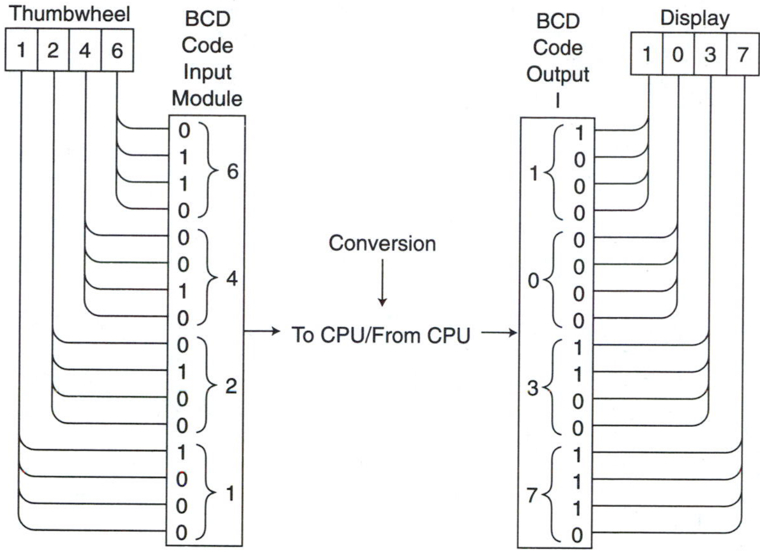

FIGURE 22–1
Analog BCD Input and Output Systems

strain gauges, and straight electrical signals. The more complicated analog system, the PID output system, will be discussed in chapter 23.

Note that the PLC handles continuous analog systems in discrete steps. The continuously varying input signal is not strictly continuous when it reaches the PLC CPU. As we divide up the input signal into more steps, it more nearly approaches the exact duplication of the input signal. Figure 22–3 shows how the input signal is divided into more parts for increased accuracy. As the number of input divisions goes up from 8 to 16, the digital signal more nearly describes the actual input signal. Some normally used PLC divisions are 1024 and 4096. As the number of divisions goes up, the PLC system cost also goes up. You must have enough steps to control your process with precision, but not so many that cost becomes prohibitive.

FIGURE 22–2
Typical Analog I/O Module Ranges

2-10 mA
4-20 mA
10-50 mA
0 to + 5 Volts DC
0 to + 10 Volts DC
± 2.5 Volts DC
± 5 Volts DC
± 10 Volts DC

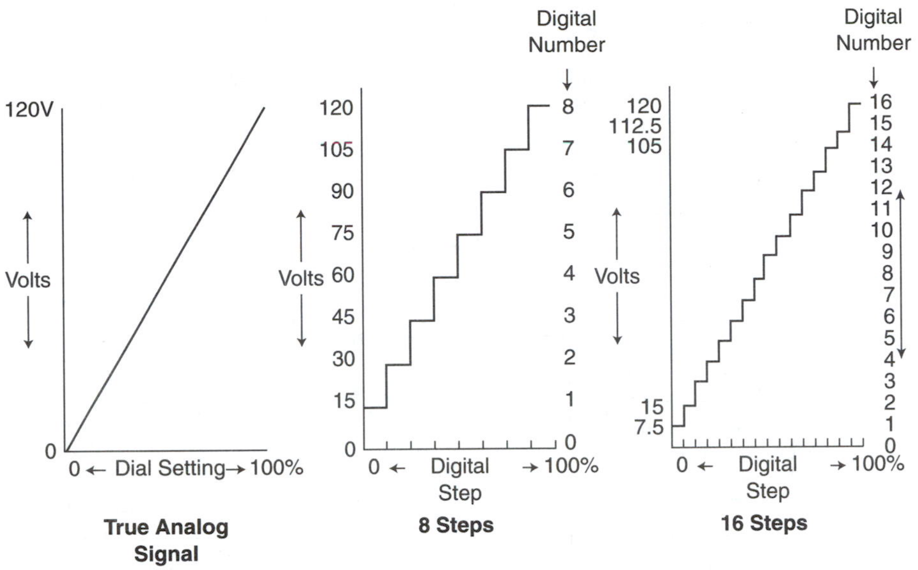

FIGURE 22–3
Analog Signal Conversion to Digital Steps

22–3 PLC ANALOG SIGNAL PROCESSING

The sensor or signaling device that feeds the input module does not usually have the same electrical range as the input module. Its lower-limit electrical value must be matched to the lower-limit electrical value of the input module. The input's upper-limit signal value must also be matched to the upper-limit electrical value of the input module by using an intermediate signal conversion. Similarly, the output module and the outputs must have their signals appropriately matched by a converter. Intermediate values must also be linearly matched by the converters for both input and output.

The input signals available have to be converted and scaled to match an available module. For example, you have a signal that varies from 0 V to 78 VAC, with 78 V representing 100 percent input voltage. You decide to use a 0–5 VDC input module. Therefore, you must convert 0–78 VAC to a linear 0–5 VDC, as shown in figure 22–4. The DC voltage fed from the converter into the module is then converted to a digital number. This digital number is sent from the analog module to an input register in the CPU, as shown in the figure.

How does the input conversion work? For illustration, trace 31 VAC. The converter analyzes the portion of 78 that 31 represents. This is 0.397. The converter, which you must design and supply, puts out a DC voltage that is this proportion of 5 VDC. This DC value, 1.987 V, is sent to the input module. Assume that the input module is an 8-bit base, which can hold a value up to 256 in decimal. The input module then takes this same proportion of 256, 102, and sends the value to a CPU input register. Which register receives the data, 102, depends on the setting of DIP switches on the module.

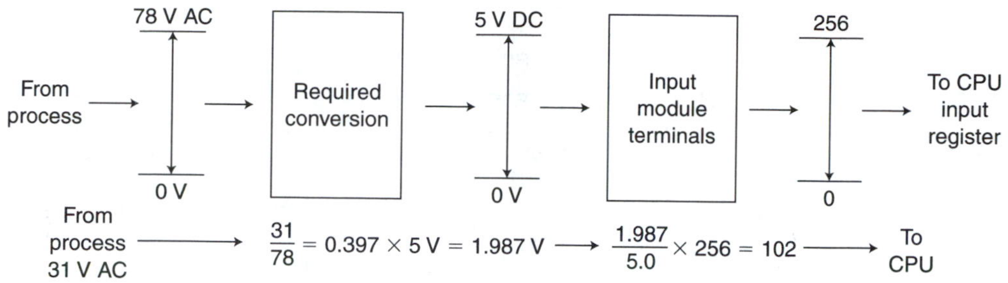

FIGURE 22–4
Analog Input Signal Path and Values

Note that the input is stepped, in 256 steps, and is not perfectly linear when the CPU receives data. The accuracy of this system is 1/256 = 0.0039, or about 0.4 percent. Other, more accurate, input modules of 10 and 12 bits can be obtained, at greater cost, if needed in your application. These would have 1024 and 4096 steps, respectively.

How does the output signal get from the CPU to an output analog device? Figure 22–5 shows an output system. For illustration it is assumed that the signal in figure 22–4 was multiplied by 2 in the CPU. The output ratio is then 0.794, as shown in figure 22–5. This would be 203 on the 256-step scale. Assume that there is an output module feeding an op amp device with a range of −10 V, to +10 V. The math shown indicates that the output would have a value of 5.9 VDC.

To illustrate how an analog system might be used, an ADD application is shown in figure 22–6. An output meter is to indicate the sum of two analog inputs. The two input values go through conversion and then through the input module. The digital values end up in IR01 and IR02, as shown. An ADD function adds the two values when the ADD function is enabled. The sum can be updated very quickly by rapid enabling of the ADD function. The sum is put into OR01. The sum is then sent to an output module and then to the indicating meter. For illustration, 17 V and 42 VAC are added and converted, giving an output of 7.8 V to the meter. The meter full scale could be set at 2 times 78, or 156 VAC, to match input and output scales.

Figure 22–7 shows a block diagram of the conversion process, along with two numerical examples. This module has 1024 steps, but it could have had 512, 4096, or some other power of 2. These are typical values for the module's divisions.

FIGURE 22–5
Analog Output Signal Path and Values

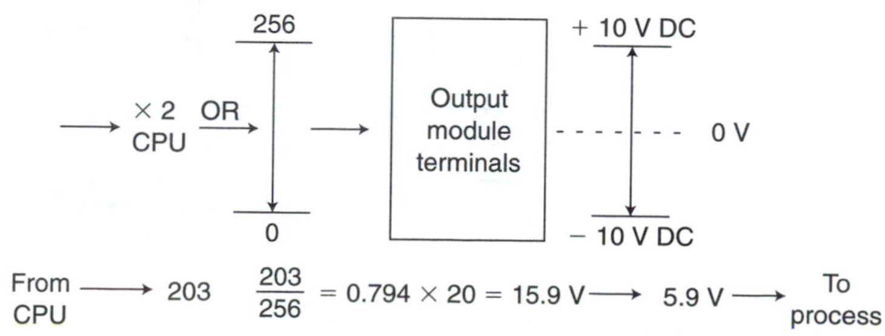

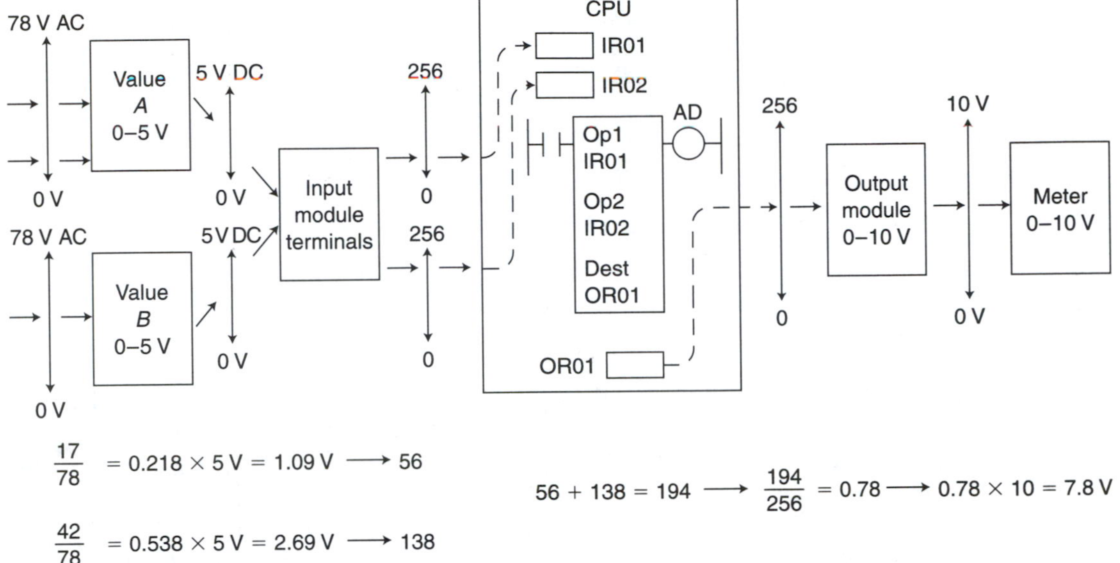

FIGURE 22–6
Analog ADD Application

Suppose you need to compare the input values with a fixed value, or with another input value for action. Say, for example, that if the sensor signal exceeds 0.5 ampere, you must turn on output 6. Further, if the sensor signal is between 0.8 and 1.1 amperes, you must turn on output 7. In this case you have an analog input and a discrete output. The signal-level processing of these electrical values is shown in figure 22–7.

A more typical situation is analog in and analog out. Figure 22–8 shows an example of an analog output system (input not shown). The upper and lower limits are shown for each block. Sixty-two percent of the total is traced through the analog output system.

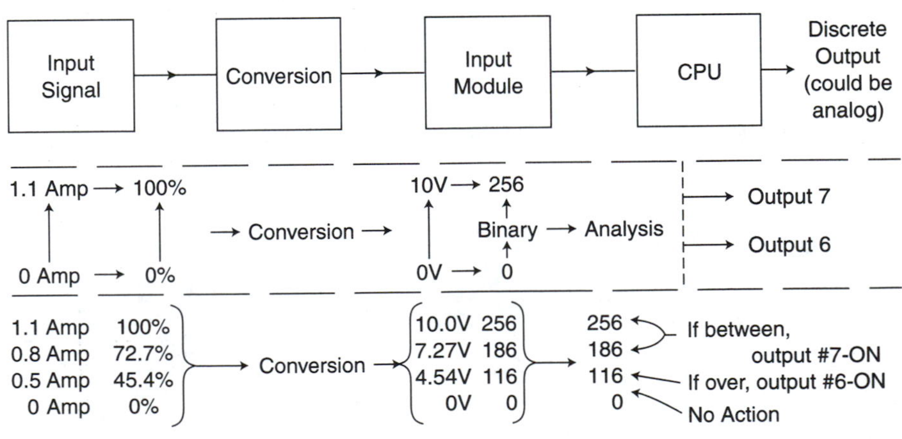

FIGURE 22–7
Analog Input System

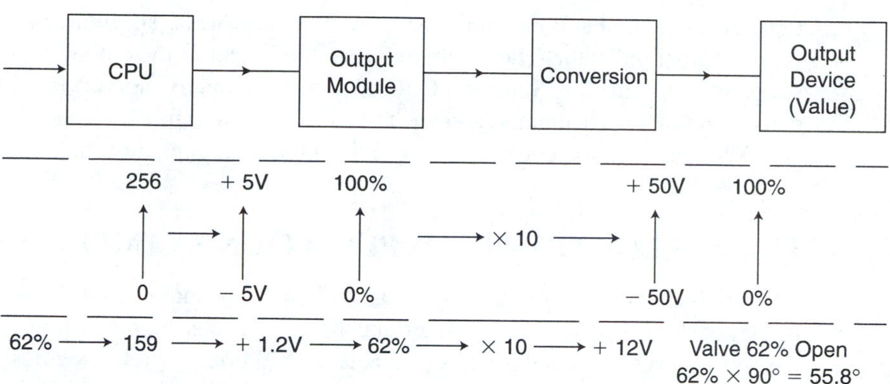

FIGURE 22–8
Analog Output System

22–4 BCD OR MULTIBIT DATA PROCESSING

BCD data is handled like analog data. Figure 22–9 shows a block diagram of how BCD devices and data are used by the PLC. The input and output devices are mathematically matched directly by the input and output modules. No conversion of values is required because the input and output devices are built to match the modules directly. In this illustra-

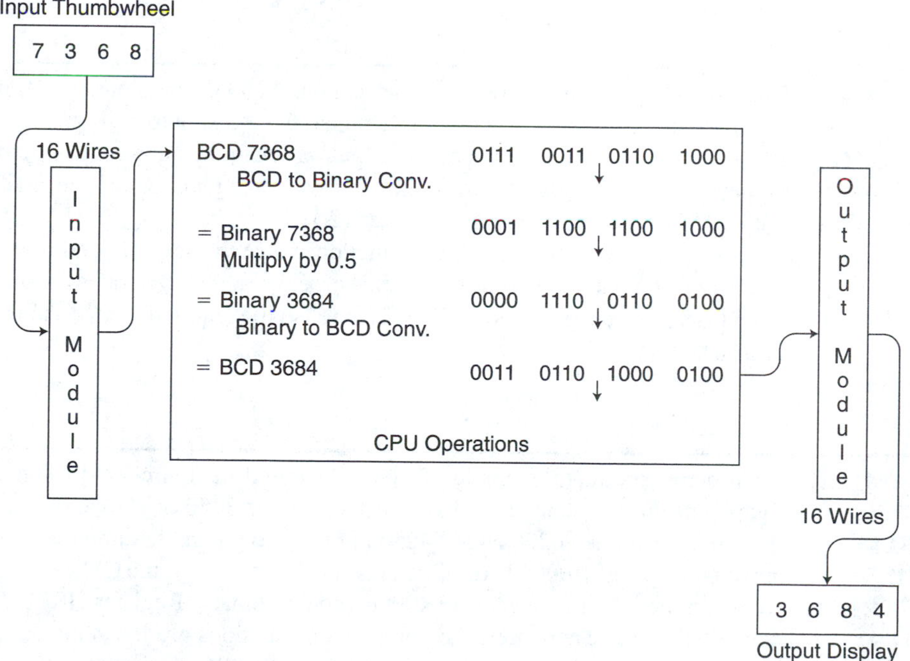

FIGURE 22–9
BCD (Multibit) Input and Output System

tion, the input number is entered directly from thumbwheels. The input data is scaled to half for the output indicator. The resulting half value is sent to the output device, a four-digit, seven-segment display. Since the CPU does math in binary, appropriate BCD and binary conversions are carried out as shown. The illustration in figure 22–9 shows multiplying by 0.5. Alternatively, you could divide by 2 and obtain the same result.

22–5 PLC ANALOG OUTPUT APPLICATION EXAMPLES

The inputs and outputs of analog systems can be straight analog or BCD, as we have shown, or they can be other types such as thermocouple, load cells, humidity transducers, and electric motor drive input and feedback signals. In addition, discrete modules may be used in combination with analog modules. Furthermore, multiple inputs and outputs can be involved. To illustrate all of these combinations would require many examples.

This section presents six examples, four of which have multiple inputs or outputs, as representative of analog PLC operations. These examples are

Example 22–1 Analog in/discrete out
Example 22–2 BCD in/discrete out
Example 22–3 Analog in/analog or BCD out
Example 22–4 BCD in/BCD or analog out
Example 22–5 Two analog in/two analog out
Example 22–6 Two BCD in/two analog out

EXAMPLE 22–1: ANALOG IN AND DISCRETE OUT

The example given in figure 22–7 was analog in and discrete out. The figure's values further illustrate the example. The problem is to have one output go on when a certain level, 0.5 ampere, is reached, and another output on when the amperage is between 0.8 and 1.1 amperes. To accomplish this, use the input values with comparison functions. (The comparison functions are covered in chapter 12.)

For the first output, use a GE function. Since the second condition has two limits, it needs two comparison functions—in this case, greater than and less than. Figure 22–10 illustrates how to program the PLC to accomplish the required comparisons and energize appropriate outputs.

EXAMPLE 22–2: BCD IN AND DISCRETE OUT

A problem similar to example 22–1 is illustrated in figure 22–11. The input is a BCD thumbwheel that counts up to 9999. If the input is 3750 or above, output 6 is to go on. If the input is between 6200 and 8542, output 7 is to go on. Assume that the input data is received in register IR0006 in the CPU. As stated previously, the CPU works in binary; therefore, you must first convert the IR0006 value to binary. Register HR0045 will receive the converted value. Thereafter, the comparison functions are the same as in example 22–1. Figure 22–11 shows the PLC programming for this BCD comparison problem.

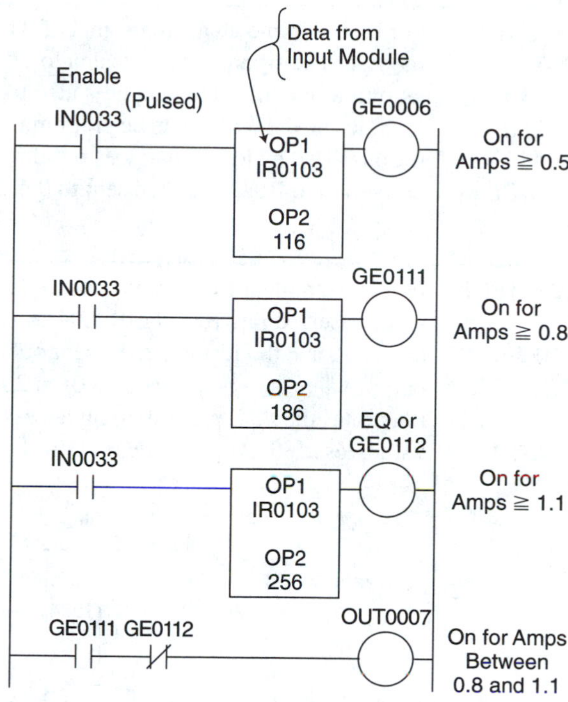

FIGURE 22–10
Example 22–1: Analog In/Discrete Out

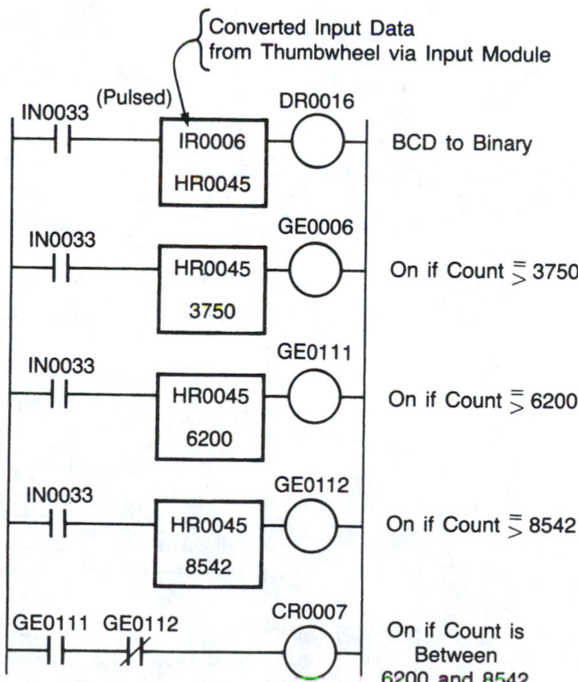

FIGURE 22–11
Example 22–2: BCD In/Discrete Out

EXAMPLE 22–3: ANALOG IN AND ANALOG OR BCD OUT

For this example an analog signal of 0 to 10 volts comes in through a converter to an input module. The signal is to be scaled to 1/5 of its value by the CPU and then sent out through an output module. The output is also to be sent to a BCD output display. Figure 22–12 illustrates how the PLC can be programmed internally to accomplish one or both output conditions. The analog signal goes out directly. The output signal to the analog output is first converted to BCD and then sent to the BCD display.

EXAMPLE 22–4: BCD IN AND BCD OR ANALOG OUT

In this example a BCD input, 0 to 9999, is received by an input module, which places the value received into register IR0004. A fixed value of 180 is to be subtracted from the value received, and the result is to be sent out to a 0–9999 BCD output display. Additionally, the output value is to be placed in a 0- to 20-milliampere analog output module. Figure 22–13 illustrates the PLC programming necessary to accomplish the transfer of the original input value, less 180. Appropriate BCD-to-binary conversions are included in the program.

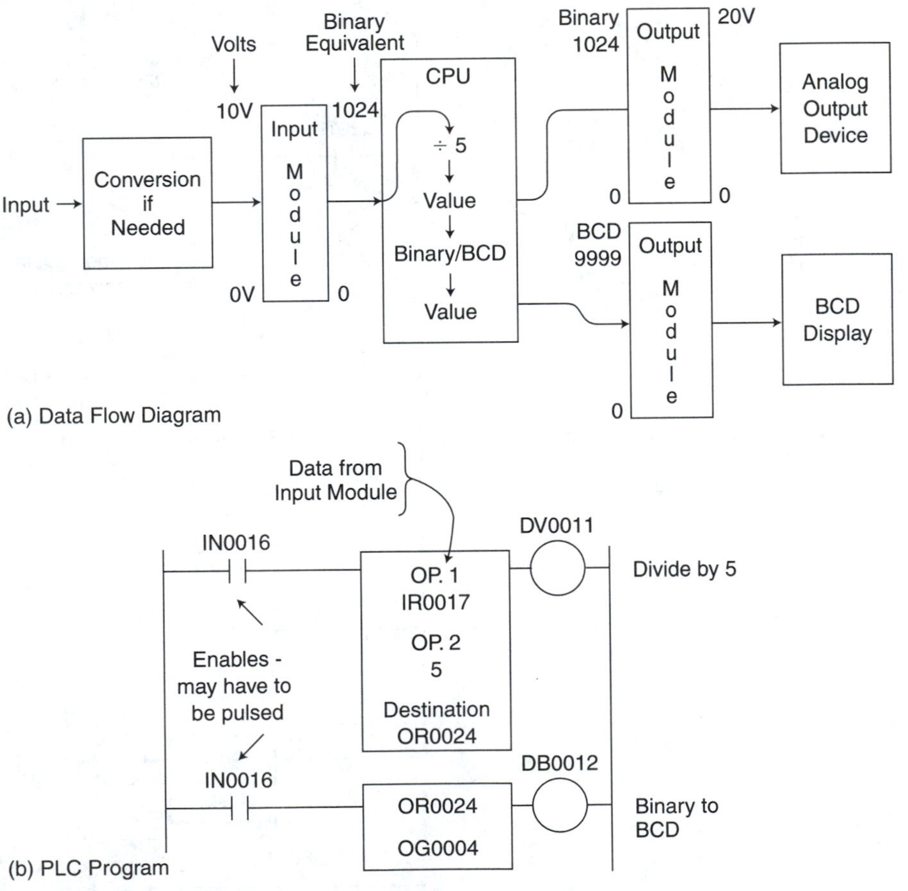

FIGURE 22–12
Example 22–3: Analog In/Analog or BCD Out

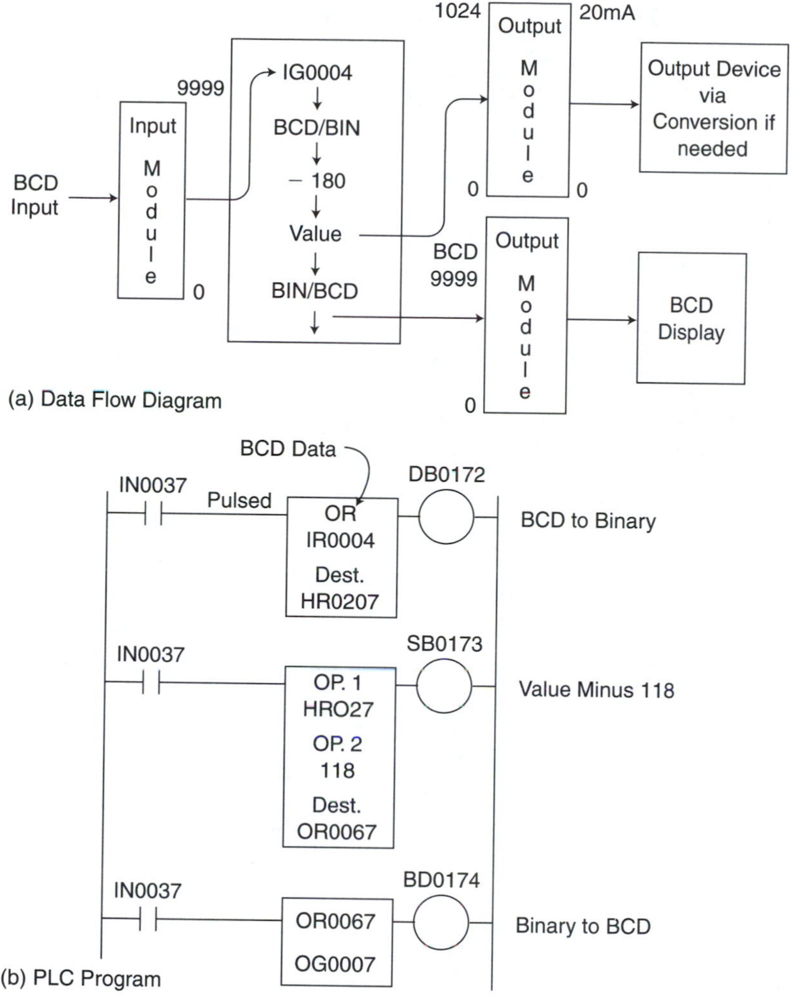

FIGURE 22–13
Example 22–4: BCD In/BCD or Analog Out

EXAMPLE 22–5: TWO ANALOG IN, MATHEMATICAL MANIPULATION, TWO ANALOG OUT

Examples 22–5 and 22–6 both have multiple inputs and outputs. They could have had more than two inputs and a single output, but in this illustration both examples use two inputs and two outputs.

The example shown in figure 22–14 has two analog inputs whose values are manipulated in the CPU. For illustration, the values are both added and subtracted. The sum is output to one analog output, and the difference is sent out to another. The internal programming to perform the mathematical manipulations is also shown.

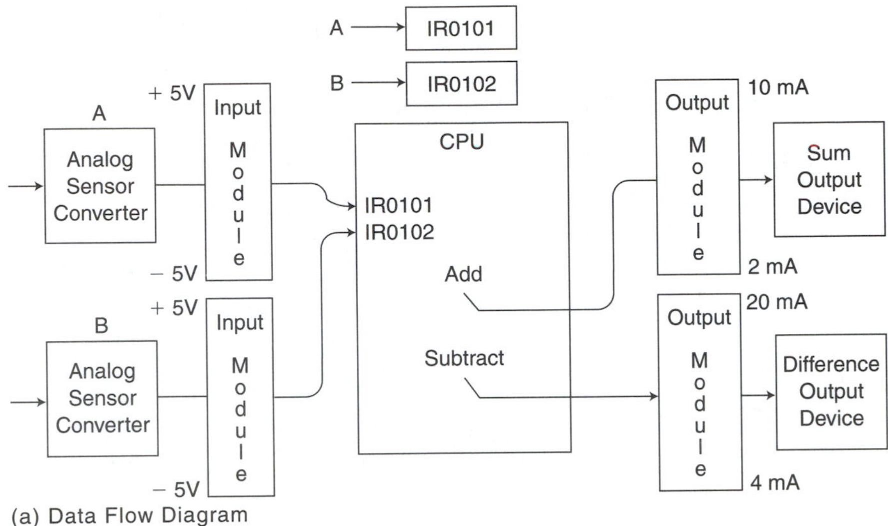

(a) Data Flow Diagram

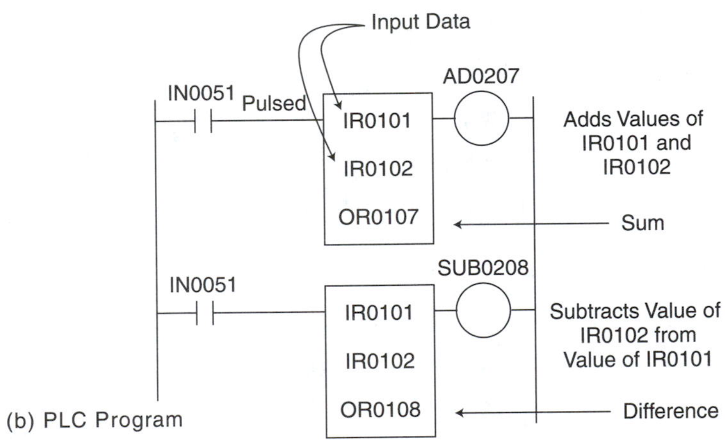

(b) PLC Program

FIGURE 22–14
Example 22–5: Two Analog In, Add and Subtract, Two Analog Out

EXAMPLE 22–6: TWO BCD IN, MATHEMATICAL MANIPULATION, TWO ANALOG OUT

Example 22–6, shown in figure 22–15, is similar to example 22–5. It has two BCD inputs and two numerical analog outputs. You could also mix and match analog and BCD inputs and outputs with little difficulty. In example 22–6, the first output is the product of the inputs and the second output is the sum, minus 155. The product is read on readout X and the sum, minus 155, on readout Y. Again, the PLC program to perform the math functions is shown.

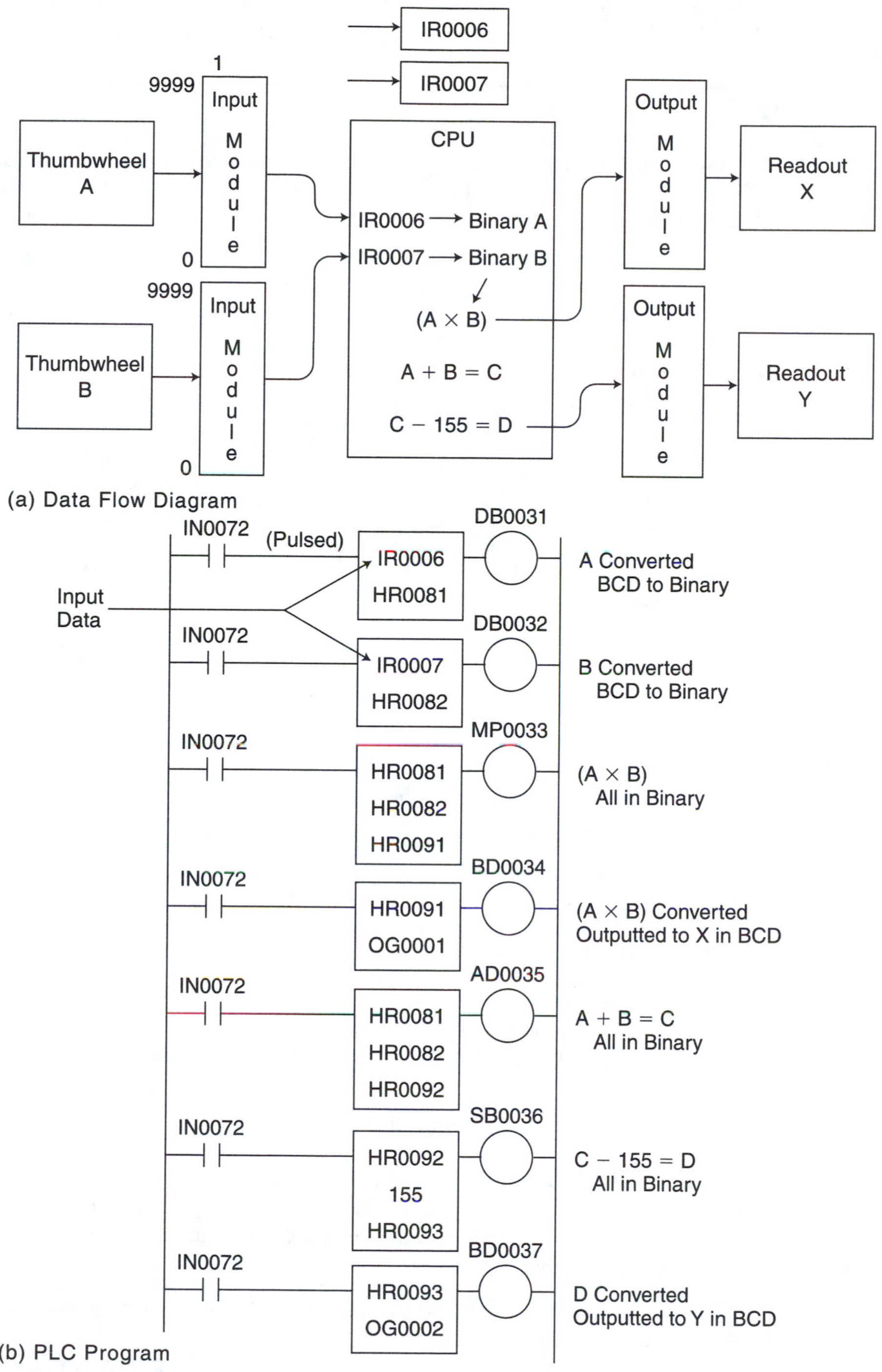

FIGURE 22–15
Example 22–6: Two BCD in, Multiplication, Addition, Subtraction, Two Analog Out

EXERCISES

1. Draw an output graph for a 32-step (5-bit) output similar to those shown in figure 22–3. 144 volts is 100 percent. The input configuration is the same as figure 22–3. Determine the output digital-step-indicated voltage (range) for dial settings of 23, 45, 46, and 78.5 percent.
2. If the input graph curve in figure 22–7 were nonlinear, as shown in figure 22–16, would the output be linear with respect to dial setting? Explain.

FIGURE 22–16
Diagram for Exercise 2

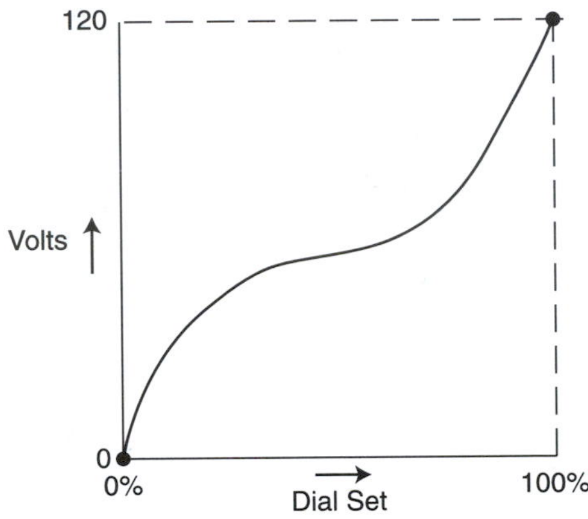

3. Refer to figure 22–4. Trace 61.5 volts through the system.
4. Refer to figure 22–5. Suppose that multiplication were by 0.5 instead of 2. The digital number would be 51. Trace 51 through the system.
5. Refer to figure 22–6. Trace for inputs of 22 VAC for valve A and 31 VAC for valve B.

For exercises 6 through 8, assume

- □ Input volts 0 to 80.
- □ Input module volts 0 to 5.
- □ Binary has 128 steps.

For exercises 6 through 8, include the following in each answer.

- □ Draw a block flow diagram as in figure 22–7.
- □ Draw the required PLC ladder program as in the chapter example.
- □ Trace a number, if requested, through the computational system similar to the tracing carried out in figure 22–7.

6. The linear input of 0 to 80 volts is to be displayed on a 9999-maximum-count BCD output. Trace 32 volts through the system.
7. Repeat exercise 6, changing the output to a linear 0 to 21 volts. Trace 53 volts input through the system.
8. Two linear input signals of 0 to 4 volts are to be multiplied and the result put out on a linear output of 0 to 150 volts. Trace the numbers if the inputs are 2.85 and 3.45 volts.

9. Two BCD numbers are to be inputted. The first is to be divided by the second. The result is to be shown on an output BCD display. Trace the computation if A is 458 and B is 35.

10. There are three BCD inputs, A, B, and C. The output is to be A plus B minus C on a BCD display. Trace the computation for an A, B, and C of 425, 283, and 63, respectively.

11. There are two BCD inputs. If A exceeds 355, output F is to go on; if B exceeds 187, output G is to go on; if both exceed their listed numbers, output H is to go on; otherwise, no outputs are to be on.

23

PID Control of Continuous Processes

OUTLINE

23–1 Introduction □ **23–2** PID Principles □ **23–3** Typical Continuous Process Control Curves □ **23–4** PID Modules □ **23–5** Typical PID Functions

OBJECTIVES

At the end of this chapter, you will be able to

- □ Briefly describe proportional, integral, and derivative control.
- □ Describe how the above three control systems are combined for effective process control.
- □ List and sketch the response curves of ineffective process control.
- □ Show and sketch the response curve of a good process control system.
- □ Explain the characteristics of a PID module.
- □ Generally describe the PID function.

23–1 INTRODUCTION

All of the processes shown in the book so far have not been of the continuous type; that is, these process examples either are on or off, or travel linearly between two points. By continuous process, we mean one in which the output is a continuous flow. Examples are a chemical process, a refining process for gasoline, or a paper machine with continuous output of paper onto rolls. Process control for these continuous processes cannot be accomplished fast enough by PLC on-off control. Furthermore, analog PLC control is also not effective or fast enough by PLC on-off control. Furthermore, analog PLC control is also not effective or fast enough. The control system most often used in continuous processes is PID (proportional–integral–derivative) control. PID control can be accomplished by mechanical, pneumatic, hydraulic, or electronic control systems as well as by PLCs.

Many medium-size PLCs and all large PLCs have PID control functions, which are able to accomplish process control effectively. In this chapter, we discuss the basic principles of PID control. We then explain the effectiveness of PID control by using typical process response curves and show some typical loop control and PID functions.

Loop and PID control are designations used interchangeably by different manufacturers. Actually, some loop controls are not strictly the PID type. However, assume they are the same.

23–2 PID PRINCIPLES

PID (proportional–integral–derivative) is an effective control system for continuous processes that performs two control tasks. First, PID control keeps the output at a set level even though varying process parameters may tend to cause the output to vary from the desired set point. Second, PID promptly and accurately changes the process level from one set point level to another set point level. For background, we briefly discuss the characteristics of each of the PID control components: proportional, integral, and derivative.

Proportional control, also known as ratio control, is a control system that corrects the deviation of a process from the set level back toward the set point. The correction is proportional to the amount of error. For example, suppose that we have a set point of 575 cubic feet per minute (CFM) in an airflow system. If the flow rises to 580 CFM, a corrective signal is applied to the controlling air vent damper to reduce the flow back to 575 CFM. If the flow somehow rises to 585 CFM, twice the deviation from set point, a corrective signal of four times the magnitude would be applied for correction. The larger corrective signal theoretically gives a faster return to 575 CFM. In actuality, the fast correction is not precise. You return to a new set point at the end of the correction, for example, 576.5 CFM, not 575 CFM. Proportional control does not usually work effectively by itself, resulting in an offset error.

To return the flow to the original set point, integral control, also known as reset control, is added. Note that integral control cannot be used by itself. Remember, with proportional control only, we had an output error from our original set point. We ended up at 576.5

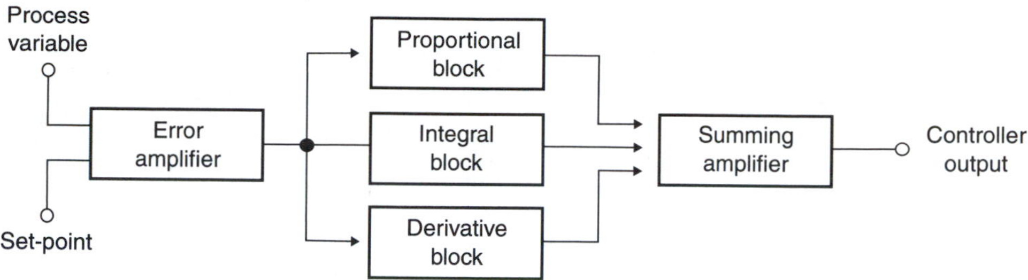

FIGURE 23–1
Block Diagram of a Typical PID Controller

CFM, not 575 CFM. Integral control senses the product of the error, 1.5 CFM, and the time the error has persisted. A signal is developed from this product. Integral control then uses this product signal to return to the original set point. An integral control signal can be used in conjunction with the proportional corrective signal. In the controller, the added integral signal reduces the error signal that caused the output deviation from the set point. Therefore, over a period of time, the process deviation from our original 575 CFM is reduced to minimum. However, this correction takes a relatively long period of time.

To speed up the return to the process control, point, derivative control is added to the proportional–integral system. Derivative control, also known as rate control, produces a corrective signal based on the rate of change of the signal. The faster the change from the set point, the larger the corrective signal. The derivative signal is added to the proportional–integral system. This gives us faster action than the proportional–integral system signal alone. A typical PID control system is shown in block diagram form in figure 23–1. This configuration is the commonly used parallel type. The controller output signal of figure 23–1 is utilized through a control system to return the process variable to the set point.

An illustration of a system using PID control is shown in figure 23–2. In this system, we need a precise oil output flow rate. The flow rate is controlled by pump motor speed. The pump motor speed is controlled through a control panel consisting of a variable-speed drive. In turn, the drive's speed control output is controlled by an electronic controller. The electronic controller output to the drive is determined by two factors. The first factor is the set point determined by a dial setting (or equivalent device). Second, a flow sensor feeds back the actual output flow rate to the electronic controller. The controller compares the set point and the actual flow. If they differ for some reason, a corrective signal change is sent to the motor controller. The motor controller changes motor speed accordingly by changing the voltage applied to the motor. For example, if the output oil flow rate goes below the set point, a signal to speed up the motor is sent. The controller then uses PID control to make the correction promptly and accurately to return to the set point flow. If the dial is changed to a new setting, the function of the PID system is to reach the new set point as quickly and accurately as possible.

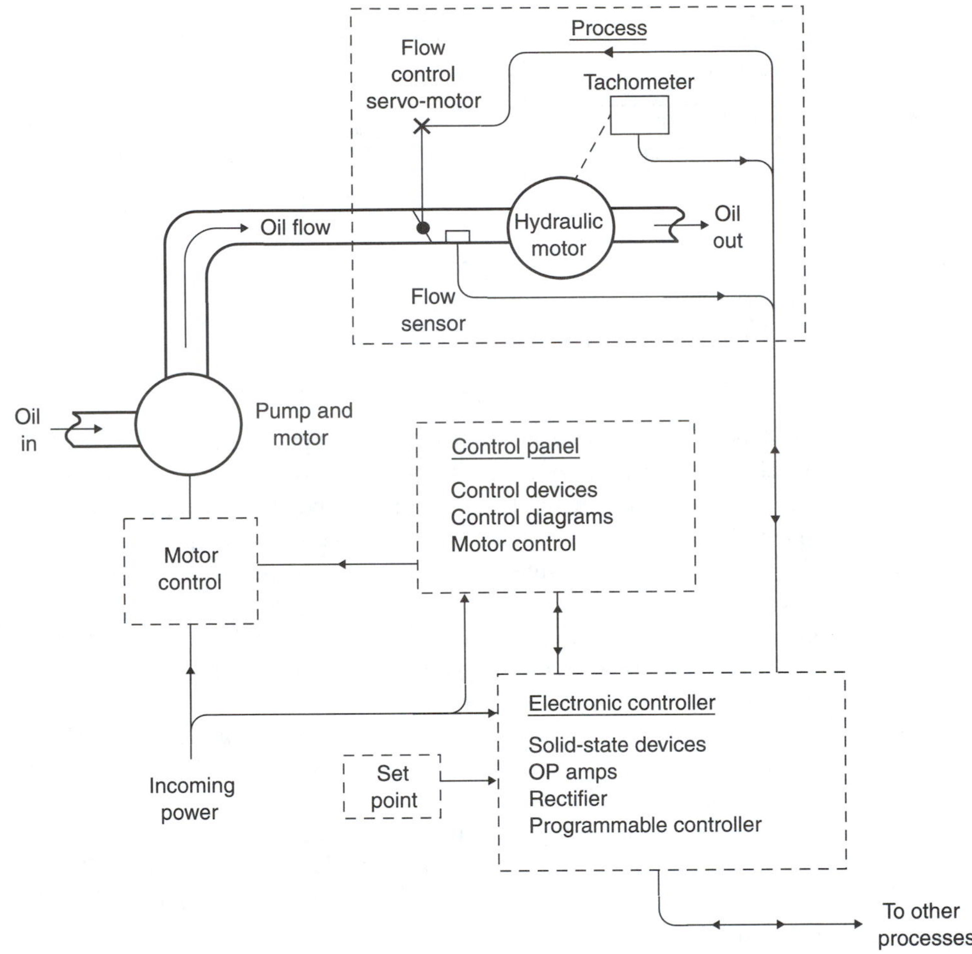

FIGURE 23–2
General Control System Diagram—Hydraulic Pump

23–3 TYPICAL CONTINUOUS PROCESS CONTROL CURVES

To illustrate some of the possible system response curves for process control systems, we will use the electromechanical system shown in figure 23–3. By response curves, in this example, we mean output position versus time. The curves to be shown are for various types of control, including PID.

Figure 23–3 shows a control system with a feedback loop, which can be PID. The dial is set to a position in degrees, and the output device is to take the position set on the dial. The output is to follow quickly and accurately any change from one dial setting to another.

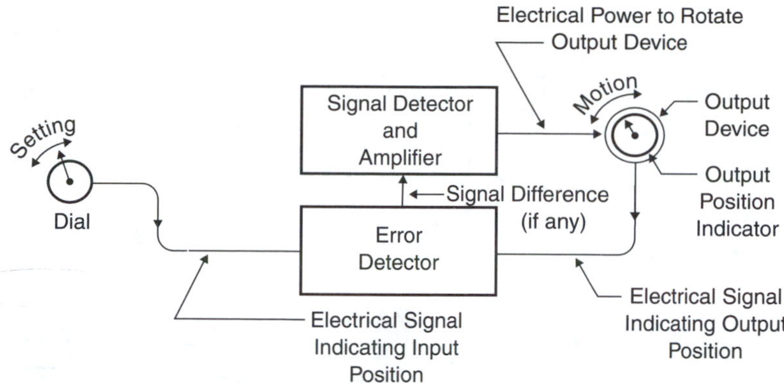

FIGURE 23–3
Position Indicator with PID Control

Furthermore, the output position should not drift out of position over a period of time. Another factor is that the indicator can have two different weights, depending on the application. These are 5 pounds and 20 pounds. Obviously, the output drive will tend to operate more slowly for 20 pounds than for 5 pounds, unless a proper PID control is set up to compensate for weight differences.

For illustration we very quickly turn the dial from 0 degrees to 108 degrees at 3 seconds after time base 0. Ideally, the position indicator should instantaneously reach 108 degrees, as shown in figure 23–4. Obviously, this does not happen in actual practice.

Figure 23–5 shows five possible curves for different types of control. A is an idealized movement but takes 4 seconds. B undershoots or overshoots the mark. C shows cyclic response and reaches an angular point near the set position but oscillates for a few seconds before reaching the proper position. D shows damped response and reaches the new position exponentially but takes a long time. E reaches the new position but continually oscillates about the final setting. None of these curves shows an acceptable control characteristic for accurate and prompt operation.

By comparison, PID control obtains the most ideal response possible—not perfect, but the best we can do. A curve for this control is shown in figure 23–6.

FIGURE 23–4
Ideal Position Control Positioning Curve

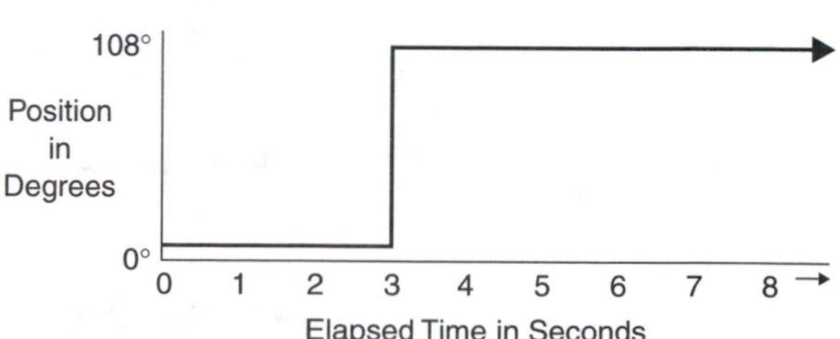

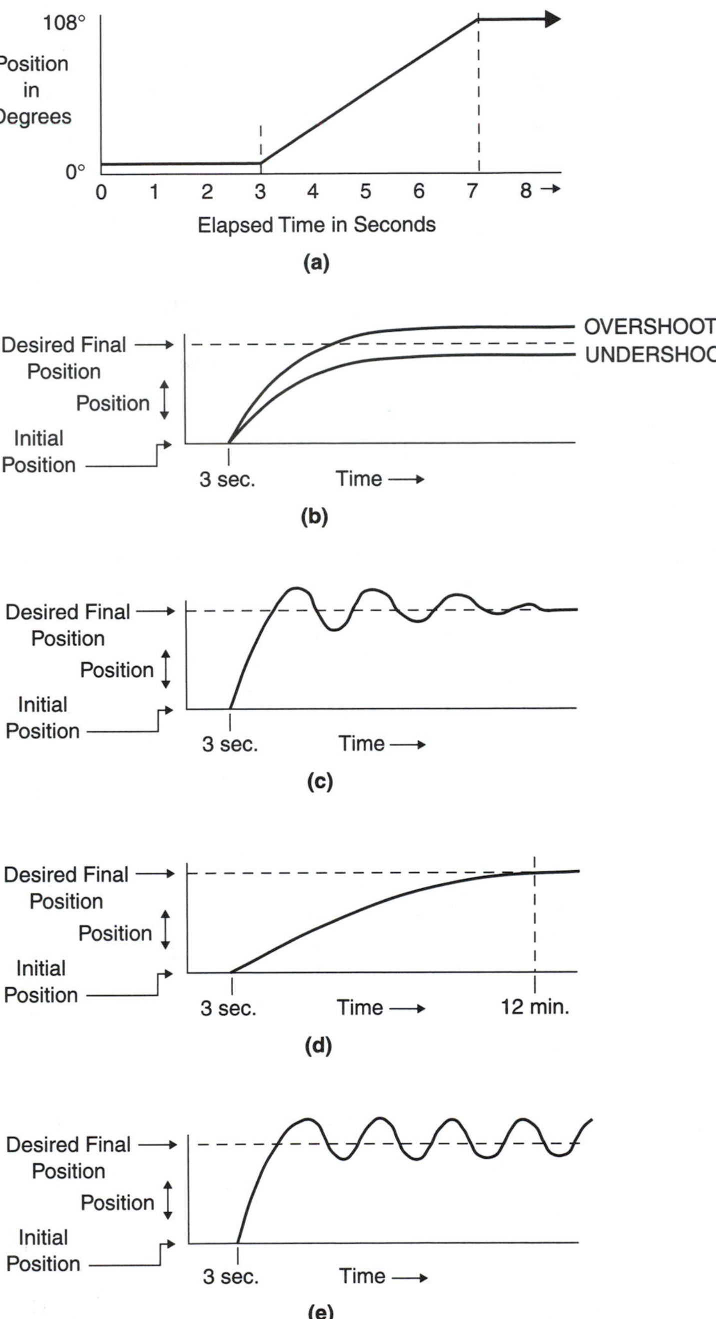

FIGURE 23–5
Typical Response Curves

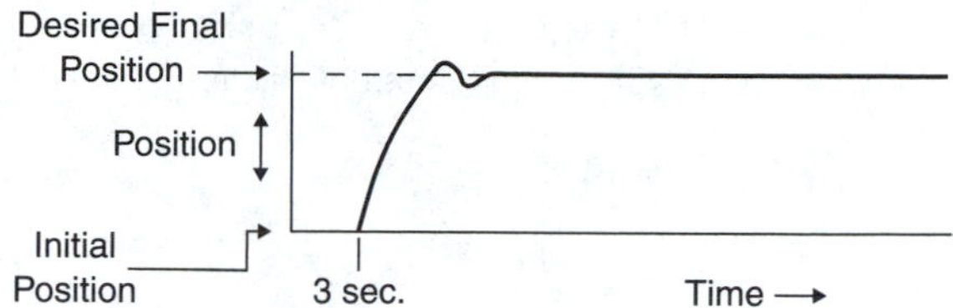

FIGURE 23–6
Ideal PID Position Control

23–4 PID MODULES

PLCs often come equipped with PID modules, used to process data obtained by feedback circuitry. Most such modules contain their own microprocessor. Since the algorithms needed to generate the PID functions are rather complex, the PID microprocessor relieves the CPU of having to carry out these time-consuming operations.

To understand the PID module, refer to figure 23–7. The PLC sends a set-point signal to the PID module. The module is made up of three elements: the proportional, integral, and derivative circuits. The *proportional circuit* creates an output signal proportional to the difference between the measurement taken and the setpoint entered in the PLC. The *inte-*

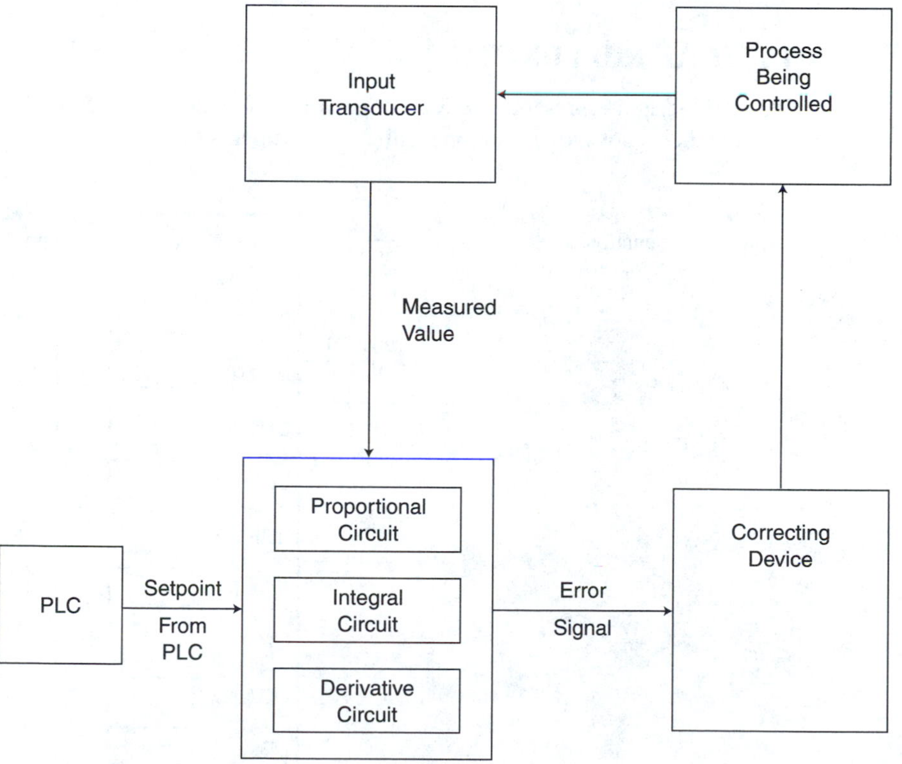

FIGURE 23–7
Block diagram of PID Module

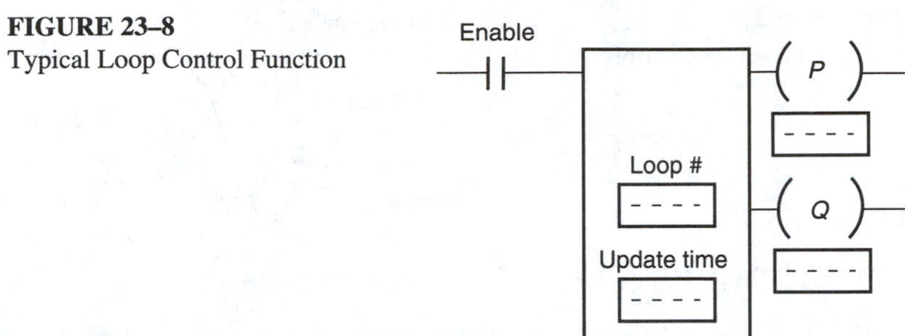

FIGURE 23–8
Typical Loop Control Function

gral circuit produces an output proportional to the length and amount of time the error signal is present. The *derivative circuit* creates an output signal proportional to the rate of change of the error signal.

The input transducer generates an output signal from the process being controlled and feeds the measured value to the PID module. The difference between the set point coming from the PLC and the measured value coming from the input transducer is the error signal. Some sort of correcting device, such as a motor control, valve control, or amplifier, takes the error signal and uses it to control the correction sent to the process being controlled.

23–5 TYPICAL PID FUNCTIONS

Figure 23–8 is a representative loop control PID function. This function controls a PID function that is not shown. A loop identifying number is specified in the block. Update time

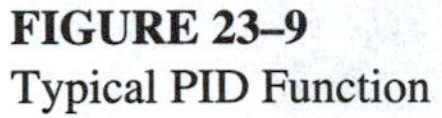

FIGURE 23–9
Typical PID Function

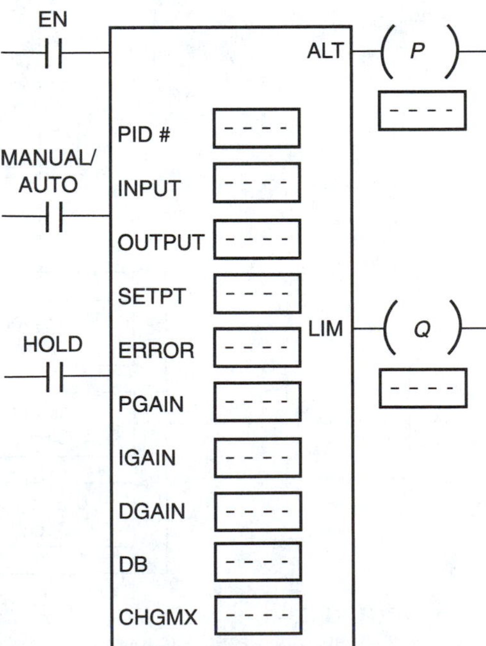

in the block is also specified. This time in seconds (i.e., 15 seconds) is the interval controlling the update procedure frequency. Coil P goes on when the function is enabled. Coil Q goes on when the update time is reached for one scan.

A typical PID function of intermediate complexity is shown in figure 23–9. Some PLCs combine the previously shown loop control function with the PID function. This example is for the PID function alone.

The functions of the inputs, outputs, and functional descriptions are

Inputs

EN: the usual function Enable line
MANUAL/AUTO: manual or automatic mode
HOLD: used for "clamping" and for logic transition control

Outputs

P: the coil number assigned
Q: an output limit coil used in the logic

Functional Descriptions and Values

PID number: the PID block identification number
INPUT: the register in which the process variable is stored
OUTPUT: the register in which the output algorithm is stored
SETPT: the register in which the set point is stored
ERROR: the register in which the value of ERROR = (SETPT – INPUT) is stored
DB: the register for the deadband value
CHGMX: the register in which the maximum allowable rate of change is stored
PGAIN: the register in which proportional gain is stored
IGAIN: the register in which the integral term is stored
DGAIN: the register in which the proportional term is stored

Most of the functions in the block are written as a percentage of the set point. The block values may be programmed as constants or moved in from other registers. DB, deadband, is effectively the tolerance you can live with for the process (in percent). The last three functional block inputs are adjusted for tuning the system in operation for optimum process control.

A more complex PID function is shown in figure 23–10. A worksheet (instruction set) for this function is shown in figure 23–11. The PID function is designated Loop control in this format. This function, when properly enabled by input contacts, performs PID analysis and control. The enabled function compares the set point with the process variable. If they differ by more than a preset value, an appropriate correcting signal is sent from the function output to the controlling device's controller. The input parameters are put into 32 consecutive registers ending with the register specified as "loop table end." The number of registers required will vary by model and manufacturer.

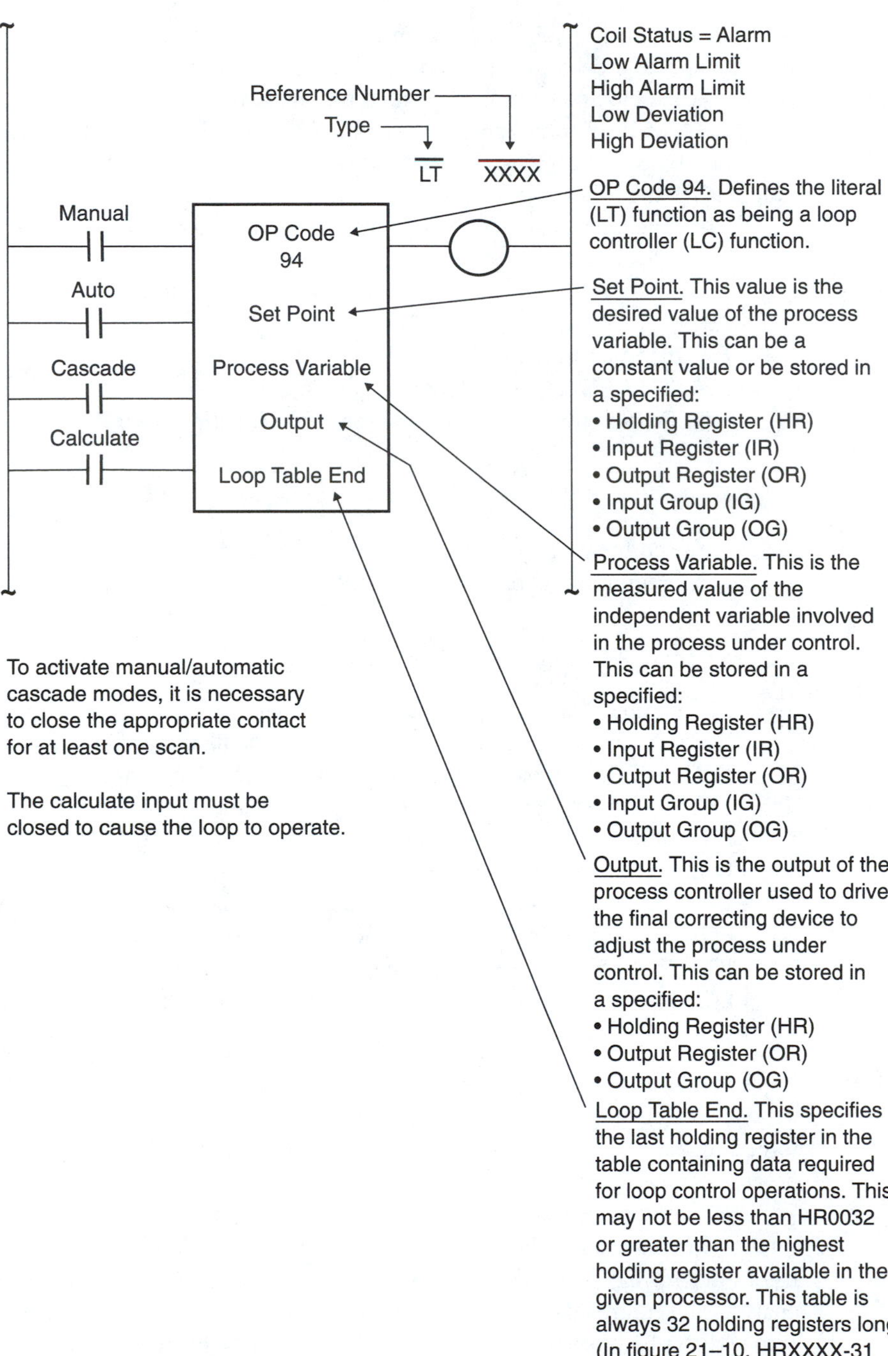

FIGURE 23–10
PID PLC Function

Loop Table Register Positions	Loop Table Actual HR Assignment	Quantity		Value/Remarks
HRXXXX-31		Proportional Term (±32,767)	C	
HRXXXX-30		Integral Term (±32,767)	C	
HRXXXX-29		Derivative Term (±32,767)	C	
HRXXXX-28		SP_n — Set Point This Sample	C	
HRXXXX-27		PV_n — Process Variable This Sample	C	
HRXXXX-26		Time Counter — Elapsed Sample Time	C	
HRXXXX-25		SP_{n-1} — Set Point Previous Sample	C	
HRXXXX-24		PV_{n-1} — Process Variable Previous Sample	C	
HRXXXX-23		E_{n-1} — Error Previous Sample	C	
HRXXXX-22		Bias (0 to Maximum Output)	C	
HRXXXX-21		RESERVED		FUTURE — DO NOT USE
HRXXXX-20		Configuration Input Word (See Below)	U	
HRXXXX-19		RESERVED		FUTURE — DO NOT USE
HRXXXX-18		RESERVED		FUTURE — DO NOT USE
HRXXXX-17		Integral Sum (±32,767)	C	
HRXXXX-16		E_n — Error This Sample	U	
HRXXXX-15		T_d — Derivative Time (0 — 327.67 Min.)	U	
HRXXXX-14		T_i — Integral Time (0 — 327.67 Min.)	U	
HRXXXX-13		T_s — Sample Time (0 — 3276.7 Sec.)	U	
HRXXXX-12		K_c — Proportional Gain (.01 — 99.99)	U	
HRXXXX-11		Inner Loop Pointer (Loop Table End)	U	
HRXXXX-10		Outer Loop Pointer (Loop Table End)	U	
HRXXXX-9		Alarm Deadband (0 — Max PV)	U	
HRXXXX-8		Batch Unit Preload (0 — Max Output)	U	
HRXXXX-7		Batch Unit Hi Limit (0 — Max Output)	U	
HRXXXX-6		Neg. Slew Limit (Max — Δ Output/Sample)	U	
HRXXXX-5		Pos. Slew Limit (Max + Δ Output/Sample)	U	
HRXXXX-4		Low Deviation Alarm Limit (0 — Max PV)	U	
HRXXXX-3		High Deviation Alarm Limit (0 — Max PV)	U	
HRXXXX-2		Low Alarm Limit (0 — Max PV)	U	
HRXXXX-1		High Alarm Limit (0— Max PV)	U	
HRXXXX		Output Status Word	C	

C = Calculated by Processor
U = User-Entered

Configuration Input Word (HRXXXX-20)

16	15	14	13	12	11	10	9	8	7	6	5	4	3	2	1

Bit Number	Definition	Status	Bit Number	Definition	Status
1	1 = Proportional Mode Selected		9	1 = Derivative on PV Selected 0 = Derivative on Error Selected	
2	1 = Integral Mode Selected		10	1 = Batch Unit Selected	
3	1 = Derivative Mode Selected		11	RESERVED FOR CONTROLLER USE	
4	1 = Deviation Alarms Selected		12	0 = Anti Reset Windup When Slew Limit Occurs	
5	1 = Error Deadband Selected		13	RESERVED FOR FUTURE USE	
6	1 = Error Squared Control Selected		14	RESERVED FOR FUTURE USE	
7	1 = Slew Limiting Selected		15	RESERVED FOR FUTURE USE	
8	1 = Reverse Action Selected 0 = Direct Action Selected		16	RESERVED FOR FUTURE USE	

FIGURE 23–11
PID PLC Instruction Set

On the worksheet, not all register functions are used for all processes. Key registers are 13 through 16 for system tuning, 27 and 28 for set point and allowable variance, and 29 and 30 for time sets. Register 20 is for "configuration word," as specified in detail in the lower table.

EXERCISES

1. The chapter does not discuss on–off discrete control. Would on–off control be effective in the control of rapid, continuous processes? Why or why not?
2. A paper mill process is run by proportional control. The process involves paper being wound onto a roll at a speed of 62 feet/second (ft/sec). At 64 ft/sec, the correction signal to the drive motor is −1.75 volts. What would be the correction signal for 66 ft/sec, 68 ft/sec, and 60 ft/sec?
3. How does proportional–integral process control improve control compared to proportional control only?
4. How does proportional–integral–derivative control improve process control compared to proportional–integral control?
5. Discuss the curves of figures 23–4, 23–5, and 23–6 in terms of a heater control process. The control system is used to maintain a constant temperature of 105°F for an oil tank.
6. What is meant by *tuning* a PID system?
7. Obtain manuals from a number of PLC manufacturers. Determine the similarities and differences in their PID functions.

24

Networking PLCs

OUTLINE

OBJECTIVES

At the end of this chapter, you will be able to

- □ List the five levels of industrial control.
- □ Define where PLC fits into these five levels.
- □ Define CIM and show how it is used.
- □ Define plant controllers, area controllers, cell controllers, and microcontrollers.
- □ Describe OSI networks.
- □ Differentiate between wide area networks (WANs) and local area networks (LANs).
- □ Show how a cell controller PLC can be used in a work cell.

24–1 INTRODUCTION

In this chapter we discuss the role of the PLC in network systems. Industrial control and where the PLC fits in are covered first. Next, computer-integrated manufacturing (CIM), a network-based manufacturing system, is discussed. A general discussion of network communication, including presently available technologies, follows. A specific PLC controller is shown next. Finally, industrial examples of network systems with PLCs are provided.

24–2 LEVELS OF INDUSTRIAL CONTROL

Figure 24–1 is a triangle showing the general levels of control of an industrial factory. At the bottom, only human control is involved. At the top, very involved computer analysis is used. Most smaller industrial operations go up through level 3. Larger factories are increasingly at level 4. Very large, multiplant operations are generally at level 5. A brief description of each level of control follows.

Level 1 is the machine level. An example is a lathe with manual controls for moving

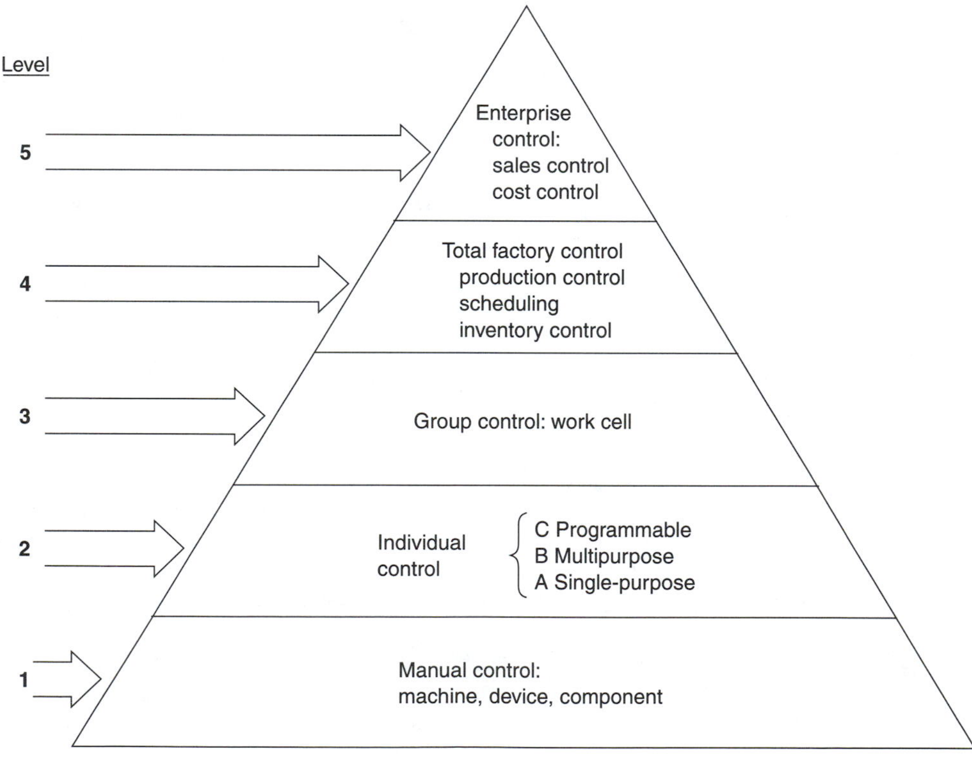

FIGURE 24–1
Control System Level Triangle

the cutter through its path. Control is manual with cranks. The lathe may have power assists for more cutting power, but control is manual.

Level 2 is reached when electrical or other controls are added. Level 2 can be divided into three sublevels as shown in figure 24–1. Suppose an electronic, computer-based control operated the lathe. Such a control for automatic feed rate for cutting the metal would be level 2A. This level 2A rate would be automatically set for each part machined. Level

med for machining more than one part. The con-
a master control as required. The master control
he third sublevel is 2C, programmable control. At
part is programmed in by an operator. When a
ch step and motion is recorded as the process pro-
d in memory. The next time the same part is to be
is recalled from memory. This recall procedure
t is to be made. Machining patterns for many dif-
needed. Note that as you go from A to C of level
ngly. The cost of more control versus the possible
ng economic feasibility.
s to be involved. Level 3 involves connecting the
es or devices to work together. An example of this
as well as controlling the lathe's operation. A mas-
e two individual controllers of the robot and the
s more than two devices are coordinated. An ex-
bly line. Conveyors, positioners, robotic welders,
vith a master computer. Such groups of machines

ves a number of work cells hooked up to a master,
l. The entire factory is under the control of a large
akes an order from sales input, checks for raw ma-
terial availability in inventory, and prepares a production plan. It then causes the required parts to be made, by running them through the appropriate operations in the plant. The master computer does such other chores as reordering an appropriate amount of raw material as it anticipates the need. The master computer also carries out such tasks as scheduling part manufacturing in a given work cell for maximum machine utilization. One result of proper control at this level is reduced amounts of raw material and in-process inventory. Inventory takes up factory floor space as well as adding to costs.

Level 5 is one more step into sophisticated manufacturing. At this level, a computer looks at past demand for each product and predicts the number of items that will be needed at each time of the month or year. Sales forecasts are also factored in. The product to be manufactured is then scheduled and made accordingly. This level gets into an area called *artificial intelligence*, at a high cost. An analysis has to be made to determine whether people can make the same predictions at lower cost and with comparable accuracy. One result of proper use of level 5 would be reduced finished material inventory, if sales and production are matched properly. Another result would be timely availability of a product for prompt delivery. Other considerations, such as the time to buy raw material at lowest cost, would be included in the computer program at level 5.

24–3 TYPES OF NETWORKING

Computer-integrated manufacturing (CIM) is a philosophy for integrating hardware and software in such a way as to achieve total automation. Although each company has its own idea of what CIM really means, most follow a pattern similar to that in figure 24–2. In this diagram, dedicated processing tasks are shown distributed around a factory. As computers are further removed from the actual manufacturing area, their function shifts from real-time control toward supervision.

It is generally agreed that at least three levels of computer integration are required for CIM to work: the *cell level*, the *area level*, and the *plant level*. Each level has certain tasks within its range of responsibility. Cell controllers, for example, are generally responsible for data acquisition and direct machine control. Area controllers are assigned the tasks of machine and tool management, maintenance tracking, material handling and tracking, and computer-assisted simulation and design facilities. The plant-level computers are responsible for such things as purchasing, accounting, materials management, resource planning, and report generation.

When you develop a CIM system, it is best to start at the lowest level. Refer to figure 24–2. Begin by completely developing the control for one cell. Then do the next cell, and so on. Then develop the area controllers. Only when the cell controllers function properly should you develop the plant controller. If CIM systems fail, they do so for two major reasons. First, the design of the system is started at the top with the plant controller and works its way down. This does not work—it is backward. A second reason for the system's failure to come on line properly is using different machines made by various manufacturers. For example, we have 17 individual cell devices in four different cells. Your company procures 17 machines and 5 computers to be used as cell controllers. When connected, the machines and computers do not "talk" to each other, and nothing works together. When you call the computer manufacturer and the 17 cell device manufacturers, you find out no one

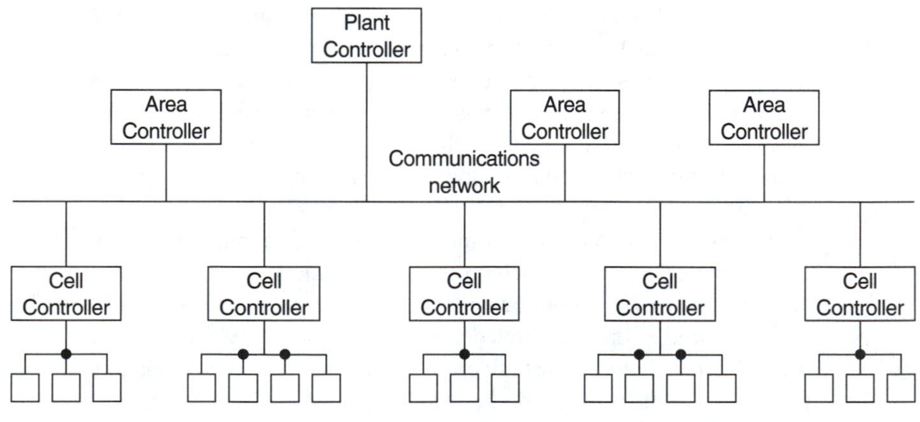

FIGURE 24–2
Distribution of Processing Tasks in a Plant According to Function

is responsible for the communication between devices. Each manufacturer is only responsible for the programs of its own devices. In a CIM project, it is necessary to ascertain communication ability between devices before buying any device. One organization or person should write or procure the overall networking programs for the CIM cell early in the project time frame.

Technically speaking, any computer, including the PLC, is capable of performing both control and supervisory tasks. The trend, however, is toward customizing computers for specific applications. These customized controllers are called *plant controllers*. It can be seen from the hierarchical structure of CIM that specialized computers would be advantageous. In fact, many companies are already producing computers to satisfy the specific requirements of every CIM level. Invariably, mainframe computers are implemented at the plant level in large facilities. In some cases, those computers were already in place to handle such tasks as inventory and payroll. To promote them to the top of the CIM pyramid, one simply selects the proper software. These computers are usually in a location remote from the actual plant floor. All the information they require is available from the host of subordinate area controllers throughout the plant.

Area controllers are usually on the plant floor and are therefore subjected to a harsh environment. A special breed of computers, called *industrial computers*, has been developed specifically for this application. The main difference between a PC and an industrial computer lies in the physical construction. Very simply, industrial computers are built to withstand higher temperatures, greater vibration, electromagnetically noisy environments, and rough handling. Since their task is still supervisory in nature, their operating hardware does not differ significantly from that of a PC. Again, it is the software that turns a generic computer into an industrial computer. The PC should not be ruled out as a strong contender for the position of area controller. Personal computers are inexpensive, and with the myriad clones available, they are plentiful. An added advantage of using PCs for area control is that there is a large software base already established for such computers.

The most specialized computers are cell controllers, those used to control work cells. A cell is defined as a group of machine tools or equipment integrated to perform a unit of the manufacturing process. Therefore, the computer that coordinates action within the cell has special hardware requirements as well as special software requirements. Specifically, such a computer must have multiple data paths (I/O ports) through which it may communicate with the various cell devices.

Traditionally, cell control has been accomplished with PLCs, which were designed for this very purpose. A CIM structure, however, requires that the cell coordinators communicate with other cell controllers, as well as with the area controller. The language of the PLC is somewhat restricting in this respect. In addition, most PLC languages do not lend themselves well to analysis and record-keeping tasks.

With these problems in mind, most PLC manufacturers are developing PLCs with greater software capability. A recent outgrowth of this effort is a computer called a *cell controller*. Cell controllers combine the software sophistication of a PC with the I/O handling capability of a PLC.

Again, the PC should not be overlooked. Many third-party manufacturers produce I/O cards designed to plug directly into the expansion slots of most PCs. These cards perform data-acquisition functions as well as data conversion and power control. A PC equipped with special software packages and I/O cards can emulate many of the popular PLCs in use today, and it can also perform the computational and communication tasks required of a cell controller.

Computers are even being implemented at the device level. That is, single-chip microcontrollers are being embedded in "intelligent" machines. Basically, a microcontroller is the marriage of a CPU to program and data memory and various I/O components, all on a single piece of silicon. The Intel 8797, for example, contains a 16-bit CPU, along with 8192 bytes of EPROM (user-programmable read-only memory), 232 bytes of random access memory (RAM), a 10-bit analog-to-digital converter, a full-duplex serial I/O port, a pulse-width modulated output, and four 8-bit I/O ports, all in a single 68-pin package. Devices of this type are, in the truest sense, complete computer systems, having all the power of their larger counterparts (PCs). Their capability is, of course, limited by the number of elements integrated onto the chip.

The concept of a microcontroller is nothing new. In fact, microcontrollers have been used in such devices as keyboards and printers for years. They may be found in applications that range from intelligent temperature transmitters to multiaxis robotic end effectors. With high-speed 32-bit microcontrollers coming on line, the future should prove very interesting.

24–4 NETWORK COMMUNICATIONS

The key to successful implementation of CIM is communication compatibility between all the computers involved in the process. This, in fact, poses quite a problem. Although communication standards do exist, each computer system may use a different standard. In addition, as the level of sophistication of communication increases, the need for more sophisticated standards arises. In anticipation of this problem, the International Standards Organization (ISO) has developed a model for what is called *open systems interconnection* (OSI). The OSI is a seven-layer model for communication network architecture (see figure 24–3). Each layer represents a different level of communication sophistication and pro-

FIGURE 24–3
Open System Interconnection (OSI) Model for Network Architecture

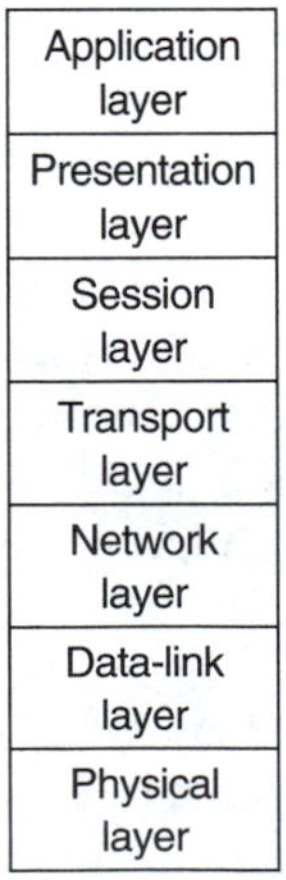

OSI model
Application layer
Presentation layer
Session layer
Transport layer
Network layer
Data-link layer
Physical layer

vides the necessary support for the layers above it. Implementing such an architecture allows the user to connect virtually any data communication device to the network and be assured of compatibility.

Buried within each OSI layer are communication protocols. A protocol is nothing more than an agreed-upon set of rules by which communication will take place. Protocols exist at all communication levels and are assigned to the appropriate OSI layer according to level of sophistication.

To illustrate the need for protocols at different levels, consider the simple action of making a telephone call. There must be some convention for how the wires are connected, the allowable bandwidth of the channel, and the signal levels used. Next, the sequence of signal exchange must be established. That is, the sequence of dial-tone transmission, dial-pulse or touch-tone detection, and ringing or busy signal transmission must be clearly defined. Finally, when connection has been made, the rules that govern human dialogue take over. There are three levels of protocols in this example: physical, transmission, and user. Note that the physical protocol provides the necessary support for the transmission protocol, which in turn supports the user protocol. Additionally, if the physical protocol is changed, the transmission and user protocols should be unaffected. That is, changes within layers should be transparent to the other layers. This principle is the essence of OSI.

To date, only layers 1, 2, and 3 of the OSI model have been clearly defined and successfully implemented. Layer 1 contains physical-link protocols—those that define such things as signal level and connection conventions. Protocols RS-232C and RS-422 are examples that would fit into this layer. Layer 2, the data-link layer, contains protocols that address such problems as circuit establishment, transmission sequence, and error control. Protocols such as SDLC, HDLC, and BISYNC belong in this layer. The network layer (layer 3) defines procedures for data routing, packet switching, and error recovery. The X.25 protocol is the most pervasive in this layer.

Several protocols can exist in each OSI layer. This ability, in fact, is what allows the network to support such a variety of different computers. A particular industrial computer may require X.25, SDLC, and RS-232C to accomplish its communication link; another may require X.25, BISYNC, and RS-422. Although the communication paths through the OSI layers would differ, both computers would use three layers, and—more important—both could use the network.

The OSI network-architecture model is by no means the only attempt at network standardization. Several other network architectures, called proprietary networks, are currently being used. International Business Machines (IBM) promotes its Systems Network Architecture (SNA), Digital Equipment Corporation (DEC) markets a system called Digital Network Architecture (DNA), and Burroughs Corporation has introduced Burroughs Network Architecture (BNA). These network architectures resemble the OSI model, differing only in the number of layers involved and, of course, the nomenclature. One disadvantage of these proprietary architectures is that they tend to lock the user into products from a single vendor. For example, SNA is designed around IBM mainframes, and DNA is designed to support DEC minicomputers. The OSI architecture, on the other hand, is designed to accommodate products from any vendor.

Generally, physical networks can be classified as *wide area networks (WANs)* or *local area networks (LANs)*. The difference lies in the transmission distance and, therefore, in the proximity of the stations to each other. Virtually all plants and factories fall into the LAN category, although different factories within the same company may communicate over a WAN.

A LAN may be characterized by its method of station access. Two access methods are in common use: *random access* and *token access*. Random access is a method by which all stations "listen" to the transmission line and wait until the line is free before attempting a transmission. If a collision occurs, all transmitting stations shut down and wait for the next opportunity to transmit. This method is called *carrier-sense multiple access with collision detection (CSMA/CD)*. Protocols such as ETHERNET (introduced by XEROX) use this type of access at the transport level of the OSI model.

Token access is a method by which possession of a token (a special code sent along the transmission line) allows a station to transmit. The token is passed from station to station until it reaches one with a message to send. That station then removes the token and replaces it with a frame of data to be transmitted. After successful transmission, the station places the token back on the line, and the token proceeds to the next station. This method allows each station an equal chance to send data. General Motors Corporation's Manufacturing Automation Protocol (MAP) is a transport protocol that uses token access.

24–5 CELL CONTROL BY PLC NETWORKS

A manufacturing cell is a group of automated programmable machine controls (programmable controllers, robots, etc.) designed to work together to perform a complete manufacturing or process-related task. The function of a cell controller is to coordinate and oversee the operation of the machine controls within the cell through its communication and information-processing capabilities. Cell control can provide your company with the unprecedented opportunity to realize flexible computer-integrated manufacturing (CIM) by applying factory floor information in real-time. The result is improved plant efficiency and profit.

A training and industrial work cell is shown in figure 24–4. The cell contains two CNC-controlled machines, a lathe and a milling machine. The abbreviation *CNC* stands for *computer numerical control*. It applies to machine tools of all types, such as lathes and milling machines. The CNC designation indicates that a machine is controlled by computer, not manually or electrically. The parts of this work cell are as follows:

1. A computer-controlled conveyor
2. A CNC lathe
3. A transverse robot with PLC control to load and unload the lathe
4. A CNC milling machine
5. A pick-and-place (PP) robot with PLL control to load and unload the mill
6. A SCARA robot for drilling or small assembly operations on the conveyor (right)
7. A master PLC to coordinate the conveyor and the other two PLCs
8. A computer-controlled storage rack (ASRS) Automatic Warehouse
9. A computer program to keep track of the location of each part in the storage rack

A

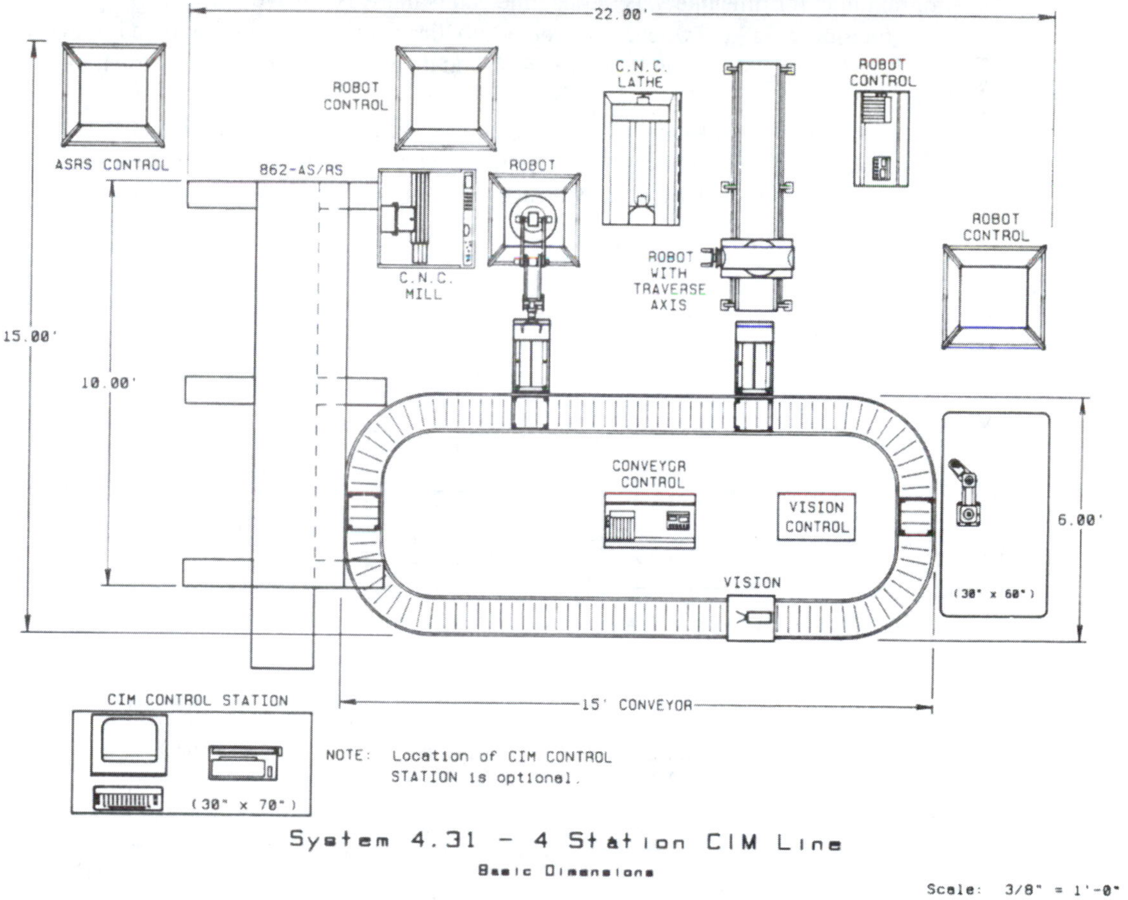

22.00'
ASRS CONTROL
ROBOT CONTROL
C.N.C. LATHE
ROBOT CONTROL
862-AS/RS
ROBOT
C.N.C. MILL
ROBOT WITH TRAVERSE AXIS
ROBOT CONTROL
15.00'
10.00'
CONVEYOR CONTROL
VISION CONTROL
6.00'
VISION
(30" x 60")
15' CONVEYOR
CIM CONTROL STATION
(30" x 70")
NOTE: Location of CIM CONTROL STATION is optional.

System 4.31 – 4 Station CIM Line

Basic Dimensions

Scale: 3/8" = 1'-0"

B

FIGURE 24–4
Industrial and Training Work Cell (Courtesy of Amatrol, Inc.)

10. A master computer to oversee the whole operation (CIM control station)
11. Safety devices and interlocks as required

A network for overall control for the cell in figure 24–4 is shown in figure 24–6. The network can be of different configurations, depending on the work cell sequence and the complexity of manufacturing the part or parts. Three major forms of possible networks are shown in figure 24–5. A is a simple star configuration in which all parts communicate directly and only with the master control. B is a "semi-star" in which groups of devices communicate with the central control. C is a commonly found bus type in which major machines and groups are connected to a common bus. The advantage of C is that individual sections can communicate with each other without going through the master computer. This gives more flexibility and faster communication times. The disadvantage of C is that programming is more complicated with a common, shared bus. Bus time sharing involves programming for time intervals of bus time and is more expensive.

Interlocks, safety switches, and other auxiliary devices are also involved in the cell. These devices are not shown in figure 24–4, and their connections are not shown in figure 24–5.

An industrial work cell is shown in figure 24–6. This large work cell is part of a manufacturing operation to fabricate large computer cabinets. The equipment shown is a multistation robotic-fabrication work cell. In this work cell, two robots are utilized for final

FIGURE 24–5
Network Configuration for CIM Cell

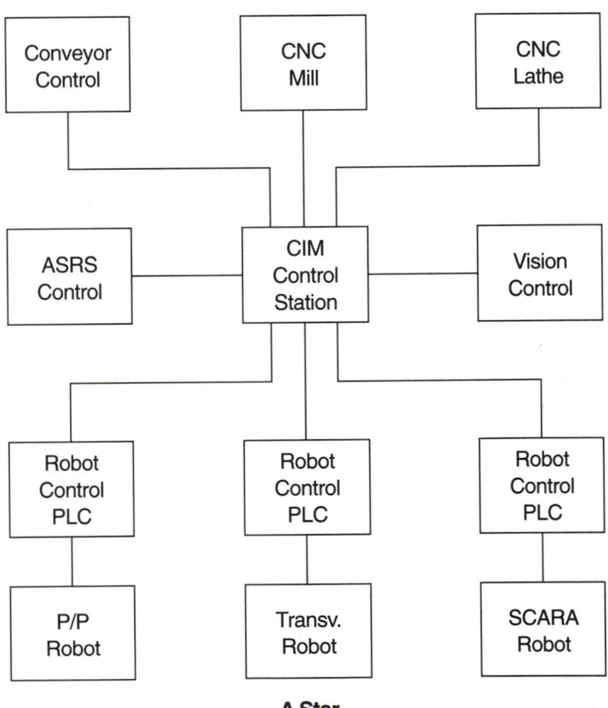

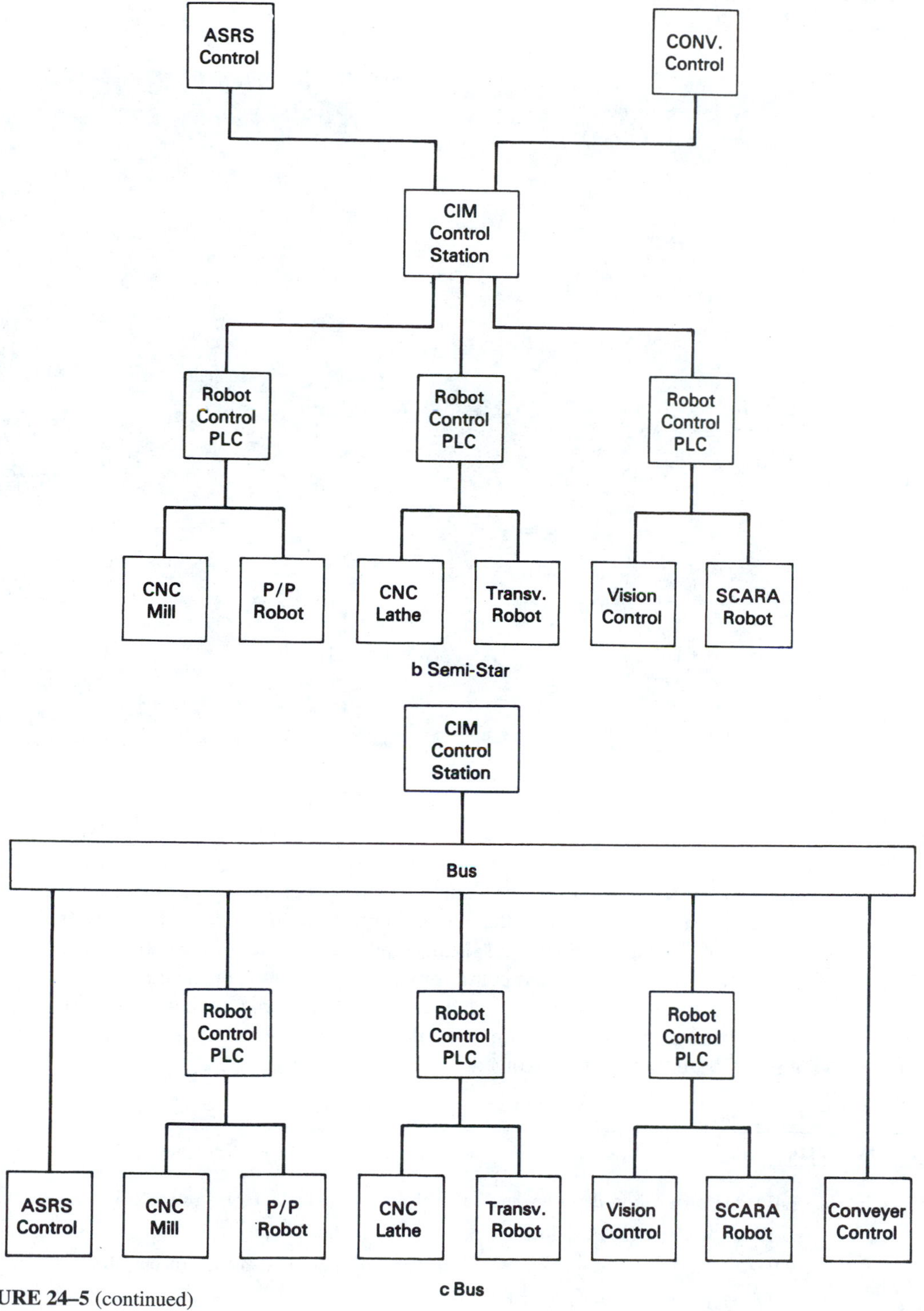

FIGURE 24–5 (continued)

FIGURE 24–6
Industrial Work Cell (Courtesy of ABB Robotics, Inc.)

weld assembly of the computer cabinets. The robots have individual controllers. Their individual controllers are under the overall control of the cell master computer. Other cell equipment, such as the conveyors, is also under the control of the cell's master computer. The robot on the left performs spot welding. The other robot, on the right, subsequently performs arc (bead) welding on the cabinets. The cell is reprogrammable for different models of computer cabinets as they come into the cell area. The network system (not shown) for this cell is quite large and involved.

EXERCISES

1. In your own words, describe the five levels of control and their interrelationship.
2. Describe where the PLC does and does not fit into the five levels.
3. Define CIM, plant controller, area controller, cell controller, and microcontroller.
4. Differentiate between WAN and LAN.

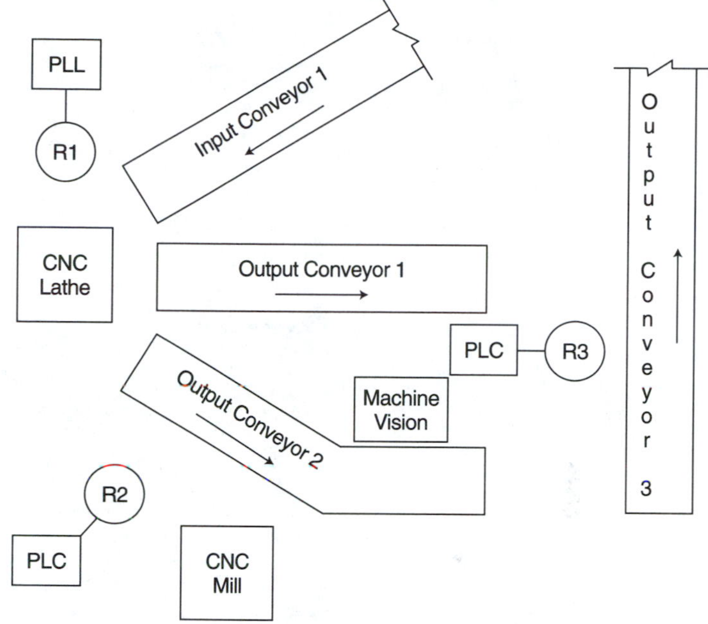

FIGURE 24–7
CIM for Exercise 5

5. Sketch star, semistar, and bus networks for the work cell shown in figure 24–7. Robot 1 takes parts from the input conveyor and puts them in the lathe. When the lathe is through turning, parts are put onto conveyor 1 or 2 by robot 1, depending on the part number. While moving on output conveyor 1, the part is checked by a vision inspection. For conveyor 2, the part is removed from the conveyor by robot 2, machined by a mill, and returned to the conveyor by robot 2. At the end of output conveyors 1 and 2, robot 3 puts finished parts in boxes on output conveyor 3.

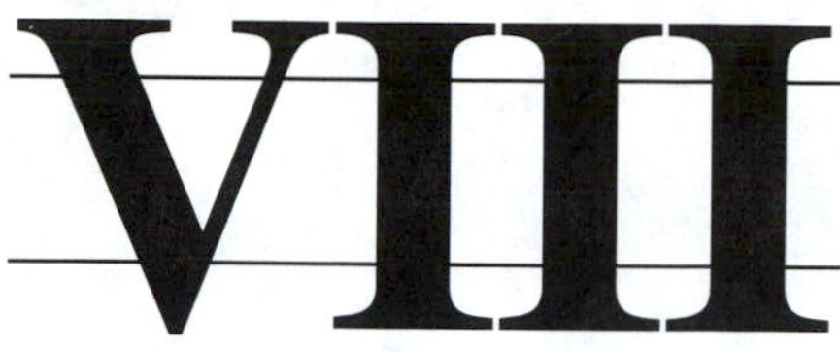

Related Topics

25

Alternative Programming Languages

OUTLINE

25–1 Introduction □ 25–2 Ladder Logic and Beyond □ 25–3 BASIC: A Step Up □ 25–4 State Languages: High-Level Programming for the PLC

OBJECTIVES

At the end of this chapter, you will be able to

- Explain why ladder logic may not be suited for more elaborate PLC machine and process control.
- Describe the advantages and limitations of BASIC as a PLC programming language.
- Define state language programming.
- Describe the four basic components of a step in a state language.
- Describe how the Quickstep State Language program works.

25–1 INTRODUCTION

Ladder logic remains the dominant programming language for today's PLCs, at least in the United States. More than 80 percent of all PLC programming in the United States is done using ladder diagrams. It is the language our electricians "speak"; it is the language they grew up on; it is the language they are comfortable with; and it is a language that worked particularly well during the transition from relays to microcomputer-based controls. Nonetheless, now that the transition is all but complete, some feel it is time to adopt a more sophisticated, higher-level programming procedure for today's and tomorrow's PLCs. Is it time to move beyond ladder logic?

With the development of more powerful PLCs, the advent of the data processing personal computer as a PLC programming tool, the demand for ever more elaborate machine and process control, the need to reduce programming time, and the necessity of communicating the programming solution to a wider spectrum of company personnel, a trend toward the use of high-level, user-friendly, graphics-based programming languages, particularly where advanced processing is concerned, is clearly discernible. Although space does not permit an in-depth review of such languages, it is nonetheless important that today's student of the PLC be introduced to alternative programming languages, languages that many feel will soon be a part of every technician's interface to the world of industrial control.

We begin this chapter with a brief look at the origins of ladder logic while examining a few of its more obvious limitations. Next, we take a step up and explore the BASIC language as a PLC in-line programming tool. Finally, we examine state languages, in particular, the Quickstep State Language, from Control Technology Corporation, as the high-level programming paradigm of choice.

25–2 LADDER LOGIC AND BEYOND

As we have seen throughout this book, ladder logic is a natural extension of older relay logic. With ladder logic, the programmer describes an imaginary relay network. If such a network were in fact real, the desired machine or process control would take place.

A problem arises, however, when we realize that real-world machine and control processes involve sequential/concurrent, or series/parallel, activity. A process performs various activities in sequence (series), one step at a time. It also works concurrently (in parallel) as the various sequences are run simultaneously.

For example, a machine may load, punch, stamp, fasten, and then eject a part, in sequence. If only one part is run, no concurrent processing takes place. However, if parts are continuously passed through the machine, where the loader, puncher, and so on, are all operating simultaneously (though on different parts), concurrent processing is also occurring.

Why should this be a problem? Because ladder logic is primarily a concurrent (parallel) processing language. True, with the use of the latching contact it also becomes a sequential (series) processing language. Yet as the ladder logic program becomes more complex, it also becomes more confusing.

What is needed, many in the field now agree, is a language that allows a programmer to specify both sequential and concurrent machine or processing activity. State languages can do that and a whole lot more. Before we examine such a language, let us look at the familiar BASIC as a PLC programming language.

25–3 BASIC: A STEP UP

The first PLCs, with their origin in the late 1960s at General Motors, were, of course, programmed in ladder logic. In the mid 1980s, the engineering side of General Motors, as well as Ford and Chrysler, started asking for BASIC or advanced BASIC programming. They wanted to program their PLCs in a higher-level, more user-friendly language.

But why choose BASIC? BASIC (Beginner's All-Purpose Symbolic Instruction Code) is a *procedural* language that is easy to use and can be developed and run on small personal or industrial computers. It consists of lines of text forming statements that tell the computer how to perform, step by step. These source code statements are then automatically compiled into object (machine) code that the computer actually executes.

For all its ease of use, however, BASIC has problems. It is relatively slow, it lacks the control structures needed for sophisticated structured programming, and it is said to produce "spaghetti" code in which it is difficult to visualize all the interlinks. When BASIC gets beyond a standard page of text, many people find it nearly impossible to read.

Nonetheless, today's compiled BASIC does overcome some of these objections, and unlike ladder logic, BASIC is primarily a sequential, not a concurrent, programming language. Since it is being used to program a number of today's PLCs, let's take a closer look.

Like all computer languages, BASIC consists of an instruction set. A partial listing of a BASIC instruction set is shown in figure 25–1. Although such instructions do not correspond exactly to ladder logic symbols, there are similarities. For example, the LET instruction would be used to assign a number value variable. Thus, if a contact is represented by the variable X, the statement LET X = 1 means that the contact is closed. The IF and THEN instructions work together to simulate the output ladder symbol. Say a ladder rung has a N.O. (normally open) contact (X) and an output (Y). When the contact is closed, the output is energized. The IF/THEN pair establishes a condition that results in a specific output. Here is the BASIC program that accomplishes this simple task.

```
10  READ  X
20  IF   X = 0  THEN  Y = 0
30  IF   X = 1  THEN  Y = 1
```

FIGURE 25–1
Partial BASIC Instruction Set

LET   INPUT   READ   DATA

IF

THEN

TIMER ON

TIMER OFF

TIMER STOP

FIGURE 25–2
Bear BASIC Program (Courtesy of Divelbiss Corp.)

```
100  INTEGER C (1)                     ´ Counters
110  INTEGER S (1)                     ´ Switch states
120  INTEGER J, K
130  C (0)=0 : C (1)=0                 ´ Initialize the counters to 0
140  FOR J=0 TO 1                      ´ Loop for both switches
150    GOSUB 300                       ´ Check for switch closure
160    IF K=0 THEN 190                 ´ Jump if no closure
170    C (J)=C (J) +1                  ´ Increment counter
180    PRINT "C" ; J;" = "; C(J)       ´ Print counter
190  NEXT J
200  GOTO 140                          ´ Loop forever
210  ´
300  ´ Subroutine to check for switch ´J´ closure. If the switch was previously
310  ´ open and is closed now, then return a 1; otherwise, return a 0.
320  K=DIN (J+1)
330  IF K=1 AND S(J) =0 THEN S(J) =K: RETURN
340  IF K=0 AND S (J) =1 THEN S(J) =K
350  K=0 : RETURN
```

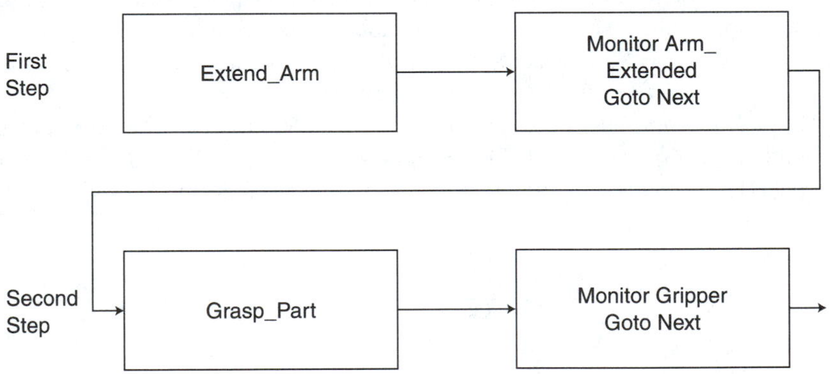

FIGURE 25–3
Components of a Simple State Language Program

Divelbiss Corporation, of Fredericktown, Ohio, has been offering a compiled BASIC, called Bear BASIC, to run with their PLC machines for some time. A sample program using their Bear BASIC is shown in figure 25–2. The program implements a system with two switches and two counters, with a counter being incremented when the corresponding switch is closed. If you have had any experience with BASIC, you should be able to figure the program out in no time.

Although BASIC and other procedural languages, such as C, are going beyond ladder logic in a number of PLC programming environments, it must be remembered that BASIC is still primarily a sequential language. Sequential PLC tasks can be handled by BASIC, but when asked to deal with parallel problems it often becomes overwhelmed.

If ladder logic is good for parallel processing and BASIC is good for series processing, obviously what is needed is a PLC language that will excel at both. We need a language that will go beyond telling the computer *how* to do something and instead tell it *what* we want done. Grafcet, which we look at next, may be an answer.

25–4 STATE LANGUAGES: HIGH-LEVEL PROGRAMMING FOR THE PLC

State languages have recently come to the forefront as superior approaches to automation programming. They came into existence to address problems that could no longer be dealt with adequately using preexisting (relay ladder logic) programming techniques. First, we see what a state language is and its main characteristics. We then look at a specific state language, Quickstep for Windows, from Control Technology Corporation.

As we have seen throughout, in the discrete manufacturing world and, to some extent, in the batch and continuous process worlds, problems often consist of a series of steps or states that a machine or process must go through to perform a series of operations. State languages attempt to exactly mimic the structure of this type of problem.

The fundamental tool of a state language is the state or *step*. A step defines the complete status of a machine or process for a finite period of time. This status, typically, consists of two components, according to Kenneth C. Crater, President of Control Technology Corporation:

1. One or more commands to create *motion* or change, thus causing a new physical state to be adopted by the machine or process.
2. One or more instructions to limit the *duration* of the step and specify the next step to proceed to upon completion of the current step.

In figure 25–3 we see the components of a simple state language program, showing motion commands (left) and instructions for proceeding to a new state (right).

In the Quickstep State Language, you are provided a natural representation of a machine control program. To start, you simply create a "sequence of events" flowchart to describe your process. (More on the flowchart in a moment.) You then convert your flowchart directly into the equivalent steps to perform the control. A Quickstep program is in turn made up of the STEPS required to reproduce the events in your flowchart. A sequence of steps is called a TASK.

There are four basic components of a step:

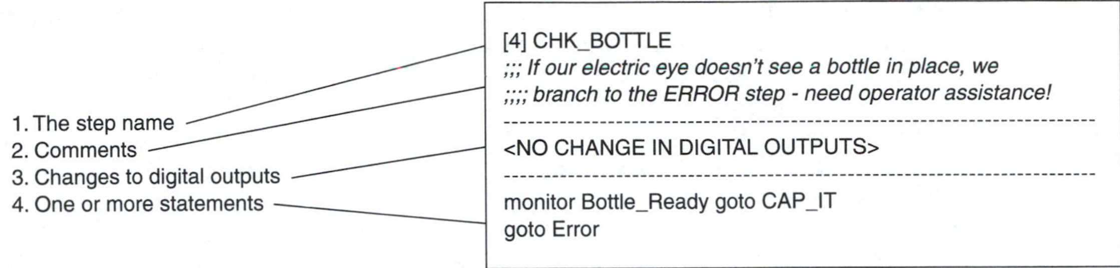

The statements are made up of Quickstep instructions which can perform many functions including:

- Mathematical operations
- Register manipulations
- Analog I/O, Display Updates
- Logical operations
- Delay operations
- Servo and Stepper profile and control functions
- Branches to other steps
- Instructions to start other tasks

FIGURE 25–4
Four Basic Components of a Step

As shown in figure 25–4, there are four basic components of a step: (1) step name, (2) comments, (3) changes to digital outputs, and (4) one or more statements. Statements are executed one by one, in the order they appear, until one of the instructions initiates a transfer of execution to another step. If none of the instructions result in a transfer to another step, each statement containing a GOto is reevaluated until a branch out of the step occurs.

As an example of a Quickstep program, we present the Automated Bottle Capper, courtesy of Control Technology Corporation. The Automated Bottle Capper program demonstrates how to use simple digital inputs and digital outputs and how to conditionally determine the next step to execute. A diagram of the process along with a list of the I/O the program uses is shown in figure 25–5.

> ***Note:*** To keep the example simple, we have assumed that we can advance each bottle perfectly into position simply by turning our servomotor a preset number of steps. We have also ignored the use of safety interlocks. However, YOU OBVIOUSLY SHOULD NOT!

In the Automated Bottle Capper Program, a servomotor is used to index each bottle into position and then the bottle is capped. The process continues indefinitely. Three digital inputs are used to detect unusual conditions. If detected, the program sounds an alarm horn and then halts execution. A detailed flowchart of the program is shown in figure 25–6. The complete program is shown in figure 25–7.

I/O and Registers Used

	Controller
Digital Outputs	**Resource**
1. Capper_Up/_Down	Output 1
2. Alarm_Horn_On/_Down	Output 2
Digital Inputs	
(1) Bottle_Ready/_Not Ready	Input 1
(2) Up_Confirm	Input 2
(3) Down_Confirm	Input 3
Servos	
[1] Conveyor	Servo 1

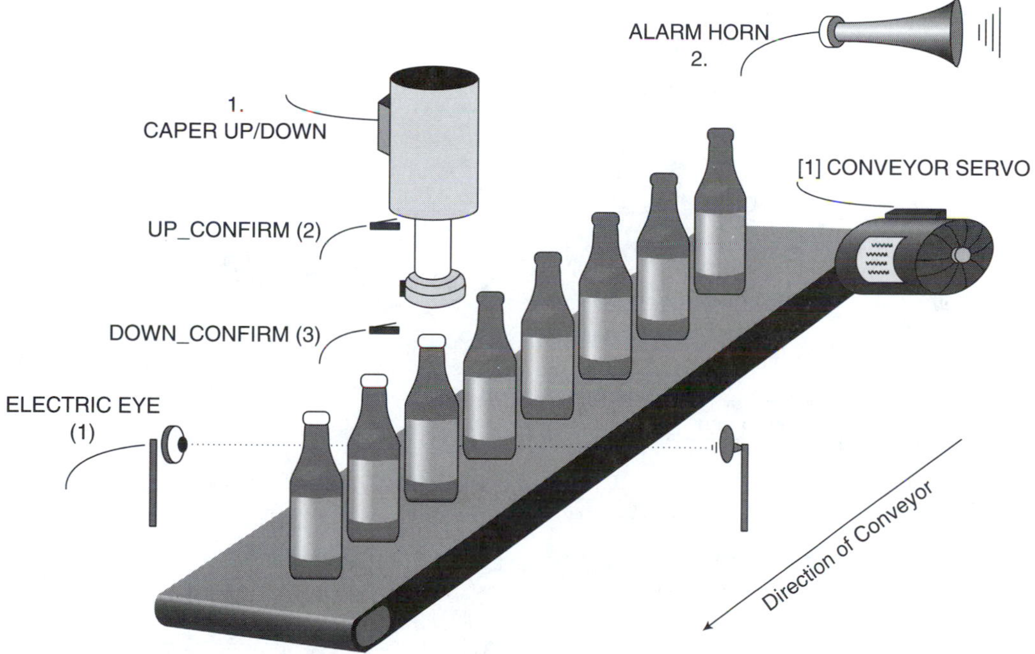

FIGURE 25–5
Automated Bottle Capper

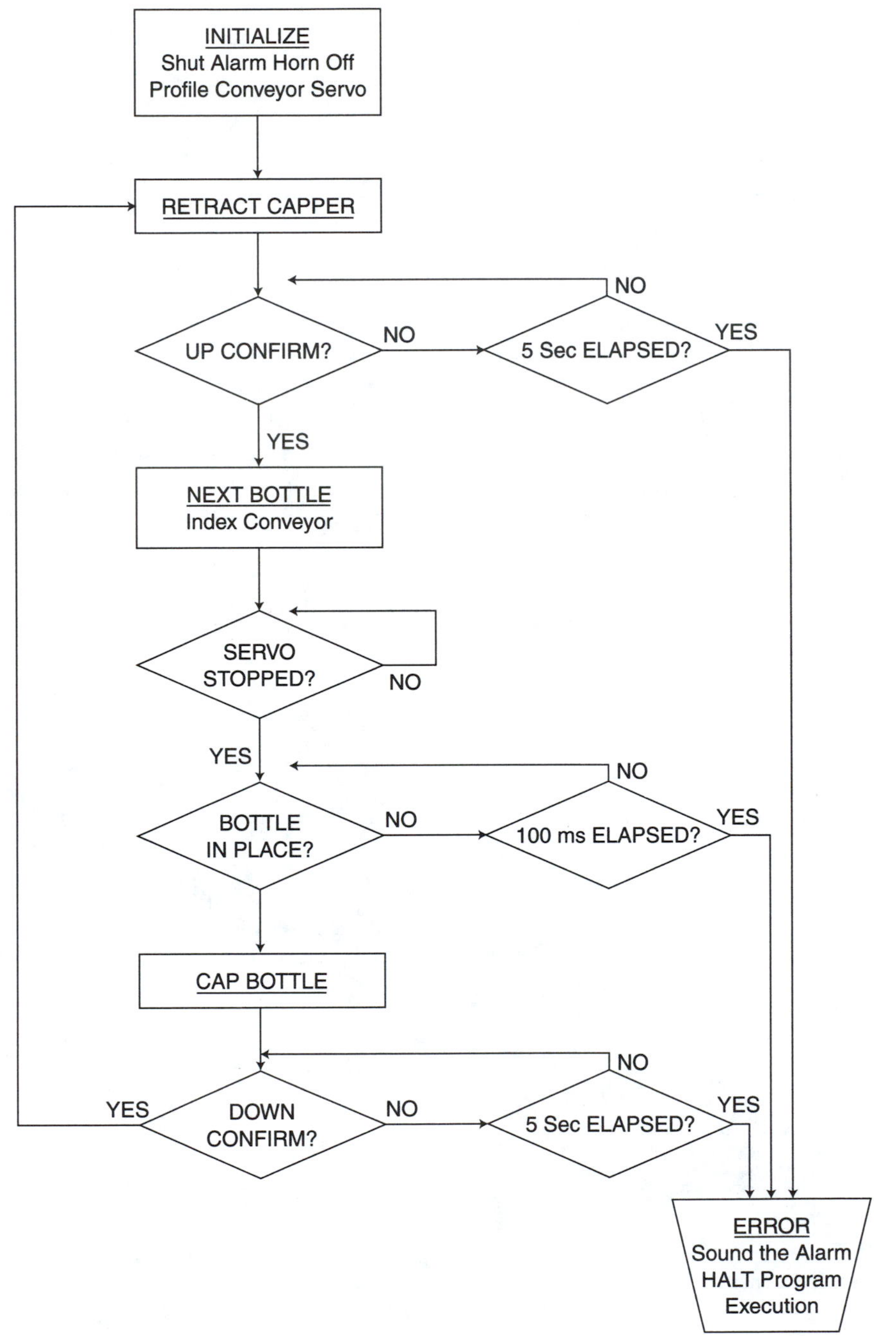

FIGURE 25–6
Flowchart for Automated Bottle Capper Program

FIGURE 25–7
Complete Program for the Automated Bottle Capper

The Automated Bottle Capper Example

A Quickstep Program For The Bottle Capper Example

```
[1]    INIT
; ; ;  Set all outputs off. Retracts the capper and ensures
; ; ;  alarm horn is off. We then go to the RETRACT_CAPPER
; ; ;  step and wait for the UP_CONFIRM limit switch.
------------------------------------------------------------
<TURN OFF ALL DIGITAL OUTPUTS>
------------------------------------------------------------
profile Conveyor servo at position maxspeed=IndexSpeed accel=RampRate
       P=PropVal I=IntegralVal D=DerivVal
goto RETRACT_CAPPER

[2]    RETRACT_CAPPER
; ; ;  Raise the capper, then monitor for the UP_CONFIRM
; ; ;  limit switch. If we don't get confirmation within
; ; ;  5 seconds, we branch to ERROR and sound the alarm.
------------------------------------------------------------
Capper_Up
------------------------------------------------------------
monitor UP_CONFIRM goto NEXT_BOTTLE
delay 5 sec goto ERROR

[3]    NEXT BOTTLE
; ; ;  Turn the servo 500 steps CW to advance the conveyor
; ; ;  to position the next bottle. After the conveyor
; ; ;  stops, we branch to the CHK_BOTTLE step.
------------------------------------------------------------
<NO CHANGE IN DIGITAL OUTPUTS>
------------------------------------------------------------
turn Conveyor cw 500 steps
monitor Conveyor:stopped goto CHK_BOTTLE

[4]    CHK_BOTTLE
; ; ;  If our electric eye doesn't see a bottle in place,
; ; ;  we branch to the ERROR step - need operator
; ; ;  assistance!
------------------------------------------------------------
<NO CHANGE IN DIGITAL OUTPUTS>
------------------------------------------------------------
monitor Bottle_Ready goto CAP_IT
delay 100 ms goto ERROR

[5]    CAP_IT
; ; ;  Cap the bottle, then branch to the RETRACT_CAPPER
; ; ;  step upon detecting the DOWN_CONFIRM limit switch.
------------------------------------------------------------
Capper_Down
------------------------------------------------------------
monitor DOWN_CONFIRM goto RETRACT_CAPPER
delay 5 sec goto ERROR

[6]    ERROR
; ; ;  This step sounds the alarm and halts the program
; ; ;  execution.
------------------------------------------------------------
Alarm_On
------------------------------------------------------------
Done
```

EXERCISES

1. Interview industry personnel to find out what they think about ladder logic versus high-level languages such as state languages.
2. Obtain a PLC program written in BASIC and analyze it, line for line.
3. Contact half a dozen PLC manufacturers and determine what programming languages, other than ladder logic, their machines are compatible with.
4. Obtain the specifications on three high-level PLC programming languages.
5. Obtain the specifications on Quickstep from Control Technology Corporation, 25 South Street, Hopkinton, MA 01748; 508-435-9595.

26

PLC Auxiliary Commands and Functions

OUTLINE

26–1 Introduction □ **26–2** MONITOR Mode Functions □ **26–3** FORCE Mode Functions □ **26–4** PRINT Functions

OBJECTIVES

At the end of this chapter, you will be able to

- □ Explain how the MONITOR mode may be called up and used for ladder diagram analysis.
- □ Explain how the FORCE mode is used for PLC program testing and analysis.
- □ List the safety precautions required when using the FORCE mode.
- □ Explain how the PRINT mode is used to print out ladder diagrams.
- □ List and explain the other major types of PLC PRINT capabilities.

26–1 INTRODUCTION

Three important PLC functions deserve a separate chapter to cover their usefulness: the MONITOR mode, the FORCE/OVERRIDE function, and the various PRINT capabilities.

After a circuit is programmed into a PLC, its operation may be watched on a screen in the MONITOR mode. The current flow from left to right as contacts open and close is indicated by a brightening of the screen pattern. Functions such as coils and timers also light up when they become energized. Other types and models of PLCs show the current flow by flashing lines and functions. Still others use a dotted pattern system. This chapter discusses the use of the MONITOR mode.

The second function to be covered in this chapter is the FORCE function. In some cases, this function may be looked upon as an override control. To use the FORCE function, first call up the contact, coil, or function to be controlled. Next, the cursor is moved to the function to be controlled. Then, the FORCE function key is depressed. Then, using the keyboard keys, the function under FORCE control may be turned on and off. The keyboard then overrides the status of the input from the outside system.

The third function to be discussed is the use of printouts to record information regarding a circuit and the status of the circuit parts. The most common printout is that of the ladder diagram. Other printouts are available on many other PLC models. These are for registers, timed status information, and other PLC equipment status.

26–2 MONITOR MODE FUNCTIONS

The MONITOR mode for ladder diagram operation is indicated on the screen in various ways. It may be indicated by a brightening of the pattern where voltage is passed through. In other cases it is indicated by the pattern changing to a dotted line or to a flashing effect. A large monitor that shows complete ladder lines normally uses the brightness enhancement effect. Smaller monitors showing a portion of a line use the other indicating systems. Figure 26–1 illustrates brightness enhancement for a standard, three-wire, motor-control, single-line ladder diagram. The figure shows the screen as the two inputs (stop and start) are energized and de-energized. The pattern changes allow us to watch the circuit operation.

The MONITOR function is especially useful for analyzing a large number of ladder lines. The MONITOR mode assists the operator in troubleshooting a large system that is malfunctioning. In some PLC models the screen is in MONITOR mode whenever it is in the EDIT mode; in other PLCs, the MONITOR mode must be called up separately.

Other system characteristics may be monitored in addition to the ladder diagram. These include register status (value), as well as individual coil and contact status. Other monitorable system parameters include a listing of the forced functions, which are discussed next. Some advanced PLC systems can also list the actual malfunctioning output devices for fast analysis.

Figure 26–2 shows how the status of four holding registers would be shown. This figure shows the register values in binary; many PLCs give you a choice of which numbering system you want used for the printout: binary, decimal, hex, octal, or ASCII.

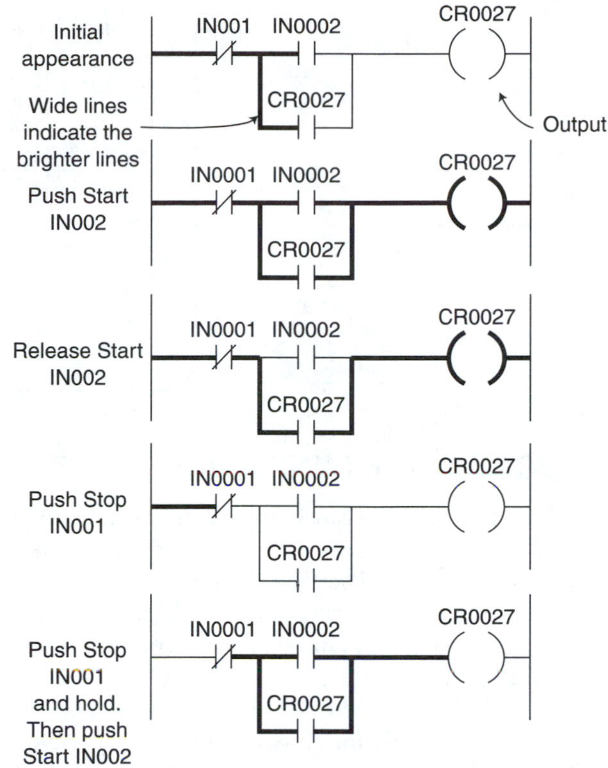

FIGURE 26–1
MONITOR Mode Example

In most PLC systems, you may call up individual coils, contacts, or both on the screen. For example, if you are looking at or in the vicinity of line 32, you may wish to see what is happening to an input contact on line 6, which is off the screen. Contacts and coils from line 6 can be inserted in a blank space by themselves and observed for on-off status. Figure 26–3 shows how these individual contacts might appear on a screen along with other PLC information, such as the ladder diagram.

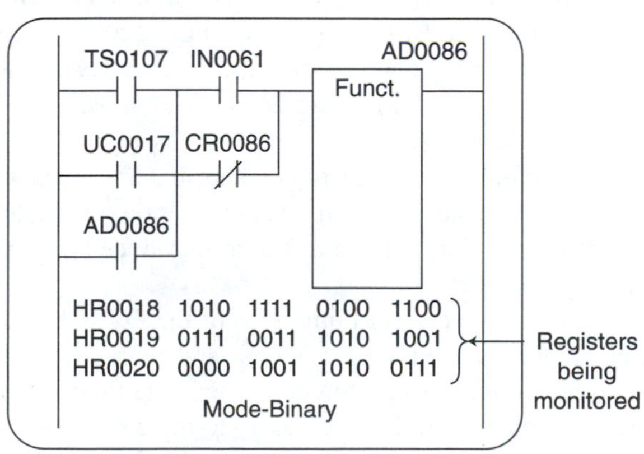

FIGURE 26–2
Individual Register Status Display

FIGURE 26–3
Contact Status Monitoring

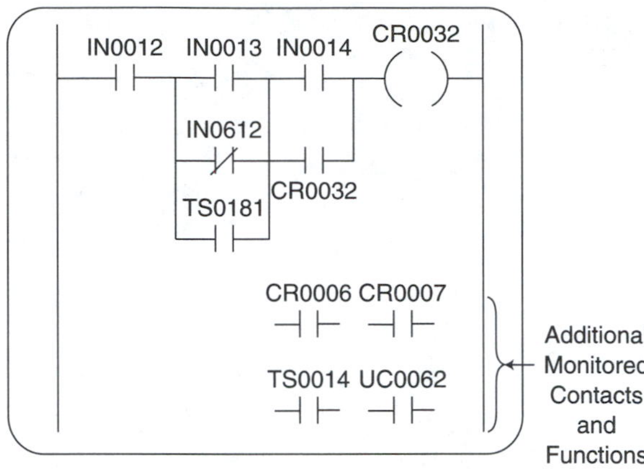

26–3 FORCE MODE FUNCTIONS

Many PLCs have the capability to carry out a FORCE function. The function is essentially an override control that enables the operator or programmer to operate the circuit from the program keyboard. The FORCE mode is useful but must be used with utmost caution in conjunction with a working process. Misuse of FORCE could lead to equipment damage and operator injury. FORCE is an override function. When turned on, it can lead to feeding the process program with incorrect information. If, for example, you were to force a safety interlock closed when the interlock is open, unsafe operation would result. It is best to use FORCE only for process malfunction troubleshooting, and then with great caution. It is, of course, useful for prerunning the process in the office to see how it works before hooking up to the actual process.

The FORCE procedure is normally carried out in the MONITOR mode. First, place the cursor over the contact, coil, or function you wish to force. Then carry out the specific keyboard procedures for forcing. Turning FORCE on changes the status of the contact or coil under the cursor. If it is a normally open contact, it will close (turn on). If it is a normally closed contact, it will open (turn off). If you force a coil or function, it will go on when forced, regardless of external commands in effect.

In most cases, forcing any contact of a relay, CR, will turn the relay coil on, as well as forcing all other contacts of that CR number. An example of a display showing a FORCE procedure is shown in figure 26–4. To remove the FORCE function, turn it off and then press the Clear key.

The individual coils or contacts that are forced on may be left in the forced state permanently by entering them, usually by pressing Return. This permanent-entry procedure must be done carefully so as not to introduce a permanent unwanted change in an operational sequence.

There are certain limitations of the FORCE mode. Not all functions react like coils and contacts when forced. For coils and contacts, forcing a contact causes its coil and all of its other contacts to be forced at the same time. Many other functions work in the same manner, but not all. It is necessary to review the operating procedures of a particular PLC to see how each function responds to the FORCE command.

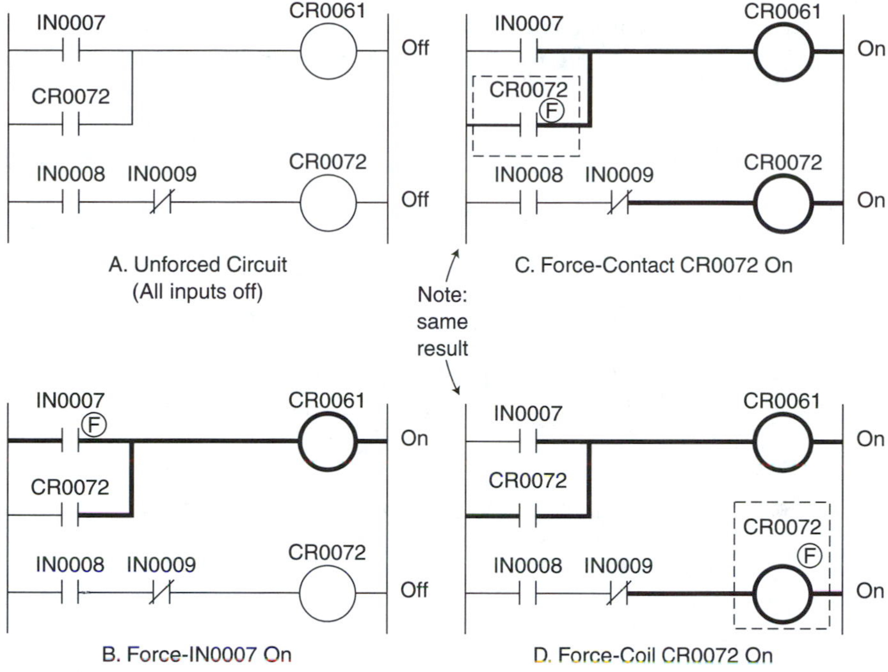

FIGURE 26–4
FORCE Procedure

For example, in some PLCs, forcing the MASTER CONTROL RELAY (MCR) function does not have the same effect as officially turning the MCR function on through its normal operating ladder program. Forcing the MCR coil does not affect its function but does turn on its associated contacts.

The same special consideration can apply to the SKIP and DR/SEQUENCER functions. See your operating manual for individual FORCE function operational characteristics.

If an industrial process is in operation, it obviously would be undesirable to insert into it periodic false signals by hooking up a keyboard to the CPU controlling the process and forcing in the false input signals. Not only would this be dangerous to equipment, someone could be injured. Do not use the FORCE function on an operating system unless all personnel in the area have been notified. It is best not to use it at all on an operating system; limit its use to simulations, if at all possible.

26–4 PRINT FUNCTIONS

Ladder diagrams on a screen cover from one to four or five rungs, depending on the PLC model. If the entire operational circuit has 20 or more rungs, for example, you may wish to see the entire circuit at once. If the PLC you are using has a PRINT mode system, the whole ladder diagram can be printed out continuously on a conventional computer printer.

```
0!                                                            !
0! IN0001  IN0002                                       CR0009!
0! --]\[-+--]  [-+------------------------------------------( )-!
1!       !      !                                             !
 !       !CR0009!                                             !
 !       +--]  [-+                                            !
 !                                                            !
 !                                                            !
 !       1, 4, #1                                             !
0!                                                            !
0! IN0003  IN0004  CR0012                               CR0011!
0! --]\[-+--]  [-+--]\[-------------------------------------( )-!
2!       !      !                                             !
 !       !CR0011!                                             !
 !       +--]  [-+                                            !
 !                                                            !
 !                                                            !
 !       2, 3, #2                                             !
0!                                                            !
0! IN0005  IN0006 CR0011                                CR0012!
0! --]\[-+--]  [-+--]\[-------------------------------------( )-!
3!       !      !                                             !
 !       !CR0012!                                             !
 !       +--]  [-+                                            !
 !                                                            !
 !                                                            !
 !       2, 3, #3                                             !
0!                                        +-----------+       !
0! CR0009                                 !           !TS0013!
0! --] [----------------------------------! PRESET    +-( )-!
4!                                        !   0005    !       !
 ! IN0007                                 !           !       !
 ! --] [----------------------------------!           !       !
 !                                        !           !       !
 !                                        ! ACTUAL    !       !
 !                                        ! HR0101    !       !
 !                                        !           !       !
 !                                        +-----------+       !
 !        5, #4                                               !
0!                                        +-----------+       !
0! TS0013                                 !           UC0014!
0! --] [----------------------------------! PRESET    +-( )-!
5!                                        !   0004    !       !
 ! IN0008                                 !           !       !
 ! --] [----------------------------------!           !       !
 !                                        !           !       !
 !                                        ! ACTUAL    !       !
 !                                        ! HR0102    !       !
 !        #5                              +-----------+       !
 !                                                            !
 !                                                       END  !
                                                              !
```

Cross Reference

FIGURE 26–5
Ladder Diagram Printout

IR0706	1010	1111	1100	0000
IR0707	0101	0010	0110	1101
IR0708	1001	0101	0011	1111
IR0709	1100	1000	1010	1101

(In Binary)

HR0062	0087
HR0063	0642
HR0064	7410
HR0065	0007

(In Decimal)

FIGURE 26–6
Register Status Printout

There are, of course, other reasons you might want a ladder printout. You might need a permanent written record, for instance. Also, in education and training, a printout is a written record of laboratory achievement (under proper controls).

One helpful extra feature of many ladder printouts is that each rung may be printed with a cross-reference system. These cross references are similar to the conventional ones used in standard ladder diagrams. Each ladder line with a coil or function is assigned a consecutive number. Then, on each line, a listing is printed of the other lines in which contacts from that line's coil or function occur.

Figure 26–5 is an illustration of a PLC ladder printout. The cross-reference system is included as numbers referencing other sections.

Other typical PRINT mode capabilities include register status, FORCE mode status, timing diagrams, input status, output status, and a listing of malfunctioning output devices. This section discusses only register status, FORCE mode status, and timing diagrams.

A register status printout is shown in figure 26–6. A user-friendly screen program lets you choose to have the status of the register, consecutive registers, or a number of nonconsecutive registers printed. In most cases, you may choose what numbering system you want the printout to display: binary, hex, octal, decimal, ASCII, or others, depending on your PLC model. The printout may be in hard copy form from the printer, or it may be printed on the monitor screen.

The status of any forced functions may also be printed out. If you have forgotten what forced contacts or functions remain in the ladder diagram, a FORCE listing printout will display them. If there are no forced contacts or coils, none will print out; otherwise, those in effect in the circuit will be printed out. Like the register status, the FORCE function listing may be put on a printer or printed on the screen.

Time	HR0307	OR0072	IR1072
0	0682	0167	6421
5	0682	0268	6421
10	0683	0167	6421
15	0683	0167	6421
20	0684	0268	6421
25	0685	0411	6421
↓	↓	↓	↓

Set for a fixed 5-second interval (in decimal)

FIGURE 26–7
Register Timing Printout

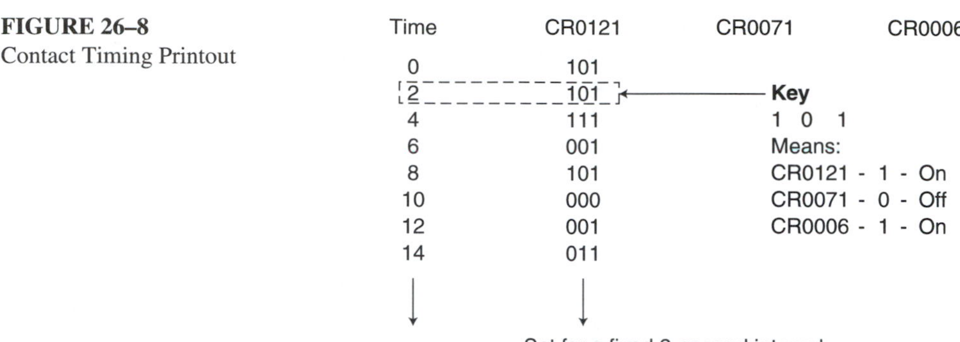

FIGURE 26–8
Contact Timing Printout

Time	CR0121	CR0071	CR0006
0	101		
2	101		
4	111		
6	001		
8	101		
10	000		
12	001		
14	011		

Timing diagrams are available on many PLC printouts. In most, you first choose the time interval you wish to use, from tenths of a second to minutes. Next, you choose the item or items to be observed, such as registers or coils and contacts. The number of items viewed is limited by the PLC's program and printer column width. Figure 26–7 illustrates the printouts for two registers being timed. Figure 26–8 shows five contacts (each with the same number as its coil) being timed. Both figures shown use a fixed, selected interval.

An alternative to fixed-time intervals is available. Exception time saves paper and the time that would be spent poring through a lot of data. Exception timing prints out only when

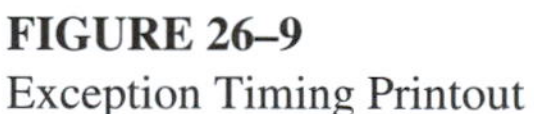

FIGURE 26–9
Exception Timing Printout

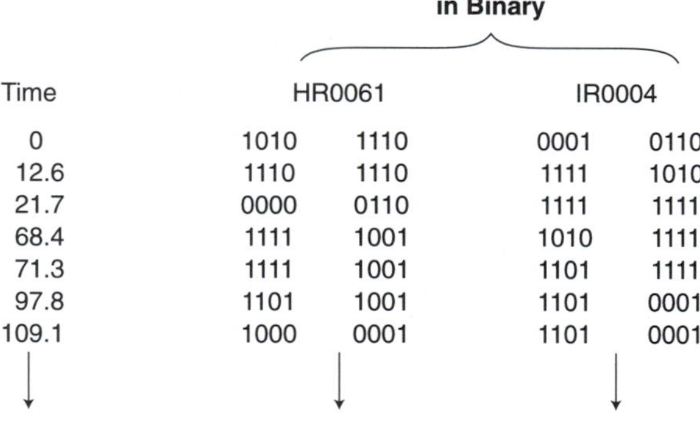

8-Bit Registers in Binary

Time	HR0061		IR0004	
0	1010	1110	0001	0110
12.6	1110	1110	1111	1010
21.7	0000	0110	1111	1111
68.4	1111	1001	1010	1111
71.3	1111	1001	1101	1111
97.8	1101	1001	1101	0001
109.1	1000	0001	1101	0001

Registers

Time	CR0702 CR0641 TS0061
0	010
6.2	010
7.2	011
7.3	111
14.1	010
14.2	111
41.3	110
116.4	000

one of the items being monitored changes status. Figure 26–9 shows how exception time works for the same registers and contacts shown in figures 26–7 and 26–8. The time of the status change is shown on the left of the printout.

Timing in intervals and by exception can be shown on the screen or printed on a printer.

EXERCISES

For exercises 1 through 4, obtain the operational manuals for one or more PLCs and review the operational procedures for MONITOR, FORCE, and PRINT.

1. Explain how the MONITOR function is made operational, how it works, and what data and functions may be observed.
2. Repeat exercise 1 for the FORCE function.
3. For the FORCE function, list how the force procedure affects each of the operational functions of the PLC, starting with contacts and coils.
4. Repeat exercise 1 for the PRINT function.

27

PLC Installation, Troubleshooting, and Maintenance

OUTLINE

27–1 Introduction □ 27–2 Consideration of the Operating Environment □ 27–3 Receiving Check, Testing, and Assembly □ 27–4 Electrical Connecting, Grounding, and Suppression □ 27–5 Circuit Protection and Wiring □ 27–6 Troubleshooting PLC Malfunctions □ 27–7 PLC Maintenance

OBJECTIVES

At the end of this chapter, you will be able to

- List and discuss the procedure for checking the parts of a PLC as received from the manufacturer.
- Describe the procedure for assembling and interconnecting the PLC system. This includes the setting of various switches in the PLC system, including the I/O switches.
- List environmental factors that may have an effect on PLC operation.
- List the reasons for PLC grounding and suppression and how they both are accomplished.
- Describe a complete testing procedure for a newly received PLC.
- Describe power line variations and disturbances that can affect PLC operation.
- Describe remedies for power line variations and disturbances.
- List and describe troubleshooting procedures for general electromechanical devices.
- List and describe specific PLC troubleshooting procedures.
- Describe corrective action and documentation for common PLC failures.
- List and describe general and preventative maintenance procedures for PLCs.

27–1 INTRODUCTION

This chapter discusses the installation and testing by the user of a new PLC. Proper installation and thorough testing before the PLC goes online ensures its dependable and continuous functioning. Included in the discussion is a check on the condition of the PLC upon receipt from the manufacturer and consideration of the environment in which the PLC is to operate. Electrical installation is covered, including grounding and suppression requirements. A master safety shutdown circuit is also discussed, as are proper testing procedures.

Power-line variations and disturbances can affect electronic devices such as the PLC. A discussion of these factors and their correction is given in this chapter.

There are universal troubleshooting procedures that apply to all electromechanical devices, PLCs included. There are also numerous unique troubleshooting procedures that apply to PLCs specifically. Both general and specific troubleshooting procedures are discussed in this chapter. Troubleshooting charts and procedures typical of PLC manuals are also listed and discussed, along with some procedures for general and preventative PLC maintenance. All PLC manufacturers have complete installation manuals for their product. These include installation and operational manuals, mechanical installation plans, and installing to codes—electrical, safety, and environmental.

27–2 CONSIDERATION OF THE OPERATING ENVIRONMENT

The factors in this section should be considered to ensure continuous, reliable operation of the PLC system after installation:

Enclosure. The PLC can be installed in the open; more often, however, it is installed in an enclosed, NEMA-type metal enclosure. NEMA, the National Electrical Manufacturers Association, sets standards for the sizes of enclosures to meet installation codes. The NEMA enclosure must be be planned to allow adequate room for the incoming control wires and power wiring and easy access to all parts and wires for installation, future alterations, and troubleshooting. Appropriate racks are needed to support groups of wires throughout the enclosure. The enclosure should be large enough to allow for future expansion.

Temperature. The PLC has upper and lower temperature operating limits, normally 0°C (32°F) and 60°C (140°F). These limits must not be exceeded during plant operation or during seasonal temperature changes affecting the PLC's ambient temperature. For example, a PLC installed over an annealing oven can soon develop operational glitches due to excessive heat, especially during the summer.

Moisture, Dust, and Corrosive Atmosphere. A PLC may be required to operate in an area of high humidity. Consideration must be given to the level of moisture, which, if too high, can cause electrical and electronic malfunctions. Dust can clog cooling ports and create paths for electrical shorts. In corrosive atmospheres, which could occur in such operations as chemical plants or refineries where oxidizing fumes may be present, electrical connection points can fail due to the buildup of oxides on the wires and terminals. In all three cases, suitable protected enclosures must be used, as specified in the National Electrical Code according to corrosion type.

Vibration. If the CPU is subjected to excessive vibration, transmitted from nearby vibrating equipment, it can malfunction. Vibration can also cause early CPU failures and reduce the life of the PLC equipment. Vibration effects can be reduced by shock-prevention mountings.

27–3 RECEIVING CHECK, TESTING, AND ASSEMBLY

There are numerous important procedures to follow before and during the assembly of a PLC system. These include checking it when received for configuration and freedom from damage, testing it out, and assembling the system. The first process is to check the system as it is received.

When you receive the PLC system from the manufacturer, inspect the packing boxes for any obvious damage. If the boxes are damaged, take a picture of them before opening, in case the parts inside the packages are also damaged. Then, inventory the parts and manuals received against the packing list provided. Also, review the purchase order. Its listing of parts ordered may differ from the parts received. Record any discrepancies between the packing list, purchase order, and parts received. If equipment is damaged, broken, or missing, the supplier should be notified. In industrial organizations, the notification is made through the company purchasing department. A disposition to rework, replace, or return must be mutually agreed upon.

When completely assembled, the PLC is ready for testing. Testing may be accomplished in any one of three modes. First, the PLC can be tested "as is," without attaching any wiring to the I/O modules. Second, it may be tested with a simulator (illustrated in appendix B). Third, it may be tested after it is hooked up to the system it is to operate.

In any testing procedure, all tests and their results should be documented in writing for later reference. More important, a review of the documentation will assure that all necessary tests have been run.

When the PLC is tested with no wiring attached to it, electrical jumpers must be used to energize the inputs. A jumper is moved around from input to input to check for correct operation. Output operations, which then are program energized, are indicated by the operation of the corresponding indicating lights on the output module. Alternately, the FORCE mode (see chapter 26) may be used to check for correct input operation. Instead of moving jumper wires around for the actuation of inputs, the keyboard is utilized. The disadvantage in using FORCE as a simulation is that input module operation is not checked, because the inputs are only simulated. The FORCE mode is described in detail shortly.

All testing should be performed in the MONITOR mode (see chapter 26). In this mode you can observe the ladder program operation on the PLC screen, which gives a better view of the PLC internal operation.

The second test method, which uses a simulator, is performed similarly to the first method; the jumper wires, however, are not necessary. A switch is attached for each input, and indicating lights are attached for each output.

The third method of PLC testing is to connect the PLC with the factory operational system it is to control. This method has one major disadvantage, however. If the PLC equipment is malfunctioning, the equipment being used for testing can be damaged. Furthermore, an operator or programmer error can cause sequence problems or even damage to equipment. Personnel in the area under the PLC's control could be injured during any malfunction.

The FORCE mode (discussed in detail in chapter 26) is often used to test the PLC for proper operations for any of these three test methods. The FORCE mode requires turning inputs and outputs on or off from the keyboard. This overrides the system's normal operation through the input module. Therefore, using FORCE mode could be dangerous to equipment or personnel. Some piece of equipment or component could be unintentionally turned on through the keyboard.

So far, the PLCs we have discussed have discrete on–off input and output modules and systems. Some PLCs have analog capabilities, which means that inputs and outputs have continuously varying values. Testing these analog PLCs requires special equipment and procedures. A special simulator is mandatory for testing a system with other than discrete modules. For these nondiscrete analog modules, the on-off light indicators are of no help in determining proper CPU and output operation. An on-off light does not indicate a variable value. Analog simulators are available as illustrated in appendix B.

Testing of peripherals such as printers, disk drives, and tape drives also takes special testing procedures. Each peripheral device should be completely tested in all possible operating modes. For example, if a printer can print five different types of information, all five modes of operation should be tested.

A complete test of the PLC system and CPU involves checking every function, not just the ones to be used in the immediate process application. Each input and output should be checked—all 200 if there are 200. Also, every function (for example, TIMER, COUNTER, MASTER CONTROL RELAY) should be operated and observed. Check each function, even if it is not to be used initially. Once the PLC warranty runs out, it is probably too late to have a malfunction fixed at no cost. Furthermore, an untested function that is later found to be faulty could cause delays until it is fixed.

Many electronic parts and assemblies are easily damaged by small charges of static electricity. To protect these parts from such damaging static discharges, the manufacturer will normally ship them in antistatic bags. When removed from the bag, these parts require special handling. Units and modules received in these bags must first be inspected for damage. If the bags are damaged, return them to the supplier for replacement or for recheck. The parts should then be removed from the bag in a static-free environment. When the modules are installed into the system, the same precautions are necessary. Portable grounding kits are available from some PLC manufacturers to prevent static damage of parts during handling.

Practically all PLCs, even lower-priced ones, have backup battery systems. Some battery systems use a common 1.5- or 9-volt long-life battery. Others use various types of batteries with special voltage ratings. Some PLCs use a rechargeable battery that is trickle-charged by a small power supply in the CPU. Not all batteries are connected when shipped; they may be in place but insulated from the battery clips by two spacers that are removed for PLC operation. In other cases, separately shipped batteries are installed according to the manufacturer's instructions. Special precautions during installation might include removal of some modules or wires to prevent static damage or electrical surges to some part of the

CPU. In all cases, battery voltage should be checked for compliance to voltage specifications listed in the PLC manual before installation.

All PLC systems have at least one fuse; many have a number of different fuses. These may be in place when the PLC is shipped. If not, the fuses must be installed according to startup instructions in the manual.

27–4 ELECTRICAL CONNECTING, GROUNDING, AND SUPPRESSION

Once you have installed all the parts, you can plug in the line cord for the CPU. Check the CPU for proper operation as you turn the keyswitch or master switch from position to position. Check to see that all operating pilot lights come on at the proper time.

If the PLC CPU does not operate properly at this point, internal visual checks are in order. The faceplate or an appropriate panel may be removed. You can then make a visual check for any loose connections.

Assembly of the input and output modules is now required on larger units. The individual modules are placed on the racks furnished by the manufacturer. The racks are not only for mechanical support, they also have interconnecting electrical wires and connections. Take care that modules are put precisely in place.

The input and output modules are next connected to the CPU with the proper cables. Care must be taken that the connecting ribbon cables are not too twisted or pulled during installation. The power should be off for this procedure. Module switches are then set as described in chapter 2. Next, the wires from the external devices and switches are attached to the I/O terminals. The incoming wires from the input and output devices must be securely fastened to the terminals. The standard practice of "hand tight" is normally followed.

Peripheral devices such as printers, disk drives, and tape drives may now be interconnected to the system by means of their cables. Remote stations and busses to other PLCs and computers should not be connected until the individual PLC checkout is completed.

Proper electrical grounding of the wiring of the equipment and cabinets is essential for personnel safety and to assure proper equipment operation. An ungrounded or improperly grounded wire or part could become shorted electrically to a metal cabinet or rack, presenting an electrical shock hazard to users. In addition, the PLC is computer based, and computers need a proper and solid grounding system for consistently trouble-free operation. A typical wiring scheme for grounding is shown in figure 27–1.

Electrical disturbances from devices outside the PLC system can cause program operation malfunctions. Solenoids, starter coils, motors, and certain other devices are electrically inductive in nature. When these inductive devices are energized or deenergized, they can cause an electrical pulse to be back fed into the PLC system. The back-fed pulse, when entering the PLC system, can be mistaken by the PLC for a computer pulse. It takes only one false pulse to create a malfunction of the orderly flow of PLC operational sequences.

Electrical disturbances in the air also can create false pulses to the PLC. These disturbances can be reduced or eliminated by the use of shielded interconnecting cables. A stranded copper outer sheath around the shielded cable prevents the disturbance pulse from

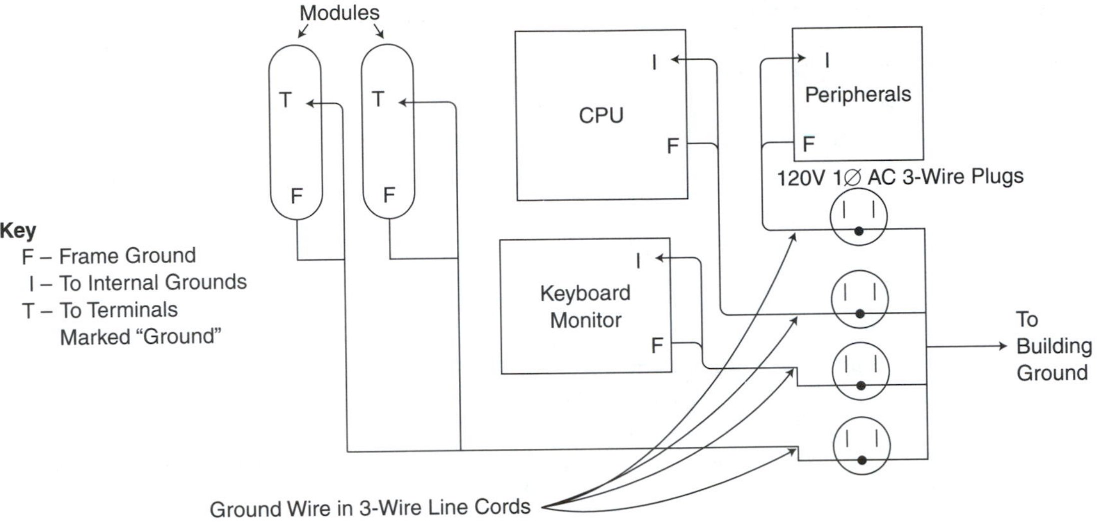

FIGURE 27–1
PLC System Grounding Scheme

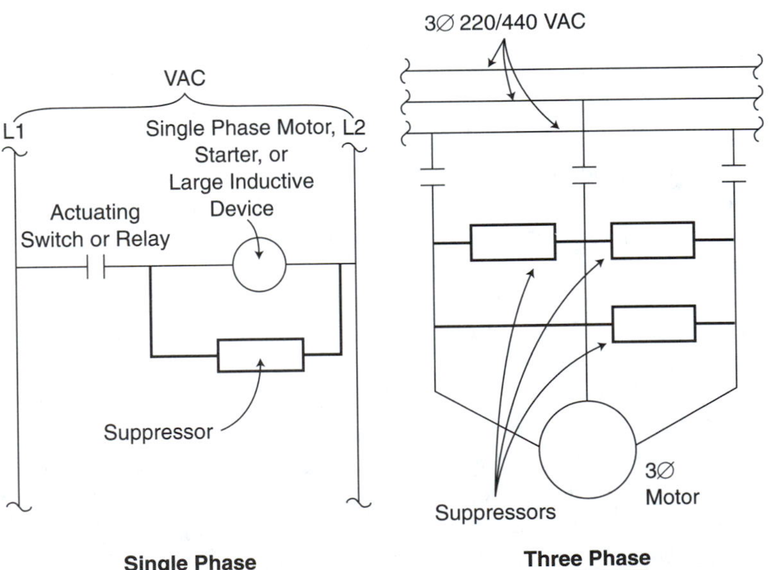

FIGURE 27–2
I/O Suppression Techniques

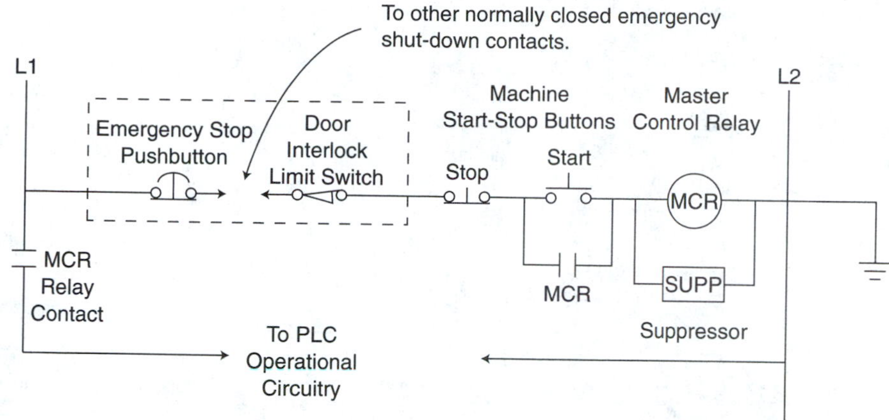

FIGURE 27–3
Master Control Safety Shutdown Scheme

getting to the cable wires inside. The disturbance can also enter through the wires themselves. These direct disturbances can be reduced or eliminated through suppression techniques. Some of these reduction methods are shown in figure 27–2. Essentially, the suppressor absorbs the inductive-caused electrical disturbances. Therefore, no disturbance signal remains that can be sent back into the PLC.

Many systems have a master control relay system for safety shutdown of the PLC operation. This is an override of the whole PLC program. When on, the safety shutdown allows the PLC to operate. When the override is deenergized, the PLC will not operate. A typical master shutdown system is shown in figure 27–3. If control power fails, the PLC operation is shut down.

27–5 CIRCUIT PROTECTION AND WIRING

Electrical circuit protection can be carried out by fuses, circuit breakers, or device overload protection. Electrical circuit protection is installed to protect the wiring and distribution system and/or the device connected to the system. Protecting the PLC requires adequate fusing and proper wiring in its feeder circuit. Inadequate fusing can lead to PLC damage, and inadequate wiring can lead to improper operation. Figure 27–4 illustrates some of the various types of fuses available.

Some of the major considerations in choosing a fuse are

1. Rated current for melting or blow
2. Rated current of the PLC
3. Interrupt capability. (A 20-ampere fuse will not stop 20,000 amperes. The current will arc over.)
4. Temperature of the environment in which the fuse will operate. (Higher temperature means faster action and blowing below rated value.)

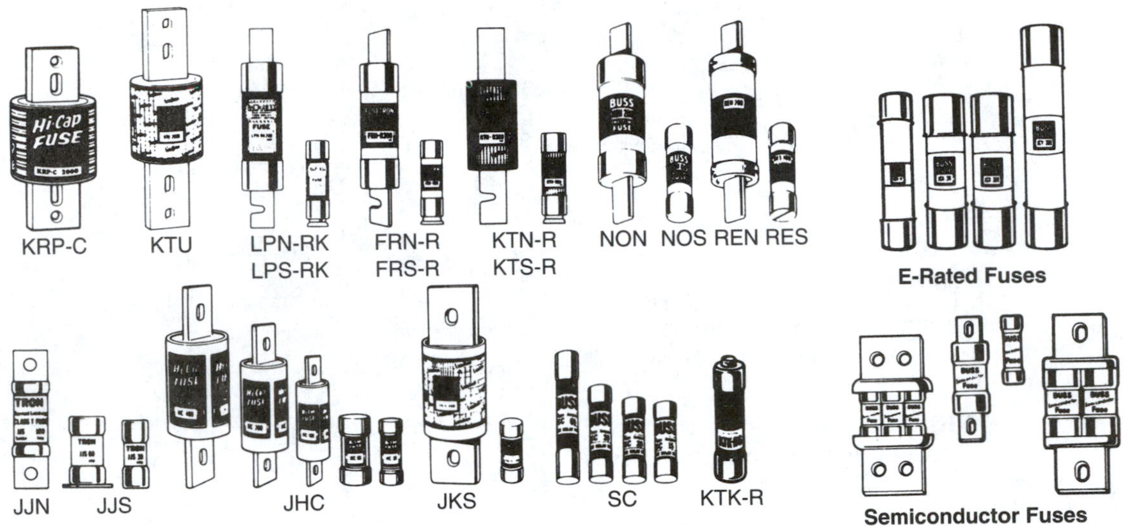

FIGURE 27–4
Various Types of Fuses (Courtesy of Bussmann Division, Cooper Industries)

5. Type of mounting
6. Replaceable link or one-time operation
7. Time delay or regular—do not use time-delay fuses for PLCs; use regular fuses
8. Other special requirements

Consult with a fuse specialist if you have difficulty choosing the correct fuse type.

Circuit breakers have advantages and disadvantages over fuses. Circuit breakers can be reset quickly and do not need to be replaced. However, fuses blow faster, giving more protection to the PLC. The considerations for choice of a circuit breaker type are similar to the considerations for fuses listed previously.

Overload protection is accomplished by internal, specially calibrated overload relays in the PLC itself. They are reset by pushing a red button on the unit. Their disadvantage is cost. Their advantage is that they can be calibrated to the PLC-rated current (for example, 16.3 amps), whereas fuses come in discrete steps (for example, 10, 15, 20, and 30.)

Wiring must be adequate. The wire size must be large enough to prevent excess line loss when a current is drawn by the PLC. Line loss is caused by voltage drop due to the feeder wire's resistance. Figure 27–5 shows wire resistance values for various wire sizes. For example, a 100-foot run of No. 28 wire has 0.6385 ohms of resistance. If the current is 15 amperes, the line drop is $I \cdot R = 15 \cdot 0.6385 = 9.6$ volts (by Ohm's law). Line voltage would then be $120 - 9.6 = 110.4$ volts. This lower voltage to the PLC can cause erratic PLC operation. The National Electrical Code and other code systems give calculations for voltage drop and choosing the proper wire size.

American Wire Gage (AWG) sizes for solid round copper.

AWG #	Area (CM)	Ω/1000 ft at 20°C	AWG #	Area (CM)	Ω/1000 ft at 20°C
0000	211,600	0.0490	19	1,288.1	8.051
000	167,810	0.0618	20	1,021.5	10.15
00	133,080	0.0780	21	810.10	12.80
0	105,530	0.0983	22	642.40	16.14
1	83,694	0.1240	23	509.45	20.36
2	66,373	0.1563	24	404.01	25.67
3	52,634	0.1970	25	320.40	32.37
4	41,742	0.2485	26	254.10	40.81
5	33,102	0.3133	27	201.50	51.47
6	26,250	0.3951	28	159.79	64.90
7	20,816	0.4982	29	126.72	81.83
8	16,509	0.6282	30	100.50	103.2
9	13,094	0.7921	31	79.70	130.1
10	10,381	0.9989	32	63.21	164.1
11	8,234.0	1.260	33	50.13	206.9
12	6,529.0	1.588	34	39.75	260.9
13	5,178.4	2.003	35	31.52	329.0
14	4,106.8	2.525	36	25.00	414.8
15	3,256.7	3.184	37	19.83	523.1
16	2,582.9	4.016	38	15.72	659.6
17	2,048.2	5.064	39	12.47	831.8
18	1,624.3	6.385	40	9.89	1049.0

FIGURE 27–5
Wire Table

A word of caution: It is best to connect the PLC and other electronic equipment to separate power feeder lines. For example, it would be poor practice to connect the PLC to the same line as a 5-horsepower motor. When the motor starts, it draws considerable current, causing line loss and line disturbance.

Electronic equipment, including PLCs, can be affected by power waveform distortion. These distortions can be caused by one of the following:

1. Surges from lightning hitting power lines
2. Disturbances from adjacent buildings or factories
3. Surges from switching action by the power company or in your plant
4. Internal factory power waveform disturbances

Some of these possible waveform distortions are shown in figure 27–6.

Correction of these waveform distortions can be accomplished by the use of a line purifier unit (figure 27–7). These units not only can purify waveforms, but they also can perform two other functions (at added cost). The purifier can take over when the line voltage is too low or too high and maintain proper voltage. Also, the unit can produce AC power from a battery standby system when power fails completely. The switchover is instantaneous, so no data or operational capability is lost.

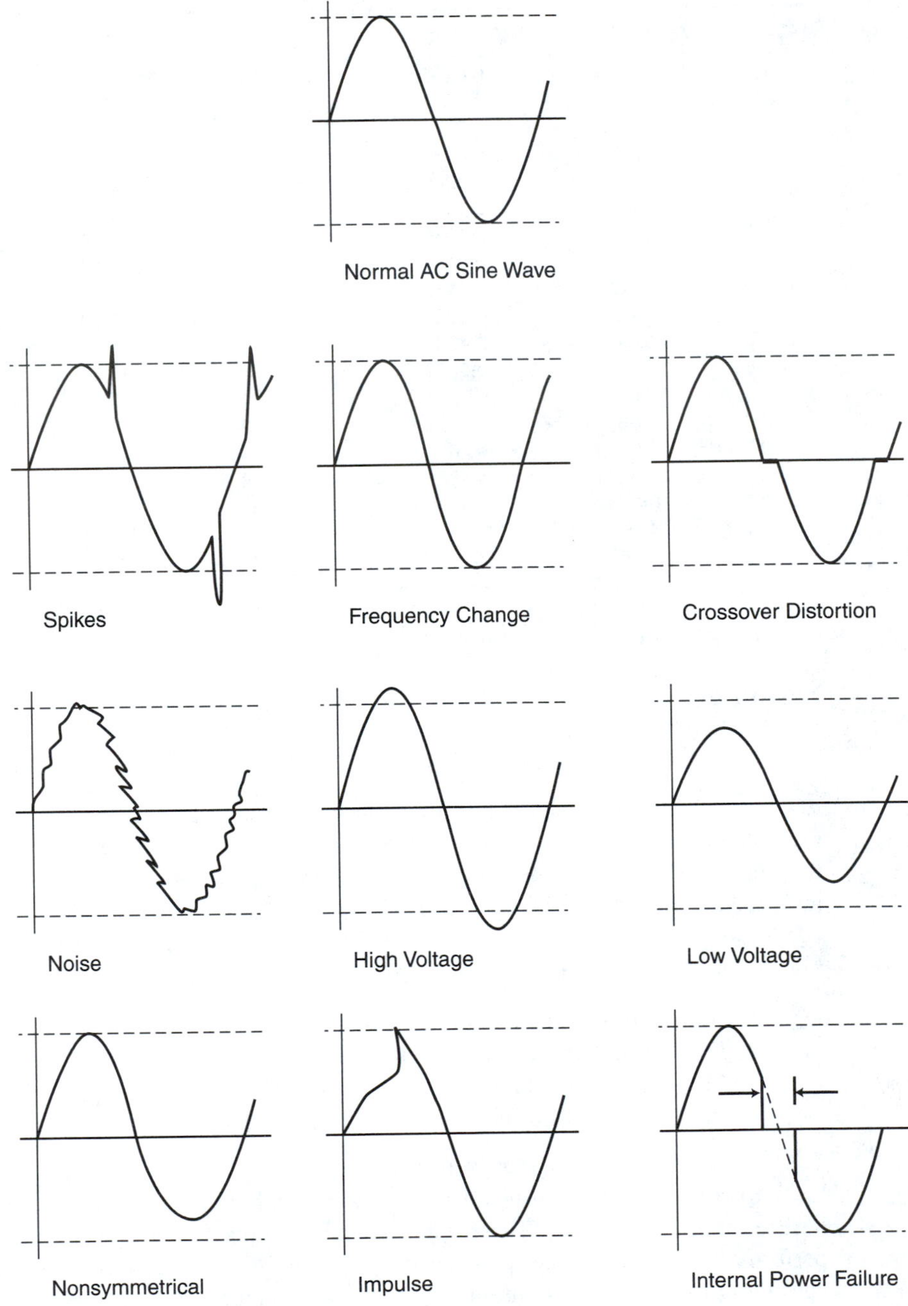

FIGURE 27–6
Possible AC Waveform Distortion

FIGURE 27–7
Power Purifier, or Conditioner, Units (Courtesy of Best Power Technology, Inc.)

27–6 TROUBLESHOOTING PLC MALFUNCTIONS

Some troubleshooting procedures have general applications to all electromechanical devices, and PLCs are no exception.

Safety is the primary consideration during troubleshooting procedures. Will the process under control by the PLC come on unintentionally during the equipment checkout? If so, the process must be operationally locked out in some manner. Turn off the power when working inside of the device unless power is necessary for the analysis. Wear safety glasses during circuit checking in case of an electrical flash or small electrical explosion. Appropriate insulated electrical tools should be used. Most important, work deliberately and consider the consequences of each step of the checking procedure.

Once general safety precautions have been taken, you are ready to proceed into malfunction analysis. One major step is to understand exactly what is malfunctioning. Try to get more than one opinion on the problem, if possible. Document the problem in detail, including the date, time, severity, and circumstances of the malfunction, and include a second opinion if one can be obtained. Then, verify that the reported problem actually exists before trying to solve it. Once the malfunction is determined, begin checking. Check for and replace any blown fuses. Next, a complete visual inspection may reveal the cause of the malfunction. Is there power to all control circuits? Are there any broken wires? Are switches set in the proper positions? Have any undocumented alterations to parts been made? A review of terminal wiring diagrams may reveal that an unrecorded change is causing the trouble.

As stated earlier, some troubleshooting procedures are exclusive to PLCs. In this section we describe some of these unique procedures.

First, establish that the system malfunction is not caused by an external part or system. Then, check the PLC itself for proper operation in each applicable mode. Larger PLCs have an available screen readout for CPU status. Call up the status and refer to the operating manual to determine what the CPU display should look like to indicate proper operational conditions. Any portion not meeting the norm shows an operational problem area. In some cases, just clearing the CPU memory and reprogramming the PLC will eliminate the malfunction.

One way to analyze a PLC problem is by subunit substitution. If you have another CPU available, replace the one in service with it. Reprogram the newly inserted CPU with the program the old CPU was using. If the problem is corrected, the removed CPU could be the culprit. If you determine that the original CPU is not the problem, analyze other PLC subparts by substitution in a similar manner.

Use of the MONITOR mode is helpful throughout the troubleshooting analysis. By observing the program ladder operation on the screen, any misoperation may be discovered. As in testing, the FORCE mode is useful in simulating operating conditions. *A word of caution, however:* A portion of the program may operate properly in the FORCE mode but not during actual operation. For example, input IN0045 may operate correctly in the FORCE mode, but not in actual operation. This would indicate that the input, IN0045, is malfunctioning—for internal or external reasons.

In addition to the CPU status screen display, many PLCs have available a fault indication register display. The fault display may appear automatically or may have to be called up. A typical fault display is shown in figure 27–8. The figure includes a display and an interpretation sheet, which is found in the operating manual. Other displays with more specific information are available on some PLCs. A message on one of these PLCs might say "OUTPUT 0024 IS SHORTED" or "REGISTER 043 IS NOT WORKING," for example. More sophisticated PLC systems have messages that also tell which external devices are not working and why (for example, "MOTOR NUMBER 45 IS OVERHEATING").

Another PLC troubleshooting aid is sequential listings of troubleshooting steps in an operating manual. The lists can be in consecutive form, as shown in this example:

1. Turn on operating switch number 7.
2. Key switch to run.

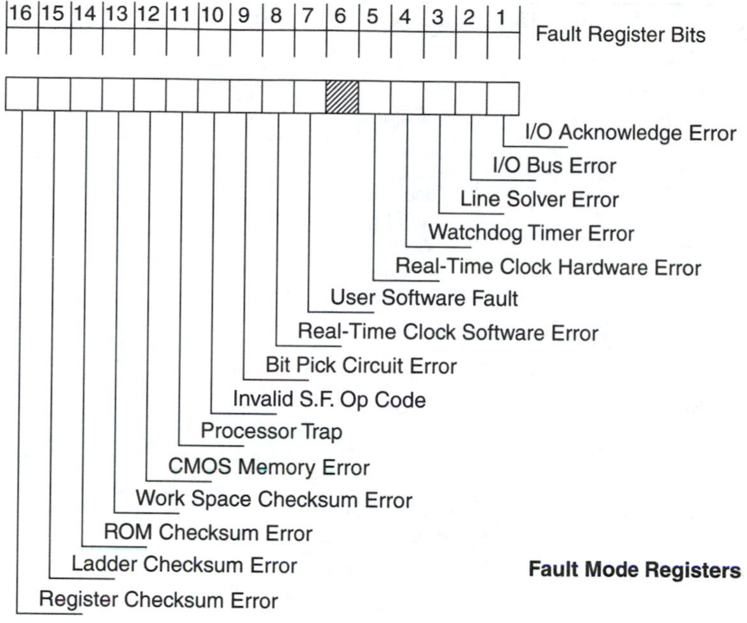

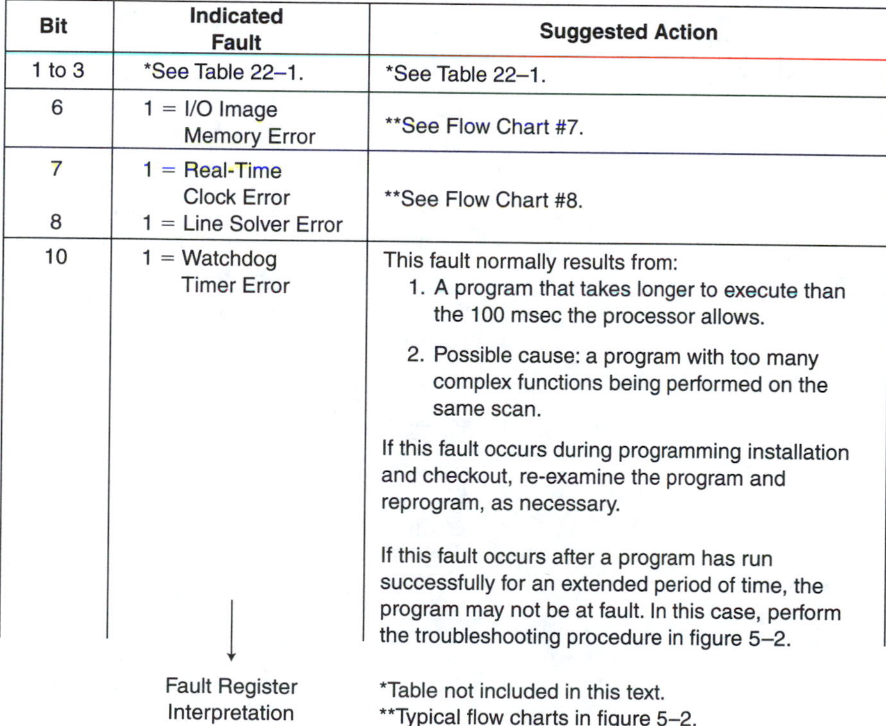

Bit	Indicated Fault	Suggested Action
1 to 3	*See Table 22–1.	*See Table 22–1.
6	1 = I/O Image Memory Error	**See Flow Chart #7.
7 8	1 = Real-Time Clock Error 1 = Line Solver Error	**See Flow Chart #8.
10	1 = Watchdog Timer Error	This fault normally results from: 1. A program that takes longer to execute than the 100 msec the processor allows. 2. Possible cause: a program with too many complex functions being performed on the same scan. If this fault occurs during programming installation and checkout, re-examine the program and reprogram, as necessary. If this fault occurs after a program has run successfully for an extended period of time, the program may not be at fault. In this case, perform the troubleshooting procedure in figure 5–2.

Fault Register Interpretation

*Table not included in this text.
**Typical flow charts in figure 5–2.

FIGURE 27–8
CPU Fault Mode Register Display and Interpretation (Courtesy of Westinghouse)

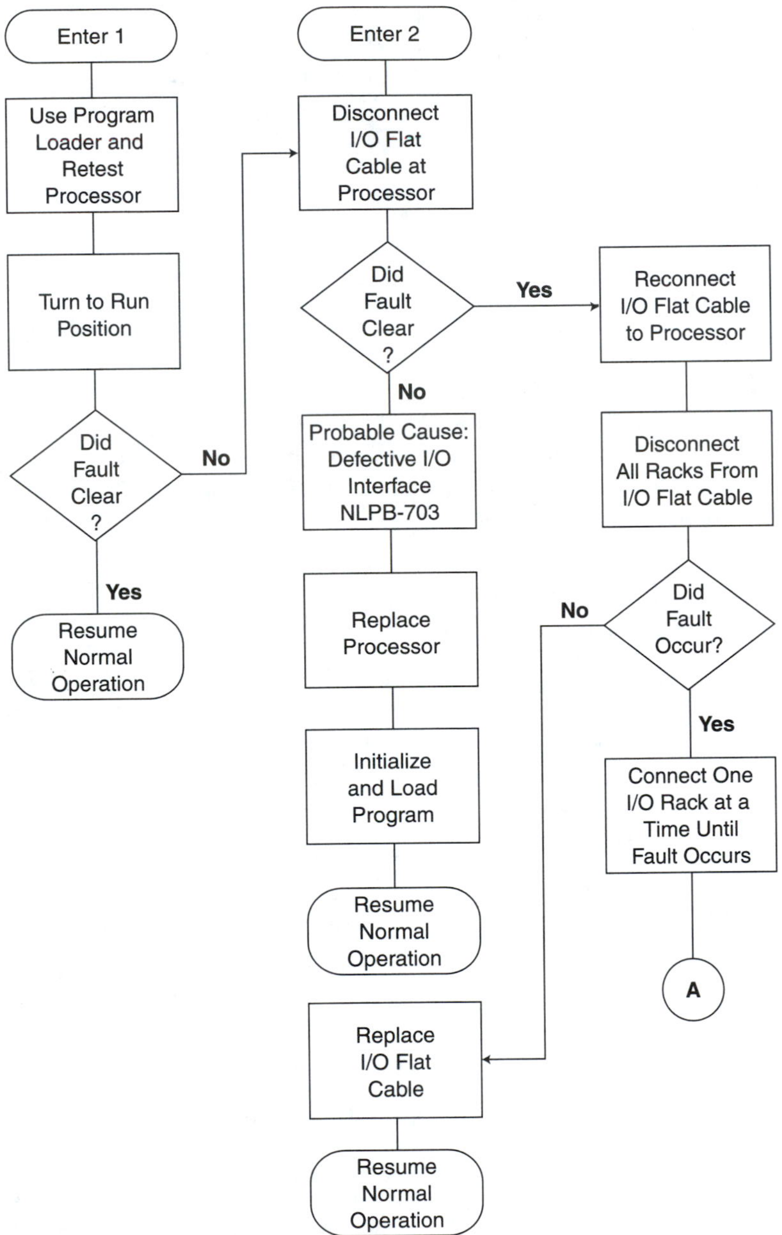

FIGURE 27–9
PLC Troubleshooting Flowchart (Courtesy of Westinghouse)

3. Light number 4 must light.
4. If light number 4 does not light, check bulb.
5. If bulb is OK, put all 1's in Address 909, etc.

Other sequential listings are in computer flow diagram form. See figure 27–9 for an example. By following the flow diagram, the subsystem may be checked for correct operation or determination of a malfunction.

Upon determining the cause of the PLC malfunction, corrective action is taken. Usually, corrective action involves replacing a faulty part with a new part. The faulty part is usually a printed circuit board or a single electronic component. Boards are usually replaced in their entirety in these cases. A PLC printed circuit board must be tested extensively to analyze failures caused by the board's faulty components; these tests are possible only in the manufacturer's facilities.

If the system still does not work when a faulty part is replaced, there are three possible reasons. First, the part used for replacement may also be faulty. A replacement part recheck is required. Second, there may be another part or parts in the unit that are also faulty. Further analysis is needed. Third, the replacement part fails when it is installed because the original defective part had failed due to an overload caused by some other malfunction of the system. The replacement part has been caused to fail by the same problem. If you have replaced a fuse twice, there isn't much loss; if you have replaced a $1500 circuit board, a second failure becomes quite expensive.

For expensive parts, some type of pretest is in order before replacing a failed part; for example, a quick precheck before a circuit board replacement to determine if its supply voltage is correct.

A written log of failures and corrections for each PLC should be kept. The log will show if there is any failure pattern after a number of entries are made. Any pattern of failures will show which PLC problems may be anticipated. More spare parts for a particular failure mode may have to be kept on hand. Another reason for the log is historical in nature. If a failure occurs a second time, reviewing the log will enable the troubleshooter to benefit from past experience. The log should include a description of the failure and the corrective action taken, as well as the date and shift (first, second, or third) of occurrence.

27–7 PLC MAINTENANCE

Maintenance and preventive maintenance of a PLC include the following:

1. Periodically check the tightness of I/O module terminal screws. They can become loose over a period of time.
2. Periodically check for corrosion of connecting terminals. Moisture and corrosive atmospheres can cause poor electrical connections. Internally, end connectors of printed circuit boards also may become corroded.
3. Make sure that components are free of dust. Properly cooling the PLC through a layer of dust is impossible.
4. Stock commonly needed spare parts. Input and output modules are the PLC components that fail most often. Stocking is especially essential if there is no conve-

nient manufacturer's service station and parts depot. Maintaining proper levels of spare parts inventory is a trade-off between costly inventory and prolonged downtime without parts.

5. Keep a duplicate record of operating programs being used. These may be recorded on paper, tape, or computer disk. These records should be kept in a plant location away from the PLC operational area. Copies of long, expensive programs should be kept off the premises to prevent their loss in case of fire or theft.
6. Replace the PLC backup batteries more often than their usable life would indicate. The lithium batteries used for backup in case of lost electrical service are usable for 3 to 5 years. However, many companies replace them yearly on a written maintenance procedure. The cost of the batteries is a small price to pay to prevent the loss of a lengthy PLC program.
7. Have a written checklist control sheet for each PLC. The sheet should have the dates on which the work is performed and the dates for the next preventive maintenance due. The next due date should be entered on a future worksheet listing to make sure it is done on the date due.
8. Keep a log sheet on maintenance for each PLC in addition to the check sheet. Records of what, who, and when should be kept. This log is often combined with the troubleshooting log discussed in section 27–6. The combined log sheet gives a valuable history record and is a guide to buying future PLCs based on performance or nonperformance.

EXERCISES

1. Choose a local industry process where a PLC might be installed. List and describe what environmental factors must be considered before installing the PLC system for control of this process.
2. Obtain a manual from a PLC manufacturer that includes an installation and testing section. List in order the operational steps to install the equipment properly.
3. From the manual you used in exercise 2, draw a block diagram showing how grounding is accomplished for all PLC system components.
4. Do electrical resistance heaters controlled by a PLC require electrical suppression? Why or why not?
5. From the manual you used for exercise 2, explain the PLC's safety shutdown or MCR system.
6. From information from the manual you used for exercise 2, list in order the necessary steps for completely testing the PLC after installation. *Extra credit:* Make up a checklist for carrying out the testing and for recording the results.
7. Describe why adequate fusing and wiring is needed for a PLC installation.
8. Describe how line transients can occur and affect PLC operation.
9. You have 200 feet of No. 16 wire (2×100 feet). If you draw 17.5 amperes, what is the voltage drop?
10. List and describe the major safety rules for PLC troubleshooting.

11. List and describe some general troubleshooting rules for electromechanical devices.
12. List and describe specific troubleshooting procedures with specific application to PLCs.
13. Set up an effective log system to record PLC failures and corrections.
14. List and describe the major areas for PLC maintenance and preventive maintenance.

Note: All exercises can be answered more fully by reviewing the applicable sections of operating manuals for PLCs from one or two different manufacturers.

PLC Manufacturers

ABB Process Automation, Inc.
650 Ackerman Rd.
Columbus, OH 43202
(614) 261-2000

ABB Kent Taylor, Inc.
95 Ames St.
Rochester, NY 14611
(800) 235-5353

Adatek
700 Airport Way
Sandpoint, ID 83864
(208) 263-1471

Allen-Bradley/Rockwell Automation
1201 2nd St.
Milwaukee, WI 53204
(414) 382-2000

Analogic Corporation
8 Centennial Dr.
Peabody, MA 01961
(508) 977-9220

Automatic Timing & Controls
203 S. Gulph Rd.
King of Prussia, PA 19406
(800) 441-8245

Basicon, Inc.
11895 N.W. Cornell Rd.
Portland, OR 97229
(503) 626-1012

Bristol Babcock, Inc.
1100 Buckingham St.
Watertown, CT 06795
(203) 575-3000

B&R Industrial Automation Corp.
1325 Northmeadow Pkwy., S-130
Roswell, GA 30076
(404) 772-0400

Cincinnati Milacron, Inc.
4701 Marburg Ave.
Cincinnati, OH 45209
(513) 841-8100

Control Technology Corp.
25 South St.
Hopkinton, MA 01748
(508) 435-9595

Digitronics Sixnet
P.O. Box 767
Clifton Park, NY 12065
(518) 877-5173

Divelbiss Corp.
9776 Mt. Gilead Rd.
Fredericktown, OH 43019
(614) 694-9015

Eagle Signal Controls
8004 Cameron Rd.
Austin, TX 78753
(512) 837-8300

Eaton Corp.
5340 Alla Rd.
Los Angeles, CA 90066
(213) 822-3061

Entertron Industries, Inc.
3857 Orangeport Rd.
Gasport, NY 14067
(716) 772-7216

Festo Corp.
395 Moreland Rd.
Hauppauge, NY 11788
(516) 435-0800

Furnas
1000 Mckee St.
Batavia, IL 60510
(708) 879-6000

GE Fanuc
PO Box 4248
Lynchburg, Va. 24502
(800) 648-2001

Giddings & Lewis, Inc.
142 Doty St.
Fond du Lac, WI 54935
(414) 921-4100

Grayhill, Inc.
561 Hillgrove Ave.
La Grange, IL 60525
(708) 354-1040

HMW Enterprises, Inc.
207 N. Franklin St.
Waynesboro, PA 17268
(717) 938-4691

Honeywell, Inc.
435A W. Philadelphia St.
York, PA 17404
(717) 848-1151

Horner Electric, Inc.
1521 Washington St.
Indianapolis, IN 46201-3848
(317) 639-4261

Icon Corp.
26 Conn St.
Woburn, MA 01801
(617) 933-9666

Idec Systems and Controls Corp.
1213 Elko Dr.
Sunnyvale, CA 94089
(408) 747-0550

International Parallel Machines, Inc.
700 Pleasant St.
New Bedford, MA 02740
(508) 990-2977

Jumo Process Control, Inc.
37 Woodbine St.
Bergenfield, NJ 07621
(201) 501-0060

Klockner-Moeller Corp.
25 Forge Pkwy.
Franklin, MA 02038
(508) 520-7080

Keyence Corp of America
1717 Route 208 North
Fair Lawn, NJ 07410
(201) 791-8811

Mitsubishi International Corp.
875 Surprime Dr.
Bensenville, IL 60106
(708) 860-4210

Modicon, Inc./Schneider Automation
1 High St.
North Andover, MA 01845
(508) 975-9529

Nolatron, Inc.
P.O. Box 4042
Harrisburg, PA 17111
(717) 564-3398

Omega Engineering, Inc.
P.O. Box 4047
Stamford, CT 06907
(800) 826-6342

Omron Electronics, Inc.
One E. Commerce Park, Dept T.
Schanumburg, IL 60173
(800) 626-6766

Opto 22
15461 Springdale St.
Huntington Beach, CA 92649
(714) 891-5861

Phoenix Contact
P.O. Box 4100
Harrisburg, PA 17111
(717) 944-1300

Pro-Log Corp.
2560 Garden Rd.
Monterey, CA 93940
(408) 372-4593

Pyramid Industries, Inc.
3706 36th Ave.
P.O. Box 23169
Phoenix, AZ 85063
(602) 269-6431

Reliance Electric
24703 Euclid Ave.
Cleveland, OH 44117
(261) 266-7000

Semix, Inc.
4160 Technology Dr.
Fremont, CA 94538
(415) 659-8800

Siemens Industrial Automation, Inc.
100 Technology Dr.
Alpharetta, GA 30202
(404) 740-3000

Square D Co./Schneider Automation
Executive Plaza One
Palatine, IL 60106
(708) 397-2600

Tenor Controls Company
17020 West Rogers Dr.
New Berlin, WI 53151
(414) 782-3800

Triconex Corp.
15091 Baker Pkwy.
Irvine, CA 92718
(714) 768-3709

Toshiba America, Inc.
375 Park Ave.
New York, NY 10152
(212) 308-2040

Uticor Technology, Inc.
4140 Utica Ridge Rd.
Bettendorf, IA 52722-1327
(319) 359-7501

Westinghouse Electric Corp.
P.O. Box 160
Pittsburgh, PA 15230
(412) 778-5183

Yaskawa Electric America, Inc.
3160-A Mac Arthur Blvd.
North Brook, IL 60062-1917
(708) 291-2340

B

Operational Simulation and Monitoring

An external simulator is a device used for simulating the operation of a programmable logic controller to train users and test program sequences. Simulator panels with switches for inputs and lights for outputs are available from some manufacturers. In other cases you may wish to build your own discrete logic simulator, or an analog simulator for those situations involving analog (variable) PLC values. For an analog simulator, variable input voltage must be available from a potentiometer or BCD device, and analog output indicators must be variable value indicators, not just indicating lights.

Using a PLC system for training is difficult without a simulator, because you must continually move shorting wires around to simulate input switch closure and you must move a voltmeter around to observe outputs being on or off.

Some training and simulation situations have specific applications for which a simulator with actual input and output devices is desirable. For example, simulation of a PLC used to control a process with pneumatic cylinders would need actual pneumatic cylinders for the output, and sensors that work just like those in an actual process for the input switches. Required elapsed time would then be included in the simulator operation.

PLC DISCRETE SIMULATORS

A discrete simulator is available from some manufacturers, or you can build your own I/O panel. A typical wiring scheme of a panel built in-house is shown in figure APB–1. An ad-

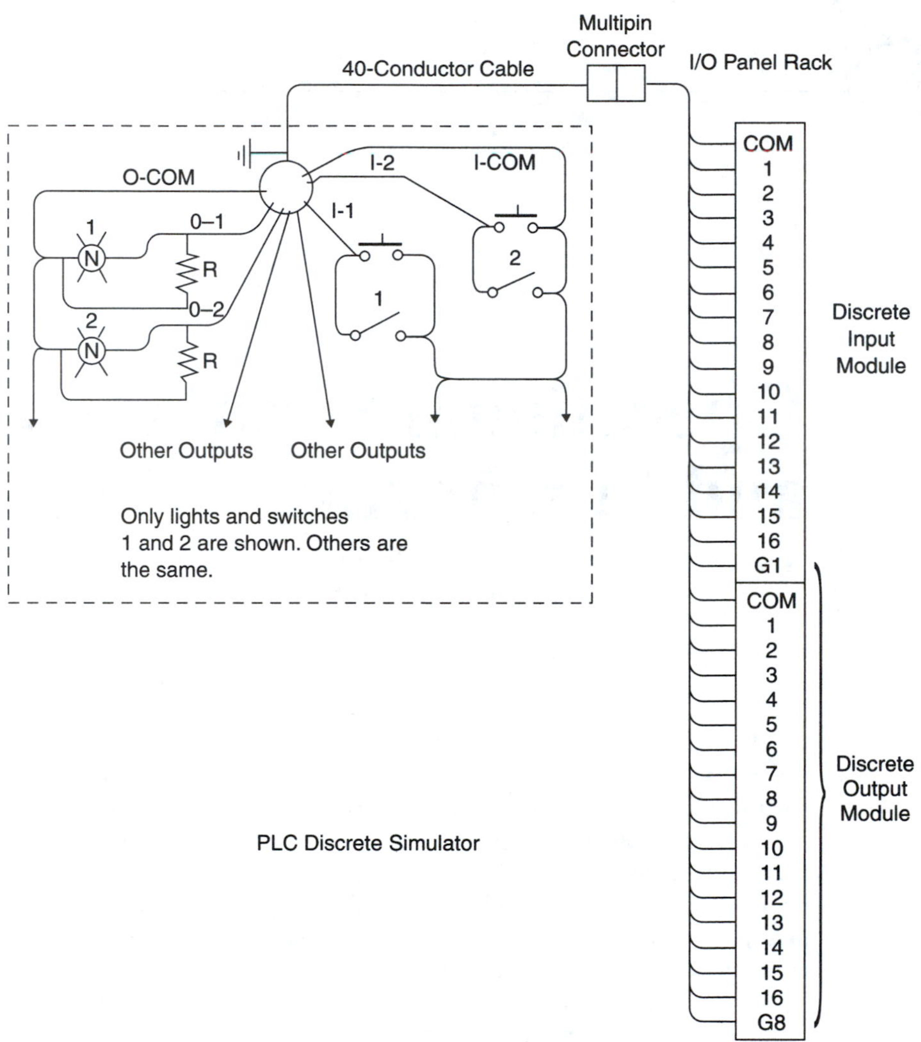

FIGURE APB–1
PLC Discrete Simulator

vantage of the panel shown is that it has both momentary and permanent contacts, both of which are found in industrial processes.

There are three precautions to be considered in building the panel: first, note that the input common connection and the output common connection are not the same polarity, must be separate, and cannot be cross connected; second, parallel resistors are needed with high-resistance neon bulbs because off-state PLC output module leakage current can keep the bulbs glowing; and third, a solid, separate safety ground wire must be run from building ground and connected to the metal chassis and cabinets.

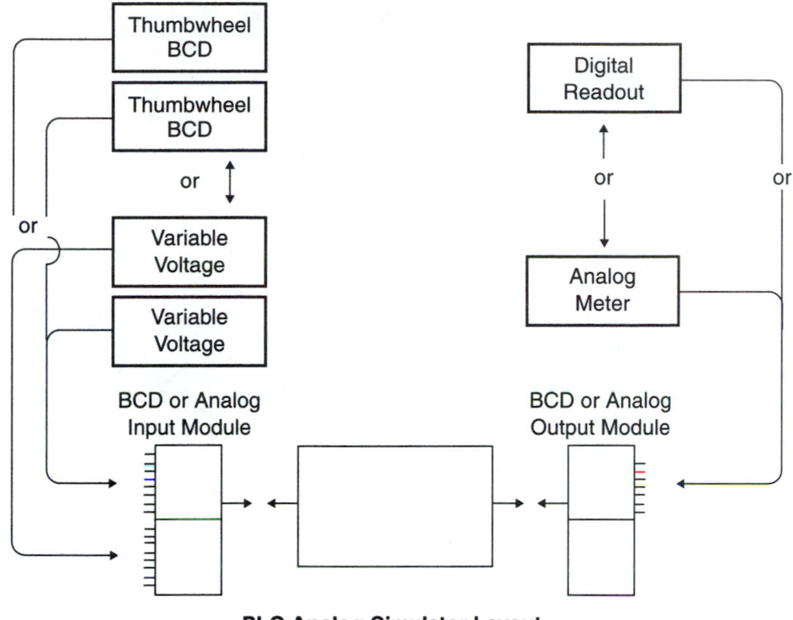

FIGURE APB–2
PLC Analog Simulator Layout

ANALOG SIMULATORS

An analog simulator is shown in figure APB–2. The previous panels shown were for on-off, discrete values. When inputs and outputs are variable, or analog, variable inputs and variable output indicators are used. Inputs may be BCD or a variable voltage device. Outputs may be a digital readout or an electrical analog device such as a meter.

Many analog simulators involve one input and one output only. The simulator should have two or more inputs to ensure the student understands the concept; otherwise, you might as well connect the input dial to the output device directly, because you get the same effect whether you go through the PLC or not. With two inputs as shown, actual CPU programming can be carried out for training and process simulation as discussed in chapter 22.

C

Commonly Used Circuit Symbols

I-Inputs

Limit Switch

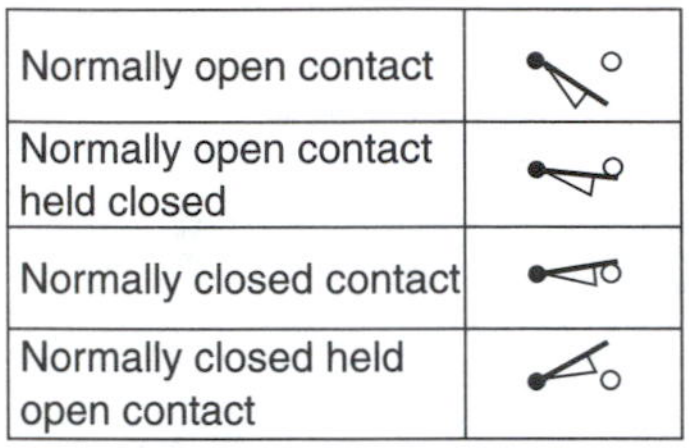

Normally open contact	
Normally open contact held closed	
Normally closed contact	
Normally closed held open contact	

Toggle Switch

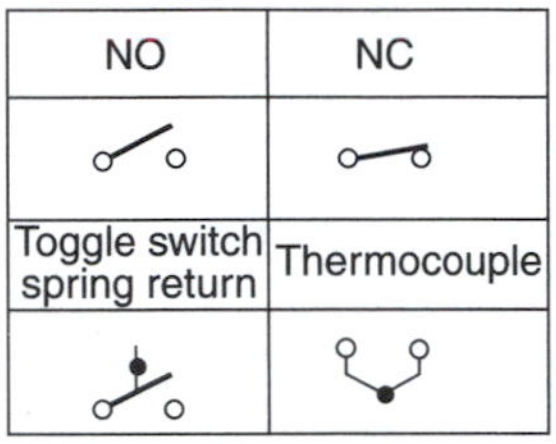

NO	NC
Toggle switch spring return	Thermocouple

Motor Centrifugal Switch

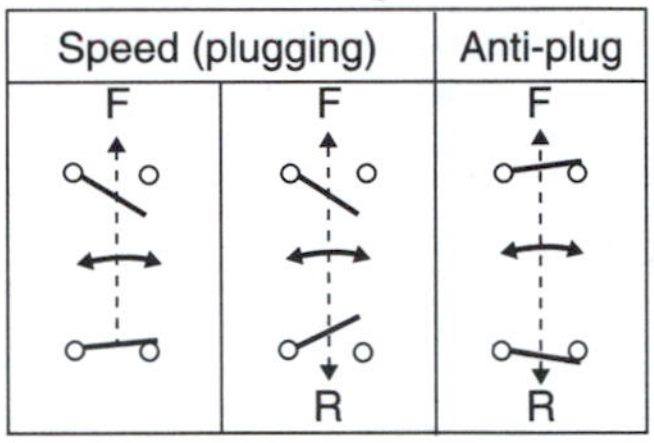

Speed (plugging)		Anti-plug
F	F R	F R

General Switches

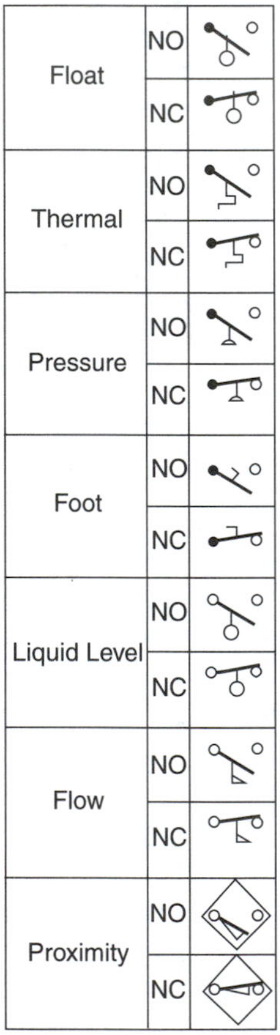

Float	NO	
	NC	
Thermal	NO	
	NC	
Pressure	NO	
	NC	
Foot	NO	
	NC	
Liquid Level	NO	
	NC	
Flow	NO	
	NC	
Proximity	NO	
	NC	

Push Buttons

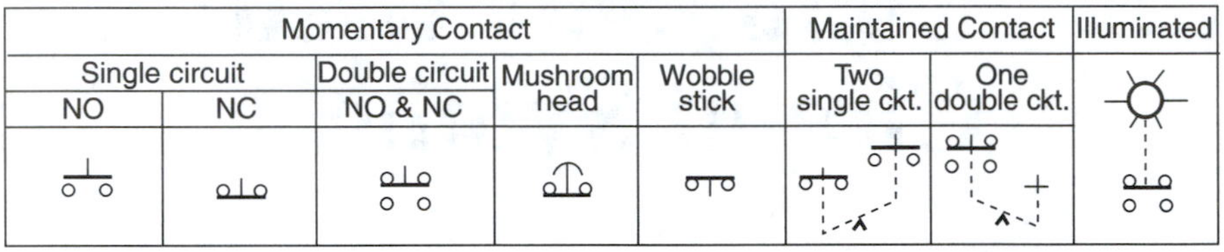

Momentary Contact					Maintained Contact		Illuminated
Single circuit		Double circuit	Mushroom head	Wobble stick	Two single ckt.	One double ckt.	
NO	NC	NO & NC					

Selector

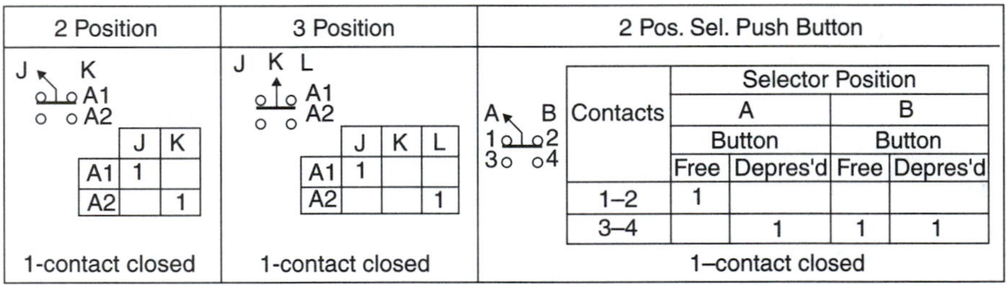

2 Position

J K A1 A2

	J	K
A1	1	
A2		1

1-contact closed

3 Position

J K L A1 A2

	J	K	L
A1	1		
A2			1

1-contact closed

2 Pos. Sel. Push Button

A B 1 2 3 4

Contacts	Selector Position			
	A		B	
	Button		Button	
	Free	Depres'd	Free	Depres'd
1–2	1			
3–4		1	1	1

1–contact closed

General Multiple Switch Convention

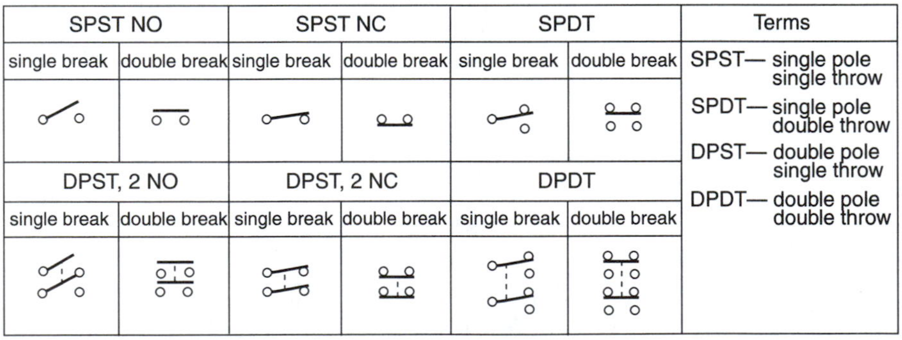

SPST NO		SPST NC		SPDT	
single break	double break	single break	double break	single break	double break
DPST, 2 NO		**DPST, 2 NC**		**DPDT**	
single break	double break	single break	double break	single break	double break

Terms

SPST— single pole single throw
SPDT— single pole double throw
DPST— double pole single throw
DPDT— double pole double throw

KEY:
NO = Normally open
NC = Normally closed

II-Outputs

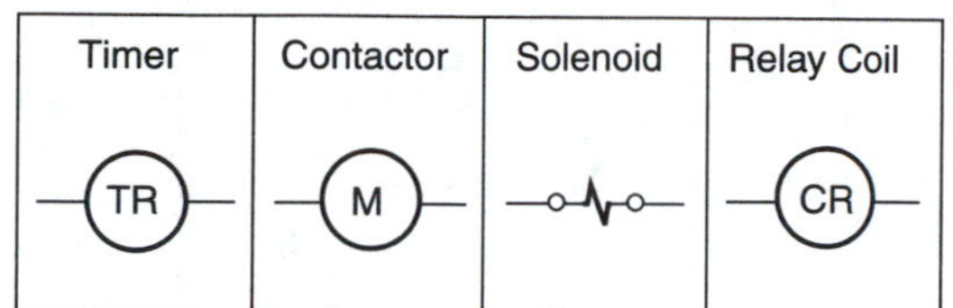

Timer	Contactor	Solenoid	Relay Coil
TR	M		CR

Annunciator	Bell	Buzzer	Horn Siren, etc.	Meter	Pilot Lights	
				Indicate type by letter	Indicate color by letter	
					Non push-to-test	Push-to-test
				VM AM	A	R

AC Motors				DC Motors			
Single phase	3-phase squirrel cage	2 phase 4 wire	Wound rotor	Armature	Shunt field	Series field	Comm. or compens. field
					(show 4 loops)	(show 3 loops)	(show 2 loops)

III – Components/Devices

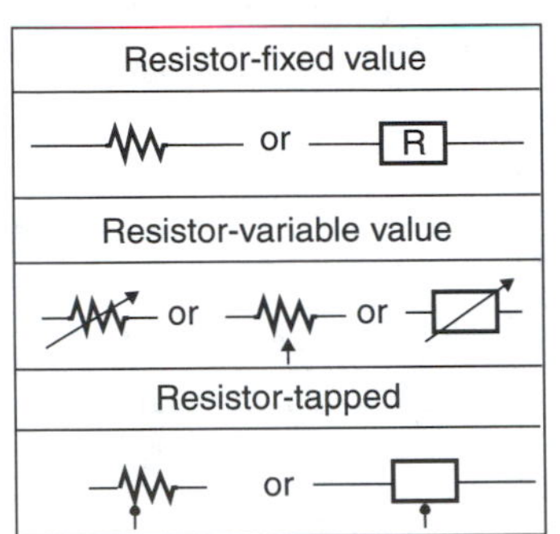

Resistor-fixed value
or R
Resistor-variable value
or or
Resistor-tapped
or

Capacitor	
fixed	variable
Inductors-fixed	
Air core	Iron core
Mechanical interlock	

Battery	AC Supply
+	

Normally open relay contact	Normally closed relay contact

Transformers

Auto	Iron core	Air core	Current	Dual voltage	Multiple output

Rectifier

Single phase half wave	Single phase full wave	Single phase bridge
	AC DC	AC DC DC AC

Symbols for static switching control devices
Static switching control is a method of switching electrical circuits without the use of contacts, primarily by solid state devices. Use the symbols shown in table above except enclosed in a diamond:

Timer

Function	Part	Symbol
On: Delay retards relay-contact action for predetermined time after coil is energized	Normally open timed closing (NOTC) contact	
	Normally closed timed opening (NCTO) contact	
Off: Delay retards relay-contact action for predetermined time after coil is de-energized	Normally open timed open (NOTO) contact	
	Normally closed timed closed (NCTC) contact	
	No instantaneous contact	
	No instantaneous contact	

IV – Electrical Protection Devices

Fuse	Circuit Breaker	
	air	oil
		CB

Overload Relays	
thermal	magnetic

Disconnect	Circuit interrupter	Circuit breaker w/thermal O.L.	Circuit breaker w/magnetic O.L.	Circuit breaker w/thermal and magnetic O.L.

V – Wiring Diagram Conventions

Not connected	Connected	Power	Control	Wiring terminal
				Ground

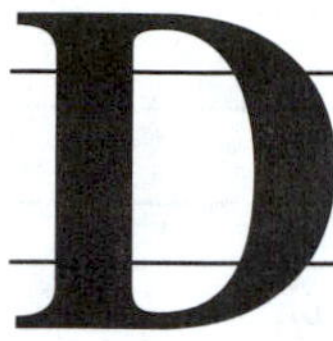

Major PLC Instruction, Function, and Word Codes by Typical Manufacturers

Instruction, function, or word code	This book	Manufacturer					
		Allen-Bradley	Modicon	GE-Fanuc	Cutler-Hammer	Omeron	IDEC
Basic							
Output	CR	OTE	O6 HEX	NO COIL	OUT	OUT	OUT
Open contact	IN –II–	X10	O8 HEX	NO CON	STR/AND	LD	LOAD
Closed contact	IN –I/I–	XIC	O9 HEX	NO CON	STN/ANN	LD NOT	NOT
Latch		OTL	OD HEX	SLAT	MCS	SET	
Unlatch		OTU		RLAT	MCR	RESET	
Leading edge differential			OA HEX		DIF	DIFV	SOT
Trailing edge differential			OB HEX		DFN	DIFD	
End of program	END					END	END
Timer/counter							
Timers	TS, TT, TH	TON, TOF, RTO	T1.0, T0.0, T0.01	ON DTR, TMR	TIM	TIM, TIMH	TIM
Count up	UC	CTU	UCTR	UPCTR	UC	CNT	CNT
Count down	DC	CTD	DCTR	DNCTR	DC		
Mathematical							*
Add	AD	ADD	ADD	ADD	ADD/ADDB	ADD	
Subtract	SB	SUB	SUB	SUB	AUB/SUBB	SUB	
Multiply	MP	MUL	MUL	MUL	MUL/MULB		
Divide	DV	DIV	DIV	DIV	DIV/DIVB		
Square root	SQ	SQR		SQRT			

Instruction, function, or word code	This book	Manufacturer					
		Allen-Bradley	Modicon	GE-Fanuc	Cutler-Hammer	Omeron	IDEC
Moves							*
Move	MV	MOV	RT	MOV	MOV		
Block move	BT	COP/FLL	BLKM	BLKMOV	BMOV		
Table-to-register	TR		T → R				
Register-to-table	RT		R → T				
Jumps						*	
Jump to a destination	JMP			JUMP/JUMPN	JUMPL/LBLL		JMP/END
Jump to a subroutine	JPS	JSR/SBR	JSR/RET	LABEL/LABELN	CALLS/RET		
Bit manipulation			*				
Bit shift right	SR	BSR		SHR	SHR	SFT	SFR
Bit shift left	SL	BSL		SHL	SHL		
Sequencer run	DR	SQO					
Sequencer compare		SQC					
Sequencer load		SQL					
Master controls			*		*		
Master control relay	MR	MCR		MCRN/END		IL/ILC	MCS/MCR
Skip	SK						
Conversions						XX	XX
Convert to BCD	BN/BCD	TOD	BCD-ON	TO-BCD4	BCD		
Convert from BCD	BCD/BN	FRD	BCD-OFF	TO-IN	BIN		
Number comparison			*		*		XX
Equal	EQ –II–	EQV		EQ		CMP	
Not equal	EQ –/I–	NEQ		NE		↓	
Less than		LES		LT			
Less than/equal		LEQ		LE			
Greater than		GTR		GT			
Greater than/equal	GE	GEQ		LE			
Limit between		LIM					
Matrix functions		*		*	*	XX	XX
And	AM						
Or	ON						
Exclusive or	XM						
Compare	CM						
Complement	CP						

[a] See manual for equivalents; XX, not normally available.

Glossary

Abort The action of terminating the progressive operation of a program or process operation.

AC *A*lternating *C*urrent. Electrical current normally alternating 60 times per second.

Address A specific location in a computer memory. Represented by a number, label, or name.

Analog Input Module A PLC module with terminals capable of receiving a continuous, varying, electrical value from an outside device or process.

Analog Output Module A PLC module with terminals capable of furnishing a continuous, varying output voltage to an output device.

Analog Signal A continuous value between two limits. Can represent position, voltage, angle, or any electrical signal with a varying value.

Analog-to-Digital Converter A circuit for converting a varying analog signal to a corresponding representative binary number.

AND (logical) A logic gate whose output is on only if all of its inputs are on.

Arithmetic Capability The ability of a computer to perform math functions.

Arithmetic/Logic Unit (ALU) A computer CPU subsystem that can perform arithmetic and logic gate operations.

ASCII *A*merican *S*tandard *C*ode for *I*nformation *I*nterchange. A 7-bit code for representing letters, numbers, and symbols appearing in written material.

BASIC *B*eginner's *A*ll-Purpose *S*ymbolic *I*nstruction *C*ode. A high-level procedural programming language, with English-like statements, that tells the computer what to do step by step.

Baud A rate of data transmission. Its rate is equal to the number of code elements per second that are transmitted.

BCD *B*inary-*C*oded *D*ecimal. A system in which each decimal digit from 0 through 9 is represented by a pattern of four binary bits.

Binary A system of counting using only 0 and 1.

Binary Number System Sometimes called *base two*, a numbering system using the digits of 1 and 0 only.

Binary Word A group of bits usually found in a single register or address location.

Bit A single binary digit. Can have a value of 1 or 0. From *B*inary dig*IT*.

Bit Pick A PLC system of choosing one bit in a register to determine a status of a PLC input or output. A bit of 1 means on and a 0 means off.

Block Format A PLC screen format with a vertical rectangle on the right. Output and internal operation data are inserted into the rectangle during programming.

Boolean Algebra A shorthand notation that expresses logic functions in equation-type expressions.

Boolean Equation Expresses the relations between logic functions in an equation written in Boolean algebra form.

Branch A parallel logic path within a user program rung.

Buffer A temporary storage area in a computer used for intermediate storage of data. Typically, receives data at one rate and then outputs the data at a different rate or in a different form.

Bus One or more conductors for transmitting data between destinations.

Byte A sequence of binary digits usually operated upon as a unit. The exact number of digits will vary with different systems, but is normally 4, 8, 16, or 32.

Cadepa A Grafcet-based PLC program from FAMIC, Inc. See *Grafcet*.

Cascading Placing two or more functions of the same kind in sequence. The purpose of cascading is to extend the number of operational steps beyond that of one function.

Cassette Recorder/Player For PLCs, a device that can transfer information between PLC memory and magnetic tape. When recording, it makes a permanent record on tape of a program or data from a processor memory. In the playback mode, the cassette recorder enters the previously recorded program or data from the tape into processor memory.

Cell Controller A specialized computer used to control a work cell through multiple paths to the various cell devices.

Central Processing Unit (CPU) The central control unit of a PLC logic system.

Chaining See *Cascading*.

Character One symbol of a set of basic symbols, such as a decimal number, a punctuation mark, or a letter.

Chip Another name for an integrated circuit. A tiny piece of layered semiconductor material mounted in a small case with terminals. Contains a large number of transistors, resistors, and capacitors in miniature.

CIM *C*omputer *I*ntegrated *M*anufacturing. A manufacturing system controlled by an easily reprogrammable computer for flexibility and speed of changeover.

Clear A command to remove data from one or more memory locations. Normally sets the memory location value to zero.

Clock A circuit that generates timed pulses to synchronize the timing of computer operations.

Coaxial Cable A special type of tubular cable having two electrical paths. One is a wire in the center; the other is circular, braided material outside tubular insulation around the center wire. May have a second, outer, braided cable for electrical shielding.

Code A system of symbols or bits for representing data, ideas, or characters.

Coil Represents the output of a programmable logic controller. In the output devices it is the electrical coil that, when energized, changes the status of its corresponding contacts.

Coil Format A PLC screen format with function coils on the right. Output and internal operational data are inserted into and around the coil.

COMPARE Function A PLC function that compares two numbers to see if they meet or do not meet the specified criterion.

Computer Interface A device that communicates between various computers or computer modules.

Contact A part in a relay with two terminals. In PLCs, it is in a conducting or nonconducting state, depending on its corresponding coil's status and the coil's initial status, whether normally open or normally closed.

Control Relay See *Relay*.

Counter A device for counting input pulses or events. Its output changes status when the preset number of counts is reached.

CPU See *Central Processing Unit*.

Cross-Reference In ladder diagrams, letters or numbers to the right of coils or functions. The letters or numbers indicate on what other ladder lines contacts of the coil or function are located. Normally closed contacts (NC) are distinguished from normally open contacts (NO) through the use of an asterisk (*) or by underlining.

CRT *C*athode *R*ay *T*ube. An electronic viewing tube on which PLC data is displayed in alphanumeric or graphic form.

Cursor An indicator that shows where the computer's action pointer is located. Any data entered through the keyboard will occur where the cursor appears on the monitor screen in the EDIT mode.

Data Processing Computer A type of computer (such as an Apple or IBM PC) used to process large amounts of data, usually in an office environment. See *Process Control Computer*.

Data Transfer A computer operation that moves data from one address or register to another.

DC *D*irect *C*urrent. Electrical current flowing continuously in the same direction, usually at a fixed rate or value.

Debug Correcting mistakes in a program through various forms of analysis.

Decimal Number System The base-10 system of counting. Digits are 0 through 9.

Diagnostic Program A computer program used to analyze faults in another computer program or in a system's operation.

Digital A system of discrete states: on or off, high or low, 1 or 0.

Digital Gate A device that analyzes the digital states of its inputs and puts out an appropriate output state.

Digital-to-Analog Converter An electrical circuit that converts binary bits to a representative, continuous analog signal.

Dip Switch A group of small in-line on–off switches. From *D*ual *I*nline *P*ackage.

Discrete Having the characteristic of being on or off.

Discrete Input Module A PLC module that processes input status information having two states only—high or low.

Discrete Output Module A PLC module that puts out only two states—on or off.

Disk Drive A device that records or reads data from a rapidly scanned flat disk.

Diskette The flat, flexible disk on which a disk drive writes and reads.

Documentation A logical, orderly recorded or written document containing software or data listings.

Double Precision The system of using two addresses or registers to display a number too large for one address or register. Allows the display of more significant figures, since twice as many bits are used.

Down Counter A counter that starts from a specified number and increments downward to zero.

Download Loading data from a master listing to a readout or another position in a computer system.

Drum Switch Synonymous with *sequencer*. Normally mechanical in nature, it operates through a multiple sequence of simultaneous on–off states.

Dump Recording the contents of a computer memory or of computer data on a tape, disk, etc. Normally done for backup in case of computer malfunction or the loss or distortion of computer data.

EBCDIC *E*xtended *B*inary-*C*oded *D*ecimal *I*nterchange *C*ode. A code similar to ASCII for representing letters, numbers and symbols, except that it uses 8 bits instead of 7.

EEPROM *E*lectrically *E*rasable *P*rogrammable *R*ead-*O*nly *M*emory. A programmable IC chip. The program can be erased after use (all bits reset to zero) by applying an electrical current to two of its terminals.

Element A part of a PLC program. Coils, contacts, timers, and the ADD function are examples of PLC program elements.

Enable To allow a function to operate by energizing a PLC ladder line. If not enabled, the PLC function will not be active.

Encoder A rotating device that transmits a coded feedback signal indicating its various positions.

EPROM *E*rasable *P*rogrammable *R*ead-*O*nly *M*emory. Same as EEPROM, except that resetting is accomplished by exposing a small section under a "window" to ultraviolet light.

Examine Off An instruction that is true only if the examined or addressed bit is off or 0.

Examine On An instruction that is true only if the examined or addressed bit is on or 1.

EXOR *EX*clusive *OR* gate. A digital gate in which the output is on only when one of its two inputs, not both, are on.

Fail-Safe A control situation that discontinues the operation of a process when the control power source fails. A non-fail-safe operation requires the application of control power to turn it off.

False Prescribed conditions are not met and the logic is disabled or remains disabled.

Feedback In analog systems, a correcting signal received from the output or an output monitor. The correcting signal is fed to the controller for process correction.

File A set of logically arranged data. A file is stored in various computer locations, normally consecutively.

Floppy Disk A recording disk used with a computer disk drive for recording data. The disk is flexible, not rigid.

FORCE Function A keyboard function used to turn elements of a PLC ladder diagram on and off. It overrides the input function status received through the input module.

Format Refers to the language arrangement and layout for a given type of PLC system.

Fuse A device that rapidly interrupts electrical current. The interruption is achieved by the melting of a thin strip of metal inside the fuse.

Gate Electronically, a device that makes logic decisions, depending on its input statuses; the gate's output is turned on or off, accordingly.

Grafcet A graphics-based, high-level computer language used to program a PLC automatically.

Gray Code A special digital code similar to the binary code, except that only one of its bits changes status when going sequentially from one number to the next.

Ground An electrical connection made for safety from a unit case to ground potential. Also sometimes defined as an undesirable connection from an electrical device's electrical system to its case, earth electrical potential, or to the computer's zero-potential level.

Hand-Held A small, portable, programming keyboard, usually with a small LCD window on which portions of the entire ladder diagram may be displayed.

Hard Copy A printed copy of data, program, etc.

Hard Disk An inflexible recording disk used with a computer disk drive.

Hardware In contrast to software, the mechanical, electrical, and electronic parts of a computer.

Hex An abbreviation for hexadecimal.

Hexadecimal A numbering system with four binary bits. Represents 0 through 15 in the decimal system by using the digits 0 through 9 and the letters A through F.

High A status representation of on, 1, or true.

High-Level Language A programming language, such as BASIC, that resembles human language.

Holding Register A type of address location in the CPU for symbol or logic storage.

Host Computer A main computer that controls other computers, PLCs, or computer peripherals.

Hydraulic A system of control using a fluid.

Input Devices Devices connected to the PLC input modules for sending status information. Switches, limit switches, pushbuttons, and electrical potentiometers are examples of input devices.

Input Group Register A single register into which the CPU records the statuses of a group of 8 or 16 input registers.

Input Module An electrical unit or circuit used to connect input devices electrically to the PLC. A module sends coded signals to the CPU indicating the status of each input.

Input Register A PLC register (address) associated with input devices.

Input Scan One of three parts of the PLC scan. During the input scan, input terminals are read and the input table is updated accordingly.

Instruction A computer command or order which causes a prescribed operation to take place.

Integrated Circuit See *Chip.*

Interfacing Connecting a PLC to outside inputs or outputs, or connecting different computers to each other.

Interlock A system for preventing one element or device from turning on while another device is on. May be applied between more than two elements or devices.

I/O An abbreviation for input/output.

I/O Module The electronic assembly of a PLC system that interfaces between the PLC CPU and the "outside world."

I/O Rack See *Rack.*

I/O Update Time The time interval in milliseconds that it takes for a PLC to update the statuses of all input and output modules.

Jog A state of being momentarily on or in motion. In controls, the momentary on state is caused by depressing a spring-return switch or pushbutton. When the switch or pushbutton is released, the device returns to the off state and is not sealed on.

Jump A command in a computer program that causes the sequence to go to, or branch to, a specified point, although not necessarily the next point in the program sequence.

Keyboard The alphanumeric keypad on which the user types instructions to the PLC.

Label The means of identification of registers, addresses, contacts, and coils—normally in letters, numbers, or alphanumeric.

Ladder Diagram A system of successive horizontal lines with symbols representing the logic operation of a control system. The symbols are drawn, in relay-logic or PLC-logic form. The control contacts are to the left, and coils and functions are to the right.

LAN *L*ocal *A*rea *N*etwork. A system control network that controls devices relatively close to each other.

Language A group of letters and symbols used for intercommunication between persons, computers, or persons and computers.

Laptop Computer A small, lightweight computer. Normally, a laptop contains a lot of computing power, a CPU, an LCD screen, and a hard disk drive in one package. So called as it easily works when held on a lap during a plane or auto ride.

Latch An electronic or mechanical device that causes an energized coil to remain on after its input signal is turned off.

Latching Relay A relay with a latching-type operation and two inputs, on and off.

LCD See *Liquid-Crystal Display.*

LED *L*ight-*E*mitting *D*iode. A type of small light used in combinations to give a visual display by emitting light.

Limit Switch A mechanical device that turns a built-in electrical switch on or off. The electrical switch is actuated by depressing its protruding arm.

Liquid-Crystal Display A low-power display, working off reflected light, used in handheld PLC programmers and laptop computers.

Loop Control A control of a process or machine that uses feedback. An output status indicator modifies the input signal effect on the process control.

Low A state of being off, 0, or false.

Magnetic Tape A thin, plastic tape covered with magnetic particles. The tape stores information by becoming magnetized at specific points as it passes through a fixed location. The stored information may be read from the tape on a subsequent passthrough.

Mask A PLC, and computer, system whereby only specified numerical operations take place. The mask allows operations going through a mask with a 1 to take place and with a 0 to not take place.

Master Control Relay A PLC function when reached in a sequence will turn the next specified number of logic lines to the OFF state.

Matrix An arrangement of data or circuit elements in an X–Y, two-dimensional array. The matrix may have any dimensions but in a PLC is often 8 by 8 or 16 by 16.

Matrix Function A PLC function which performs multiple numerical operations or functions. The data for the multiple operations is in matrix array form.

Mechanical Drum Programmer See *Drum Switch.*

Memory In a PLC, the group of addresses or registers where information and programs are stored. Storage may be permanent or temporary and erasable.

Memory Map An allocation of segments of a PLC's solid-state memory that defines which areas the PLC can use for specific purposes.

Menu A list of programming choices shown on a PLC screen.

Microprocessor A computer on a chip containing functions normally found on many different chips. It has all the capabilities of the digital computer, such as ALU, memory, logic, and registers.

Microsecond One millionth of a second.

Millisecond One thousandth of a second.

Mnemonic Codes A short code for a function, usually an abbreviation or combination of key letters for easy recognition. For example, SB for SUBTRACT, EQ for EQUAL TO, and MCR for MASTER CONTROL RELAY.

Mode The functional form in which a computer is operating: for example, run or program.

Module An electronic functional device that may be attached to or plugged into a bus. The bus is connected electrically/electronically to other modules and, more often, to a main computer.

Monitor A mode of operation in which information is displayed. A screenlike device for displaying the program, the operational status of the PLC, or the status of the process the PLC is controlling.

Move A description of a class of PLC functions which moves specified data from one specified location to another, normally by duplication. Data can be individual or multiple.

NAND A digital gate whose output is off only when all its inputs are on.

Negative Logic In digital logic, a system in which the high or on state is more electrically negative than the low or off state.

NEMA Standards A set of industry standards of dimensions, specifications, and performance parameters published by the National Electrical Manufacturers Association.

Nesting In ladder diagrams, locating a series of contacts logically within another series of contacts.

Network A number of interconnected logic devices.

Node A common electrical or logic point with two or more points of the circuit or diagram connected to it.

Nonretentive Describes a PLC logic device that loses its count of increments when the input goes off or low.

NOR A digital gate whose output is off when one or more of its inputs are on.

Normally Closed Contact (NC) A contact that is conductive when its operating coil is not energized.

Normally Open Contact (NO) A contact that is nonconductive when its operating coil is not energized.

NOT A digital inverter gate. On translates to off, and off translates to on through the gate.

Octal A numbering system using three binary bits equivalent to a decimal 8. Numbers used are 0 through 7.

Off-Delay Timer A timer that initiates an action at a specified time after another action ceases.

One Shot An action that takes place once per initiation. Once started, the action lasts for its specified period regardless of further status changes of its initiating input.

Operand A number used in an arithmetic operation as an input.

Optical Isolation Electronic isolation of two parts of a circuit by using a small light beam between the two stages. One stage produces a light beam of appropriate varying intensity; the other receives and decodes the varying light's pattern.

OR A digital gate that is on if any one or more of its inputs are on.

OSI *O*pen *S*ystems *I*nterconnection. A standardized seven-layer model for network architecture.

Output An electrical signal from a PLC used to control a process device under the PLC's control.

Output Devices Devices connected to the output modules to receive status information. Relay coils, fans, lights, and motor starter coils are examples of output devices.

Output Group Register A PLC register (address) that can control multiple outputs, on or off, by the status of its individual bits. Usually controls 8 or 16 outputs.

Output Module An electrical unit or circuit used to connect the PLC to outside devices that are to be turned on or off.

Output Register A PLC register (address) associated with output devices.

Output Scan One of three parts of the PLC scan. During the output scan, data associated with the output status table are transferred to the output terminals.

Parallel Circuit An electrical or control circuit in which the opposite ends of two or more components or elements are each connected to the same nodes. A parallel circuit may make up the whole circuit or be a portion of an overall larger series parallel circuit.

Parallel Transmission A computer operation in which two or more bits of information are transmitted simultaneously.

PC *P*ersonal *C*omputer.

Peripherals In computer systems, the devices connected to or controlled by the computer's central processor.

Pick A method of selecting the state of a specific, single, register bit.

PID *P*roportional–*I*ntegral–*D*erivative. A sophisticated analog control system for accurately and speedily controlling output parameters.

Pilot Run A preproduction run of a small amount of product to work out any "bugs" in the process or program before going into full production.

PLC *P*rogrammable *L*ogic *C*ontroller.

Pneumatic A system run by air pressure.

Pneumatic Controller An industrial control device using compressed air.

Port In computers, a point of connection to a peripheral input or output device.

Positive Logic In digital logic, a system in which the high or on state voltage is more positive than the low or off state voltage.

Power Purifier An electronic package placed electrically between the power company supply and an electronic device. As input power deviates from normal, the purifier furnishes proper electrical waveform, voltage, and frequency to the electronic device. Often also includes standby power capabilities to take over during power outages.

Power Supply For computers, the device that converts line power, usually 120 volts AC, to the power type required by the computer, which can be various values of low-voltage DC.

Procedural Language A programming language, such as BASIC, that requires the programmer to specify the procedure the computer has to follow to accomplish a task.

Process Control Computer A type of computer used to control manufacturing and industrial processes. See also *Data Processing Computer*.

Program A logical sequence of computer steps carried out sequentially by the computer.

Program Scan One of three parts of the PLC scan. During the program scan, data in the input status table is applied to the user program, the program is executed, and the output status table is updated appropriately.

Programmable Logic Controller (PLC) A user-friendly computer that carries out control functions of many types and levels of complexity.

Programmer A keyboard or other device used to enter a program into the computer. Can also monitor, edit, control, and modify the program.

PROM *P*rogrammable *R*ead-*O*nly *M*emory. A ROM chip programmed at the factory for use with a given PLC. It is nonvolatile.

Protected Memory Instructions or data stored in a computer memory that cannot be erased or altered.

Proximity Device A noncontact, input-indicating device for detecting the presence of an object associated with a process. May be discrete or varying-value analog, depending on the process being controlled.

Rack A mechanical channel or chassis on which PLC input and output modules are mounted. May also include wiring channels and connectors.

RAM *R*andom *A*ccess *M*emory chip. A chip that has read and write capabilities.

Rated Voltage The electrical voltage required for proper operation of a PLC. Usually has allowable upper- and lower-range values. Example: $+5.0$ volts $\pm$ 0.2 volt.

Read/Write Memory A computer memory that can receive and store (read) information, and can be accessed for retrieval of stored information (write). Stored information can be erased or replaced. Writing out information does not change the stored information that has been accessed; the stored information is duplicated in the write destination.

Register A location in a PLC memory for storing information, usually in bit form. Essentially, a specified address.

Relay A device actuated by a voltage or signal. The actuation changes the status of discrete mechanical or electronic contacts associated with the device.

Reliability The ability of a device to perform its function correctly over a period of time or through a number of actuations or operations. Can be expressed as a decimal or a percentage.

Repeatability The ability of a device or system to operate properly and accurately on a repetitive basis.

Reset The activity of restoring a function or a value to the initial state or status.

Response Curve In control systems, a plot of position versus time, which shows the movement of a device from a given position to a newly specified position.

Retentive Timer (or Counter) A timer with two inputs. One is enable/reset, and the other actuates the timing cycle. If the timing cycle is interrupted during its interval, the accumulated time is retained. When the input is reclosed, the timer starts at its retained time. The time may be reset to its initialized value only when the enable/reset input is de-energized. A retentive counter operates similarly.

ROM *R*ead-*O*nly *M*emory. An integrated circuit chip with unalterable information.

RS-232C A standard type of computer interconnecting wiring system. Set by the Electronics Institute of America (EIA).

Rung The ladder PLC system that controls one output. May be on more than one horizontal control line.

Scan The sequential operation of a PLC that goes through the ladder diagram from top to bottom and updates all of the outputs according to input statuses. The scan usually takes place from left to right on each rung. Scan time is in microseconds for each scan and is repeated continuously in normal PLC operation.

Scan Time The time required for one complete sweep through the PLC's entire ladder program.

Schematic An electrical diagram symbolically showing components and their wiring schemes. It normally is in logic form, not in wiring connection diagram form.

SCR *S*ilicon *C*ontroller *R*ectifier. A solid-state thyristor device used to control DC power levels. Control of the levels is accomplished by varying the on or firing time within each AC pulse by pulses to the SCR gate terminal. Each single SCR operates only on one half of the sine wave input.

Scratch Pad Memory A small group of registers or addresses used for intermediate computer calculation operations.

Seal A contact paralleling an input device contact. The seal keeps the output on when the input device is turned off, since the parallel contact (seal) is controlled and turned on by the output.

Sensor A PLC input device that senses the process condition. Its status is fed to the PLC through an input module.

Sequence The order in which events take place.

Sequencer A PLC function which sequences through a fixed sequence of events or states. Also can sequence through a listing of comparisons or other operations. Can be used with a mask for selection in the operation.

Serial Operation A system of transferring data sequentially rather than simultaneously.

Series Circuit A circuit in which elements are connected end to end. May be the whole circuit or a portion of an overall series parallel circuit.

Servomechanism A control system that uses feedback for accuracy and process correction.

Significant Digit A digit used to give the accuracy required to a number or an operation.

Simulator An external device used with a PLC that represents the process to be controlled. It would contain input switches and signal devices, and discrete and analog output indicators. It is not as complicated or as costly as the process it represents.

SIP A group of small on-line on–off switches. From *S*ingle *I*n-line *P*ackage.

Skip A PLC Function which when reached in a sequence will skip the specified next number of logic lines. The skipped lines remain in their previous state.

Software The programs that control a computer.

Solenoid A magnetic coil that changes the on or off status of an output device. The change is accomplished by the movement of an iron plunger, which may be a spring-return type.

Solid-State Made of semiconductor material. May be transistors, thyristors, or complete circuits in the form of integrated circuits.

Status The condition of a device, usually on or off.

Three-Wire Control An on–off seal control system with pushbuttons. Requires three wires between the control station and the motor starter or electrical device.

Thumbwheel Switch A series of small, adjacent, numbered (0 through 9) rotary wheels that may be set to a given number. Their settings may be inputted to a PLC for process control.

Timer A device for monitoring or determining times. It begins time at one event and causes another event to occur at the end of the specified time.

Toggle Switch A small electrical switch with an extended lever for actuation. Usually panel mounted.

Transient In electrical power terms, an abrupt “spike” in the waveform that can cause malfunctions in electronic equipment.

Triac A solid-state thyristor device capable of being switched on at a given point in an AC voltage cycle. Works on both the positive and negative portion of the input sine wave. Essentially, a two-way SCR.

True Prescribed conditions are met and the logic is enabled or remains enabled.

Truth Table A yes/no (1/0) matrix indicating the status of one function and how it depends on the status of one or more other functions.

TTL *T*ransistor–*T*ransistor *L*ogic. A type of logic on a chip that includes multiple transistors and gates.

Unlatch Instruction A PLC command that turns a function off and keeps it off, overriding any other subsequent instruction to turn it on.

Up Counter An event counter that starts from 0 and increments up to the preset value.

User Friendly A term indicating that a PLC program operation can be run by a person who, with minimal training or instruction, can proceed through the program by following sequential instructions appearing on the screen.

Value The number or symbol located in the position specified.

Volatile Memory A memory of values or status that is lost when power is turned off. Memory locations are usually reset to 0 as a result.

WAN *W*ide *A*rea *N*etwork. A system control network that controls devices which are relatively far apart. This includes devices in the vicinity and in other plants.

Waveform A plot of electrical value versus time. Often repetitive.

Word A group of bits used to represent a number or symbol.

Work Cell A group of machines that works together to manufacture a product. Normally includes one or more robots. The machines are programmed to work together in appropriate sequences. Often controlled by one or more PLCs.

Write For PLCs, the insertion of information into an address or register.

Bibliography

PROGRAMMABLE LOGIC CONTROLLERS

Bertrand, *Programmable Controller Circuits.* Albany, NY: Delmar, 1996.

Bryan and Bryan, *Programmable Controllers—Collection,* 2nd ed. Chicago: Industrial Text Company, 1997.

Cox, Richard D. *Technician's Guide to Programmable Controllers,* 3rd ed. Albany, NY: Delmar, 1995.

Filer, *Programmable Controllers + Design. Seq.* Philadelphia: W.B. Saunders, 1992.

Industrial Computing. ISA Services. 67 Alexander Drive, Research Triangle Park, NC 27709.

Jones, C. T., and L. A. Bryan. *Programmable Controllers—Concepts and Applications.* Atlanta, GA: IPC/ASTEC Publishers, 1983.

Kissell, Thomas E. *Understanding and Using Programmable Controllers.* Upper Saddle River, NJ: Prentice Hall, 1986.

Oldfield, *Field-Programmable Gate Arrays.* New York: Wiley, 1995.

Petruzelka, Frank D. *Programmable Logic Controllers,* 2nd ed. Albany, NY: Glencoe, 1998.

Simpson, Colin. *Programmable Logic Controllers.* Upper Saddle River, NJ: Prentice Hall, 1994.

Stenerson, Jon. *Fundamentals of Programmable Logic Controllers, Sensors, and Communications.* Upper Saddle River, NJ: Prentice Hall, 1993.

Swainston, *Systems Approach to Programmable Control.* Albany, NY: Delmar, 1992.

Warnock, *Programmable Controllers.* Upper Saddle River, NJ: Prentice Hall, 1989.

CONTROLS

Bateson, Robert. *Introduction to Control System Technology.* Columbus, OH: Merrill Publishing Company, 1980.

Beach, P. David, and G. Roy Bridges. *Industrial Control Electronics.* Upper Saddle River, NJ: Prentice Hall, 1990.

Herman, Stephen L., and Walter N. Alerich. *Industrial Motor Control.* Albany, NY: Delmar, 1985.

Rexford, Kenneth. *Electric Controls for Machines.* Albany, NY: Delmar, 1987.

Rockis, Gary, and Glen Mazur. *Electric Motor Controls.* Alsip, IL: American Technical Publishers, 1982.

Wiring Diagrams Bulletin SM304. Milwaukee, WI: Square D Company.

INDUSTRIAL ELECTRONICS

Moloney, Timothy J. *Industrial Solid State Electronics—Devices and Systems.* Upper Saddle River, NJ: Prentice Hall, 1986.

Webb, John, and Kevin Greshock. *Industrial Control Electronics*, 2nd ed. Indianapolis, IN: Macmillan, 1990.

DIGITAL ELECTRONICS

Floyd, Thomas. *Digital Fundamentals,* 5th ed. Indianapolis, IN: Macmillan, 1994.

Reis, Ronald. *Digital Electronics Through Project Analysis.* Indianapolis, IN: Macmillan, 1991.

INDICATORS AND SENSORS

Honeycutt, Richard D. *Electromechanical Devices.* Upper Saddle River, NJ: Prentice Hall, 1986.

Seippel, Robert G. *Transducers, Sensors, and Detectors.* Reston, VA: Reston Publishing Company, 1983.

REFERENCE MANUALS FOR FUNCTIONS/INSTRUCTION SETS

Allen-Bradley, *PLC-5 Instruction Set Reference*, Publ. 1785-6.1, - February 1996.

Allen-Bradley, *SLC 500 Reference Manual*, Publ. 1747-6.15, - January 1996.

Cutler-Hammer, *D50 Programmable Controller Software Program* Publ. SA-129, June 1996.

GE-Fanuk, *Logicmaster 90-30/20 Programming Software User's Manual*, March 1993.

IDEC, *MICRO-1 User's Manual*, Publ. EM228-7U, February 1996.

Modicon, *Ladder Logic Block Library* Publ. 3000 3/97 KP.

Omeron, *SYSMAC Programmable Controller Catalog*, Publ. P15-E1-2, reprint.

Index

DEVRY INSTITUTE OF TECHNOLOGY
LIBRARY
MISSISSAUGA, ONTARIO